Summary of Contents

About the Editors xxiii
About the Authors xxv
Foreword xxix

1	Overview — *Lynn L. Bergeson and Bethami Auerbach*	1
2	U.S. Chemical Control Laws	15
2.1	Toxic Substances Control Act (TSCA) — *Lisa M. Campbell, Sheryl L. Dolan, and Kathleen M. Roberts*	17
2.2	Federal Insecticide, Fungicide, and Rodenticide Act (FIFRA) — *Lisa M. Campbell, Lisa R. Burchi, Sheryl L. Dolan, and Henry M. Jacoby, MS*	83
2.3	State Laws — *Carla N. Hutton*	143
3	Canada — *Lisa R. Burchi*	179
4	Central America and Mexico — *Michael S. Wenk*	221
5	South America — *Michael S. Wenk*	233
6	Europe	257
6.1	REACH — *Leslie S. MacDougall and Ruth C. Downes-Norriss*	259
6.2	RoHS and WEEE — *Lynn L. Bergeson and Leslie S. MacDougall*	295
6.3	Biocides — *Lisa R. Burchi*	309
7	Asia	339
7.1	China — *Leslie S. MacDougall, Andrew G. Burgess, and Lynn L. Bergeson*	341
7.2	South Korea — *Leslie S. MacDougall and Lynn L. Bergeson*	361
8	Concluding Remarks — *Lynn L. Bergeson*	377

Glossary 379
Index 411

GLOBAL CHEMICAL CONTROL HANDBOOK

A GUIDE TO
Chemical Management Programs

Lynn L. Bergeson, Editor

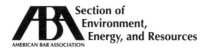

Cover by Monica Alejo/ABA Publishing.

The materials contained herein represent the opinions and views of the authors and/or the editors, and should not be construed to be the views or opinions of the law firms or companies with whom such persons are in partnership with, associated with, or employed by, nor of the American Bar Association or the Section of Environment, Energy, and Resources, unless adopted pursuant to the bylaws of the Association.

Nothing contained in this book is to be considered as the rendering of legal advice for specific cases, and readers are responsible for obtaining such advice from their own legal counsel. This book is intended for educational and informational purposes only.

© 2014 American Bar Association. All rights reserved.

No part of this publication may be reproduced, stored in a retrieval system, or transmitted in any form or by any means, electronic, mechanical, photocopying, recording, or otherwise, without the prior written permission of the publisher. For permission, contact the ABA Copyrights and Contracts Department by e-mail at copyright@americanbar.org, or complete the online request form at http://www.americanbar.org/utility/reprint.html.

Printed in the United States of America.

18 17 16 15 14 5 4 3 2 1

Library of Congress Cataloging-in-Publication Data

Global chemical control handbook : a guide to chemical management programs / edited by Lynn L. Bergeson. — First edition.
    pages cm
Includes bibliographical references and index.
    ISBN 978-1-62722-739-1 (alk. paper)
    1. Chemicals—Law and legislation. 2. Hazardous substances—Law and legislation. 3. Chemical industry—Law and legislation. 4. Chemicals—Safety regulation. 5. Chemicals—Law and legislation—United States. 6. Chemical industry—Law and legislation—United States. 7. Hazardous substances—Law and legislation—United States. I. Bergeson, Lynn L., 1953– editor.
    K3672.5.G56 2014
    344.04'24—dc23

2014041342

Discounts are available for books ordered in bulk. Special consideration is given to state bars, CLE programs, and other bar-related organizations. Inquire at Book Publishing, ABA Publishing, American Bar Association, 321 North Clark Street, Chicago, Illinois 60654-7598.

www.ShopABA.org

All the contributors to this Handbook are grateful to the extraordinary efforts of Allison MacDougall Davidson, Bergeson & Campbell, P.C. Allison's relentless pursuit of perfection, her careful eye when editing, and impeccable proofreading capabilities are unmatched, and we are all indebted to Allison for making this Handbook the great read that it is.

Contents

About the Editors	*xxiii*
About the Authors	*xxv*
Foreword	*xxix*

1　**Overview**　　1
　　Lynn L. Bergeson and Bethami Auerbach

Introduction	1
United States	2
TSCA	2
FIFRA	4
State Law	5
Canada	6
CEPA 1999	6
Central America, Mexico, and South America	7
Europe	9
Asia	11
China	12
South Korea	13
Conclusion	14

2　**U.S. Chemical Control Laws**　　15

2.1　**Toxic Substances Control Act (TSCA)**　　17
　　Lisa M. Campbell, Sheryl L. Dolan,
　　and Kathleen M. Roberts

Executive Summary	17
Statutory Overview	17
Basic Provisions of TSCA	18
Scope of TSCA	19
Scope of Substances Subject to TSCA	19
Substances Exempt from TSCA	19

Scope of Activities Subject to TSCA	23
Scope of Persons Subject to TSCA	23
Notifcation, Approval Requirements, and Compliance	25
Summary of Authority Governing the Requirements for Notification	25
TSCA Inventory and "New" and "Existing" Substances	25
TSCA Inventory—Public and Confidential Lists	25
Determining Inventory Status	26
Exemptions from New Chemical Notification Requirements	27
Exemptions Requiring EPA Approval	30
Exemptions Requiring EPA Approval— Recordkeeping	32
Standard of Review/Burden of Proof	32
Inventories or Other Listing Procedures	33
EPA Review and Response for Chemicals Added to the Inventory	35
Regulation of Substances	37
Authority to Regulate Inventory/Listed Substances	37
Prioritization and Risk Assessment	38
Use Restrictions	38
Chemical or Product Bans	40
Test Data Development	41
Authority to Require Testing/Testing Triggers	41
Chemical Substances/Products That Must Be Tested/Implementing Test Requirements	42
Data Sharing and Compensation	43
Reporting and Recordkeeping Requirements	44
Reports to Be Submitted to EPA	44
Recordkeeping	52
Risk Notification	55
Imports and Exports	62
Authority to Regulate Imports and Exports	62
Import Requirements and Exemptions	65

Confidential and Trade Secret Information	72
Authority	72
TSCA Provisions Relating to Confidential Business Information	72
Enforcement and Penalties	74
Unlawful Activities	74
Civil Penalties	74
TSCA Section 16 Settlements	74
Criminal Penalties	75
Specific Enforcement	75
Subpoena Power	76
Specific Enforcement and Seizure	76
Citizens' Petitions	77
Key Business Issues	78
Practical Tips for Commonly Encountered Issues	78
Trends	79
Resources	82

2.2 Federal Insecticide, Fungicide, and Rodenticide Act (FIFRA) — 83
Lisa M. Campbell, Lisa R. Burchi, Sheryl L. Dolan, and Henry M. Jacoby, MS

Executive Summary	83
FIFRA Overview	83
Basic Provisions	83
Scope of FIFRA	84
Scope of Substances/Products Subject to and/or Exempt from FIFRA	84
Scope of Activities Subject to FIFRA	89
Scope of Persons Subject to FIFRA	90
Notification/Registration/Approval Requirements and Compliance	92
Summary of Authority Governing the Requirements for Notification/Registration/Approval	92
Standard of Review/Burden of Proof	93

Inventories or Other Listing Procedures	96
EPA Review and Response	98
Regulation of Substances/Products	101
Authority to Regulate Inventory/Listed Substances or Products	101
Prioritization and Risk Assessments	102
Use Restrictions	104
Chemical or Product Bans	107
Test Data Development	112
Authority to Require Testing/Testing Triggers	112
Chemical Substances/Products That Must Be Tested/Implementing Test Requirements	113
Data Sharing and Compensation	114
Reporting and Recordkeeping Requirements	118
Reports to Be Submitted to EPA	118
Recordkeeping	120
Risk Notification	123
Imports and Exports	125
Authority to Regulate Imports and Exports	125
Export Requirements and Exemptions	126
Import Requirements and Exemptions	127
Confidential and Trade Secret Information	128
Authority Relating to Confidential Business Information	128
FIFRA Provisions Relating to Confidential Business Information	129
Enforcement and Penalties	130
Unlawful Activities	130
Civil Penalties	132
Criminal Penalties	133
Specific Enforcement	134
Key Business Issues	135
Commonly Encountered Issues	135
Practical Tips	136
Trends	138
Resources	141

2.3	**State Laws**	**143**
	Carla N. Hutton	

Executive Summary 143
2.3.1 California 144
2.3.1.1 Proposition 65 144
Proposition 65 Overview 144
Scope of Proposition 65 145
    Scope of Substances Subject to Proposition 65 145
    Scope of Activities Subject to Proposition 65 145
    Scope of Persons Subject to Proposition 65 146
Notification/Registration/Approval Requirements
and Compliance 147
    Summary of Authority Governing the Requirements
    for Notification/Registration/Approval 147
    Standard of Review/Burden of Proof 147
    Listing Procedures 147
    OEHHA Review and Response 148
Confidential and Trade Secret Information 149
Enforcement and Penalties 149
    Unlawful Activities 149
    Civil Penalties 150
Key Business Issues 151
    Commonly Encountered Issues 151
    Practical Tips 152
Resources 153
2.3.1.2 SCPR 153
SCPR Overview 153
Scope of SCPR 154
    Scope of Products Subject to and/or Exempt from SCPR 154
    Scope of Activities Subject to SCPR 155
    Scope of Persons Subject to SCPR 155
Notification/Registration/Approval Requirements
and Compliance 156
    Summary of Authority Governing the Requirements
    for Notification/Registration/Approval 156
    Listing Procedures 156

Regulation of Substances/Products	157
Authority to Regulate Inventory/Listed Substances or Products	157
Prioritization and Risk Assessments	157
Use Restrictions/Chemical or Product Bans	160
Test Data Development	162
Authority to Require Testing/Testing Triggers	162
Chemical Substances/Products That Must Be Tested/Implementing Test Requirements	163
Data Sharing and Compensation	163
Reporting and Recordkeeping Requirements	163
Reports to Be Submitted to DTSC	163
Recordkeeping	164
Risk Notification	164
Confidential and Trade Secret Information	165
Authority Relating to Confidential Business Information	165
SCPR Provisions Relating to Confidential Business Information	165
Enforcement and Penalties	165
Key Business Issues	166
Resources	166
2.3.2 Massachusetts	167
TURA Overview	167
Scope of TURA	167
Scope of Substances/Products Subject to and/or Exempt from TURA	167
Scope of Activities Subject to TURA	169
Scope of Persons Subject to TURA	169
Regulation of Substances/Products	171
Reporting and Recordkeeping Requirements	171
Reports to Be Submitted to the Massachusetts DEP	171
Recordkeeping	173

Confidential and Trade Secret Information	173
Enforcement and Penalties	174
Civil Penalties	174
Specific Enforcement	175
Key Business Issues	175
Commonly Encountered Issues	175
Practical Tips	176
Resources	176
2.3.3 Trends Overview	176

3   Canada   179
Lisa R. Burchi

Executive Summary	179
CEPA Overview	179
Basic Provisions of CEPA	180
Scope of CEPA 1999	181
Scope of Substances/Products Subject to and/or Exempt from CEPA 1999	181
Scope of Activities Subject to CEPA	186
Scope of Persons Subject to CEPA	187
Notification/Registration/Approval Requirements and Compliance	188
Summary of Authority Governing the Requirements for Notification/Registration/Approval	188
Standard of Review/Burden of Proof	188
Inventories or Other Listing Procedures	191
Environment Canada/Health Canada Review and Response	193
Regulation of Substances/Products	196
Authority to Regulate Inventory/Listed Substances or Products	196
Prioritization and Risk Assessments	196
Use Restrictions	199
Chemical or Product Bans	201

Test Data Development	203
Authority to Require Testing/Testing Triggers	203
Chemical Substances/Products That Must Be Tested/Implementing Test Requirements	203
Data Sharing and Compensation	204
Reporting and Recordkeeping Requirements	204
Reports to Be Submitted to Environment Canada/Health Canada	204
Recordkeeping	205
Risk Notification	205
Imports and Exports	206
Authority to Regulate Imports and Exports	206
Export Requirements and Exemptions	207
Import Requirements and Exemptions	210
Confidential and Trade Secret Information	210
Authority Relating to Confidential Business Information	210
Provisions Relating to Confidential Business Information	211
Enforcement and Penalties	212
Unlawful Activities	212
Civil Penalties	214
Criminal Penalties	215
Key Business Issues	216
Commonly Encountered Issues	216
Practical Tips	217
Trends	218
Resources	219

4 Central America and Mexico   221
Michael S. Wenk

Executive Summary	221
4.1 Costa Rica	222
Regulation of Substances/Products	223
Trends	223

4.2 Panama		224
Regulation of Substances/Products		224
Trends		225
4.3 Honduras		226
Regulation of Substances/Products		226
Use Restrictions		227
4.4 Mexico		227
Regulation of Substances/Products		228
Trends		230
5 South America		**233**
Michael S. Wenk		
Executive Summary		233
5.1 Argentina		234
Notification/Registration/Approval Requirements and Compliance		234
Resolution No. 709/98 of the Ministry of Health and Social Action Overview		234
Scope of Resolution No. 709/98		235
Regulation of Substances/Products		235
5.2 Chile		236
Notification/Registration/Approval Requirements and Compliance		236
Scope of 1968 Chilean Sanitary Code (Decree No. 725)		237
Regulation of Substances/Products		237
Reporting and Recordkeeping Requirements		238
Imports and Exports		239
5.3 Colombia		239
Regulation of Substances/Products		240
Authority to Regulate Inventory/Listed Substances or Products		240
Prioritization and Risk Assessments		240
Imports and Exports		241
5.4 Venezuela		242

Regulation of Substances/Products 243
Key Business Issues 244
   Commonly Encountered Issues 244
     Practical Tips 245
     Trends 245
5.5 Uruguay 246
Reporting and Recordkeeping Requirements 247
Enforcement and Penalties 247
5.6 Peru 248
Imports and Exports 249
Key Business Issues 249
5.7 Paraguay 249
Regulation of Substances/Products 250
Trends 251
5.8 Brazil, Fernando Tabet 251
Regulation of Substances/Products 251
   Main National Laws and Regulations on Chemicals 251
Biotechnology and the Access to Genetic Resources—Provisional Measure No. 2,186-16/2001 254
Enforcement and Penalties 255
Trends 256

6 Europe 257

6.1 REACH 259
Leslie S. MacDougall and Ruth C. Downes-Norriss

Executive Summary 259
   Overview 259
     Basic Provisions 260
   Scope of REACH 262
     Scope of Activities Subject to REACH 264
     Scope of Persons/Entities Subject to REACH 265
   Notification/Registration/Approval Requirements and Compliance 266
     Registration Requirements 266
     The Registration Dossier and Chemical Safety Report 268

	ECHA Fees and Review	270
	Regulation of Substances	271
	Authorization/Substances of Very High Concern	271
	Restrictions	273
	Test Data Development	274
	Data Sharing and Compensation	276
	Registration Software	279
	Duty to Inform ECHA of Current Information	280
	Imports and Exports	281
	Confidential and Trade Secret Information	281
	Enforcement and Penalties	283
	Unlawful Activities	284
	Civil and Criminal Penalties	284
	Key Business Issues	285
	Substance Identity	285
	Preregistrations and SIEF Issues	286
	Guidance and Interpretation Shortfalls	287
	Data-Sharing and Cost-Sharing Issues	288
	Managing Adverse SVHC Inferences	289
	Practical Tips on Supply Chain Communication	289
	Trends	291
	Resources	293
6.2	**RoHS and WEEE**	**295**
	Lynn L. Bergeson and Leslie S. MacDougall	
	Executive Summary	295
	Overview	296
	Basic Provisions	296
	Exemptions	297
	Scope and Implementation Deadlines of RoHS 2 and WEEE 2	298
	Notification/Registration/Approval Requirements and Compliance	301
	Manufacturer Obligations	301
	EEE Waste Compliance	303

Regulation of Substances/Products
under WEEE   303
  Importer Obligations   305
  Distributor Obligations   305
  Enforcement and Due Diligence   306
  CE Markings   306
  Waste EEE Containing Hazardous Substances   307
Resources   308

6.3 Biocides   309
Lisa R. Burchi

Executive Summary   309
  Overview of BPR   310
  Basic Provisions   310
Scope of BPR   310
  Scope of Substances/Products Subject to
  and/or Exempt from BPR   310
  Scope of Activities Subject to BPR   315
  Scope of Persons Subject to BPR   316
Notification/Registration/Approval
Requirements and Compliance   317
  Summary of Authority Governing the Requirements
  for Notification/Registration/Approval   317
  Standard of Review   319
  Burden of Proof   319
  Inventories or Other Listing Procedures   320
  ECHA and Respective Member State Competent
  Authorities Review and Response   320
Regulation of Substances/Products   323
  Listed Substances or Products   324
  Prioritization and/or Risk Assessments   324
  Use Restrictions   324
  Chemical or Product Bans   325
  Exclusion   325
  Substitution   326

Test Data Development	327
Chemical Substances and Products That Must Be Tested	327
Testing Triggers	328
Implementing Test Requirements	328
Data Sharing	328
Reporting and Recordkeeping Requirements	331
Imports and Exports	332
Confidential and Trade Secret Information	332
Authority Relating to Confidential Business Information	332
Provisions Relating to Confidential Business Information	333
Enforcement and Penalties	334
Commonly Encountered Issues	335
Substance Sameness and Technical Equivalence	335
Data and Cost Sharing	336
Data Protection	337
Trends	337
Resources	338

7   Asia   339

Executive Summary	339

7.1   China   341
Leslie S. MacDougall, Andrew G. Burgess, and Lynn L. Bergeson

Overview	341
Governance Framework	342
Cultural Framework	343
Scope of the Three Primary Laws	344
Decree No. 591	344
SAWS Order No. 53	347
MEP Order No. 7	348

       Notification/Registration/Approval Requirements
          Inventory of Existing Substances   348
          Chemical Substance Notifications   349
          MEP-SCC Review of New Chemical Notifications   352
          Postchemical Notification Obligations   352
          Exemptions from Notifications   353
       Risk Assessment   354
          Risk Assessment Overview   354
          Environmental Risk Assessment   354
          Human Health Assessment   354
       Test/Data Development   355
       Imports and Exports   356
       Enforcement and Penalties   356
          Unlawful Activities   357
          Guidance Documents   357
       Key Issues and Practical Tips   358
       Resources   359

7.2 South Korea   361
Leslie S. MacDougall and Lynn L. Bergeson

       Overview   361
       7.2.1 TCCA   362
       Overview   362
       Scope of TCCA   363
          Notification/Registration/Approval Requirements
          and Compliance Chemical Inventory   364
          New Chemicals Notification   364
          Exemptions from New Chemical
          Notification Requirements   365
       Test Data Development   366
       Resources   366
       7.2.2 K-REACH   366
       Overview   366
          Notification/Registration/Approval Requirements
          and Compliance Reporting of Chemical Substances   367
          Registration of Chemical Substances   368

	Evaluation	370
	Hazardous	370
	Authorization	371
	Restricted and Banned Chemicals	371
	Only Representative	372
	Supply Chain Communications	372
	Supply Chain Requirements Applicable to Product Manufacturers	373
	Enforcement Provisions	374
	Trends	374
	Resources	375
8	**Concluding Remarks** *Lynn L. Bergeson*	**377**
Glossary		*379*
Index		*411*

About the Editor

Lynn L. Bergeson is managing partner of Bergeson & Campbell, P.C. (B&C®), a law firm based in Washington, D.C., that concentrates on conventional and engineered nanoscale and biobased chemicals, pesticides, and other specialty chemical product approval, regulation, litigation, and associated business issues. Ms. Bergeson is also president of The Acta Group (Acta®), B&C's scientific and regulatory consulting arm, with offices in the United States, United Kingdom, and China, and president of B&C Consortia Management, L.L.C., which helps the chemical industry achieve shared research, testing, regulatory, and access goals. Ms. Bergeson counsels clients on a wide range of issues pertaining to chemical hazard, exposure and risk assessment, risk communication, and related legal and regulatory aspects of conventional and nanoscale chemical regulatory programs under the Federal Insecticide, Fungicide, and Rodenticide Act (FIFRA), the Toxic Substances Control Act (TSCA), and the European Union's Registration, Evaluation, Authorization and Restriction of Chemicals (REACH) regulation, and on issues pertinent to nanotechnology and other emerging transformative technologies. Ms. Bergeson is former chair of the American Bar Association (ABA) Section of Environment, Energy, and Resources (2005–2006) and current vice chair of the section's Pesticides, Chemical Regulation, and Right-to-Know Committee, and serves in other ABA leadership capacities. Ms. Bergeson is listed in *The International Who's Who of Business Lawyers* (2004–2014), and *The Chambers USA: America's Leading Lawyers for Environmental Law* (2005–2014). Ms. Bergeson is a graduate of Michigan State University (BA, magna cum laude), and the Columbus School of Law, Catholic University of America, where she was a member of the Law Review. She is admitted to the bar of the District of Columbia and several federal circuit courts.

About the Authors

Bethami Auerbach, of counsel with Bergeson & Campbell, P.C., has more than 35 years of legal experience with particular expertise in environmental law gained while in private practice, government service as a U.S. Environmental Protection Agency (EPA) attorney, and in academia. She is an expert on the Clean Air Act (CAA) and the Federal Insecticide, Fungicide, and Rodenticide Act (FIFRA), as well as many business aspects of environmental law. She received her JD from Stanford Law School, where she was on the editorial board of the Law Review.

Lisa R. Burchi is of counsel with Bergeson & Campbell, P.C. Her work involves Toxic Substances Control Act (TSCA) and FIFRA regulatory matters. She has particular expertise in data compensation matters, and also counsels on matters related to California law, including Proposition 65 and the Green Chemistry Initiative/Safer Consumer Products Regulations. She received her JD from George Washington University's National Law Center and was a clerk with the U.S. Senate Judiciary Committee.

Andrew G. Burgess is director of China regulatory affairs for The Acta Group, the consulting affiliate of Bergeson & Campbell, P.C. He has been deeply engaged in China chemical registration and support activities for the last decade and has gained unique insight and knowledge of Chinese chemical control legislation and its practical implication for business.

Lisa M. Campbell is a partner and shareholder of Bergeson & Campbell, P.C. with more than 25 years of experience in addressing pesticide and chemical product approval, defense, and regulatory issues. She offers an unparalleled breadth and depth of expertise in finding and implementing solutions to challenging pesticide and chemical product regulatory issues, whether helping clients keep products on the market or bringing new products to the market. She received her JD from Stanford Law School.

Sheryl L. Dolan, senior regulatory consultant with Bergeson & Campbell, P.C., specializes in addressing complex chemical compliance management under federal and state law. Her clients include manufacturers of pesticides, conventional and nanoscale chemicals, intergeneric microorganisms, and industrial and consumer products, and she regularly represents them in interactions with EPA's Office of Pesticide Programs (OPP) and Office of Pollution Prevention and Toxics (OPPT). She received her JD from George Washington University.

Ruth C. Downes-Norriss, a regulatory specialist with The Acta Group, the consulting affiliate of Bergeson & Campbell, P.C., focuses on the European Union (EU) Registration, Evaluation, Authorization and Restriction of Chemicals (REACH). She has been assisting clients register substances since REACH's inception in 2008, and has extensive experience with Substance Information Exchange Forums (SIEF) and consortia to coordinate Lead Registration submissions and meet REACH requirements. She graduated with honors from the University of Derby.

Carla N. Hutton is a regulatory analyst with Bergeson & Campbell, P.C., whose areas of focus include green chemistry, nanotechnology, FIFRA, TSCA, Proposition 65, and the REACH program. She graduated with honors from the University of Pennsylvania, and received a JD from Washington College of Law.

Henry M. Jacoby, MS, regulatory consultant with Bergeson & Campbell, P.C., has more than 40 years of experience in assisting pesticide, insecticide, herbicide, fungicide, antimicrobial, wood preservation, and antifouling paint manufacturers and formulators in environmental science, government guidelines, evaluations, and analysis, including 25 years working for OPP at EPA. Few consultants have a greater degree of knowledge of the workings of OPP, familiarity with personnel, and knowledge of the regulations.

Leslie S. MacDougall is director of regulatory affairs at The Acta Group, the consulting affiliate of Bergeson & Campbell, P.C. A former EPA scientist, Ms. MacDougall has over two decades of global chemical control experience in government, private industry, and consulting roles, and an exceptionally broad understanding of the strategic business and technical requirements of international chemical control schemes.

Her knowledge of the regulatory and cultural nuances of many regions (Asia in particular) makes her an in-demand speaker and presenter.

Kathleen M. Roberts is vice president of B&C® Consortia Management, L.L.C. She has more than 20 years of experience in chemical control issues related to TSCA and is highly regarded as a leader in domestic and international science and policy program management for industry groups engaged in legislative and regulatory advocacy, research, and public outreach and communications. She graduated from the University of North Carolina, Chapel Hill.

Fernando Tabet is managing partner of Tabet Advogados, an environmental law firm in São Paulo, Brazil. He holds degrees in environmental law and international law from the University of São Paulo, and a degree in civil process law from Getúlio Vargas Foundation (GVlaw). Mr. Tabet is also co-president of the Inter-American Network of Environmental Law Specialists (RIELA), which congregates independent law firms with expertise in environmental matters in all Americas. He is also coordinator of the Sustainability Commission of the France-Brazil Chamber of Commerce in São Paulo and of the Thematic Committee on the Management of Environmental Risks of the Brazilian Council for Sustainable Construction. Both Mr. Tabet and the firm are highly ranked in Brazilian environmental law by Análise Advocacia 500, The Legal 500—Latin America, Who's Who Legal, and Chambers Latin America.

Michael S. Wenk is the manager, product stewardship and regulatory affairs, Americas, for Kemira Chemicals, Inc. He has been managing regulatory affairs at Fortune Global 500 chemical and consumer product firms since 1997. He pairs extensive regulatory expertise with a well-developed understanding of corporate operations, providing a unique blend of business process knowledge and technical acumen.

Foreword

Lynn L. Bergeson

Chemicals are a global business. Perhaps in response to a growing global awareness of the need for heightened and improved management of all aspects of chemicals—production, processing, distribution, use, and disposal, based in part on heightened awareness of chemical hazards—governments around the world have been enhancing their domestic chemical management programs and ratcheting up the level of regulatory control over chemicals and, in some instances, manufactured articles containing chemicals believed to be "of concern." Additionally, chemical stakeholders are developing new and improved management standards by which to address their business operations. Often these standards are more stringent and superior to regulatory standards.

What these measures are, on what segment of the global supply chain they apply, when, and how such measures might impact the business of chemicals are increasingly topics legal practitioners, product stewards, and others engaged in the business of chemicals must be aware. The need to know is driven less by the particular standard of care the law and/or regulation of a particular country might impose on a chemical or class of chemicals than by a need to be aware of and navigate around the real-world consequences of failing to know a particular chemical might impose a trade or import restriction, for example, or invite a production requirement that was not known and now impedes commercialization plans. Lawyers, business managers, product stewards, regulatory affairs personnel, and others need to possess a general understanding of chemical governance systems globally to ensure their client's and/or businesses' interests are protected.

This book is intended to provide a broad overview of key chemical management programs in the United States, Europe, Asia, and Central and South America. Our aim is to provide an overview of the laws and their regulatory implementation in these jurisdictions to familiarize the reader with a basic understanding of the law and to identify key business issues of which practitioners should be aware. The book is not intended to provide a granular review of these authorities.

It provides sufficient information to point to other authorities and resources if a more in-depth look at a particular law or provision is needed. No book can answer all the questions and issues that will undoubtedly arise with respect to a particular law or its implications for a business. Our hope is this book will provide sufficient information for readers to know generally how a country's chemical management program works, and thus to facilitate an understanding of how the law might apply to a product line or otherwise help the reader anticipate the transactional impact and enforcement opportunities a particular regulatory program presents.

1

Overview

Lynn L. Bergeson and Bethami Auerbach

INTRODUCTION

Few, if any, environmental professionals can stay on top of the many, varied, and always challenging regulatory programs throughout the world that attempt to achieve the 21st-century goals of safe chemicals, expanded sources of publicly available information about them, and hazard communication inside and outside the workplace. The worthiest environmental professionals do all this while balancing the demands of robust risk assessment without overtesting or overspending, while working within the realities of constrained public resources. It is a tall order and a dynamic one.

This volume collects in one place both the essential background of the key chemical regulatory programs worldwide and insightful observations and commentary from a team of experts who work with these programs daily. Especially where a program is evolving, some issues have no easy resolution, but each chapter in this book can assist working professionals in identifying and understanding the issues that they may encounter. This introduction describes what the book covers, which is organized by country and/or regulatory program. This introduction also includes stand-alone sections discussing the forecast and trends.

UNITED STATES

TSCA

Enacted in 1976, the Toxic Substances Control Act (TSCA) by now can be called venerable. Along with its implementing regulations, TSCA establishes a comprehensive structure aimed at protecting human health and the environment from potentially harmful chemical substances and mixtures. Chapter 2.1 of this book provides an in-depth discussion of TSCA's subchapter 1, describing the key provisions and how they work.

These fundamental provisions include the following:

- section 4, authorizing the U.S. Environmental Protection Agency (EPA) to issue rules that require manufacturers, importers, and processors to test existing chemicals;
- section 5, authorizing EPA to regulate new chemicals before they are manufactured, imported, or processed for commercial uses as well as "significant new uses" of existing chemicals;
- sections 6 and 7, authorizing the regulation and, if necessary, the preemptive seizure of existing substances or mixtures that EPA determines pose an "unreasonable risk" to human health or the environment;
- section 8, authorizing data collection/submission, recordkeeping, and risk reporting by manufacturers and processors;
- section 12, authorizing rules to require export notification and to regulate imports; and
- sections 11, 15, 16, and 17, providing an enforcement structure, authorizing facility inspections for compliance, assessment of civil and criminal penalties for violations, and seizure of noncompliant substances or mixtures.

TSCA section 3(2)(A) defines the term "chemical substance," which identifies those substances that trigger the applicability of the various statutory requirements listed above. Section 3(2)(B) sets out a number of exclusions from the definition, including (but not limited to) pesticides and other chemical substances regulated under other laws. Although "mixtures" are excluded from the definition of "chemical substance," each chemical component included in a "mixture" is nonetheless subject to TSCA requirements. Despite the various exclusions, TSCA's reach is broad; EPA has stated that it covers microorganisms as well as traditional chemicals and that nanoscale substances that meet the definition of "chemical substance" also are covered.

The TSCA Chemical Substance Inventory (TSCA Inventory) of all chemicals manufactured or processed in the United States, which EPA has compiled as mandated by section 8(b), plays a pivotal role. Under section 5(a), no person may manufacture or import a chemical substance unless it is listed on the TSCA Inventory, subject to exemptions discussed in chapter 2.1, which also describes the "public" and "confidential" components of the Inventory. The Inventory enables the delineation of a category of "existing" chemicals—those that are listed—as opposed to "new" chemicals—those that are not. For a "new" chemical to be added to the Inventory, it must undergo EPA review, including the submission of a premanufacture notice (PMN) and, later, a notice of commencement (NOC), both discussed in chapter 2.1. Also discussed is the requirement that EPA must be notified of any intended "significant new use" of an existing chemical substance addressed through promulgation of a section 5(a) Significant New Use Rule (SNUR).

Other aspects of TSCA covered in chapter 2.1 are EPA programs to assess, prioritize, and address risks; test rules and data requirements; varied and detailed recordkeeping and reporting requirements; specifics of export, import, and customs requirements; confidential business information and trade secrets; enforcement and penalties; citizen petitions; pointers for businesses; trends; and resources.

Readers may be aware that the initiatives in recent years on Capitol Hill to update, amend, modernize, and reform TSCA all have failed because of bipartisan differences. In May 2013, S. 1009, the Chemical Safety Improvement Act, was introduced in the Senate with bipartisan sponsorship and with co-sponsors from both sides of the aisle. The unexpected joint effort drew favorable early comments from both industry and environmental organizations. Although the outcome of the current initiative is impossible to predict, some observers view this latest bill as a significant step toward updating TSCA. As of this writing, no additional legislation has been introduced in the Senate. No legislation has been formally introduced in the House, but a "discussion draft" was circulated in 2014.

FIFRA

As noted, TSCA exempts chemical substances regulated under other federal laws. Key among these substances are pesticides, which the Federal Insecticide, Fungicide, and Rodenticide Act (FIFRA) defines broadly as "any substance or mixture of substances intended for preventing, destroying, repelling, or mitigating any pest." Pesticides are regulated under FIFRA as well as under the Federal Food, Drug, and Cosmetic Act (FFDCA), as amended by the Food Quality Protection Act (FQPA). Subject to certain exemptions, EPA identifies a pesticide by a regulatory test that turns on the presence of one or more "active ingredients."

As discussed in chapter 2.2, FIFRA affects all activities relating to pesticides and pesticide-containing products; likewise, it affects anyone engaged in an activity involving pesticides. EPA's authority flows from the central prohibition in FIFRA section 3(a) that no person may sell or distribute a pesticide unless it has been registered. Registration, in turn, necessitates a determination by EPA, among other findings, that the pesticide generally will not cause "unreasonable adverse effects on the environment" (defined to include human health) when used as intended. Chapter 2.2 describes the various types of registrations that EPA may issue, as well as the statutory requirements for applications to register a pesticide, including labeling criteria and the submission or citation of data (sometimes considerable and expensive) for review. Through regulations and guidance, EPA has fleshed out significantly these and other requirements in the law. Separately, based on its authority under the FFDCA/FQPA, EPA also sets maximum permissible pesticide residue levels, or "tolerances," for food products.

As chapter 2.2 explains, risk assessment plays a major role not only for an initial registration but also beyond. For example, EPA completed a statutorily required second look at older registrations in light of fresher data. This process, called reregistration, has been replaced by "registration review," under which EPA will review a pesticide every 15 years to ensure that it continues to meet statutory standards. All of these exercises require the development or acquisition and submission of data to satisfy EPA's elaborate and extensive requirements. Where more than one registrant is involved, FIFRA provides for a system of data sharing to encourage competition and avoid unnecessary repetition of testing. FIFRA also provides data compensation, to preclude latecomers from riding on the coattails of the original data submitter without fairly compensating the latter. This system also is addressed in chapter 2.2.

Chapter 2.2 offers an in-depth discussion of regulatory control options available to EPA, from use restrictions (including appropriate label language) to product bans or the cancellation (including voluntary cancellation) or suspension of a product registration. Where EPA seeks to pursue avenues to cancel, suspend, or restrict use of a pesticide, certain procedures must be followed and certain criteria established.

Other aspects of FIFRA covered in chapter 2.2 include the following:

- recordkeeping and reporting requirements, including risk notification,
- export and import requirements and exemptions,
- confidential business information and trade secrets,
- prohibited acts, enforcement, and penalties,
- pointers for businesses,
- trends, and
- resources.

State Law

Chapter 2.3 discusses in detail two states, California and Massachusetts, which have enacted their own chemical programs. (Programs elsewhere—in Washington, Maine, and Minnesota—are described briefly in the Trends Overview subsection.) The most influential program is California's Safe Drinking Water and Toxic Enforcement Act of 1986 (Proposition 65), under which the implementing agency maintains and publishes a list of chemicals known to the state to cause cancer or reproductive toxicity, and businesses have a corresponding obligation to notify the state about lists of chemicals in products they buy or release to the environment. A more recent project is California's development of Safer Consumer Products Regulations (SCPR), under the state's Green Chemistry Initiative, which require manufacturers to evaluate the availability of safer alternative ingredients for products containing "candidate chemicals."

Chapter 2.3 also addresses the Toxics Use Reduction Act (TURA), enacted in Massachusetts. TURA requires businesses using large quantities of listed chemicals to report on their use of those chemicals annually, pay an annual toxics use fee, and prepare a submission that examines their use of the chemicals and how a toxics use reduction strategy might be implemented.

CANADA

CEPA 1999

Those with a working knowledge of TSCA can find many familiar concepts in the Canadian Environmental Protection Act, 1999 (CEPA 1999). As detailed in chapter 3, although Canada's approach to managing chemical substances reflects a different regulatory structure, different terminology, and some different policy choices, certain key elements are functionally similar to those found in TSCA. Becoming conversant with one statute and its implementing regulations can give an environmental professional a leg up in mastering the other. Those who are familiar with the European Union's approach, on the other hand, may notice that some elements found in CEPA 1999 also appear in the subsequently adopted REACH regulation (which is discussed in chapter 6).

Like TSCA, CEPA 1999 regulates and distinguishes between "existing" chemical substances (those on Canada's Domestic Substance List (DSL), analogous to the TSCA Inventory) and "new" substances (those that are not listed). CEPA 1999 also provides for a Non-Domestic Substance List (NDSL), which consists of chemicals present in commerce in the United States but not, so far, in Canada; less stringent registration and data requirements apply to these substances than for entirely new substances. Also like TSCA, a notification must be submitted and approved, following a substantive review under new substance notification (NSN) regulations, before a subject new chemical may be commercially manufactured or introduced into commerce. Additionally, CEPA 1999's Significant New Activity (SNAc) notification requirement is somewhat similar to TSCA's SNUR requirement.

As discussed in chapter 3, despite similarities in some respects to TSCA, differences in substance and implementation should not be overlooked. Key in implementing CEPA 1999 are the dual roles of Environment Canada, which has primary responsibility, and of Health Canada, which performs health reviews where needed, as for NSNs. Among the substantive differences, the term "toxicity" has a different meaning under CEPA 1999 than it does under TSCA, where it refers to the inherent toxicity of a substance to humans or the environment. Under TSCA, risk is defined by combining this "toxicity" factor with an exposure term, or the extent to which humans or the environment is exposed to the substance. In Canada, by contrast, under CEPA 1999, the term "toxic" denotes risk in the sense that it encompasses both hazard and exposure. Thus, identifying a substance as "toxic" under CEPA 1999 is significant. As discussed in the chapter, "inherently toxic" also is a term of art under CEPA 1999.

Chapter 3 addresses additional fundamental aspects of the law, including the following:

- the NSN review process;
- risk assessment and prioritization of chemicals and the creation of a Priority Substances List (PSL), which is made up of those chemicals that, based on toxicity, pose the greatest risk to human health and the environment;
- use restrictions, including the role of the SNAc concept;
- chemical and product bans, including the endpoint of virtual elimination;
- testing triggers and requirements;
- recordkeeping;
- risk notification obligations;
- exports and imports;
- confidential business information and trade secrets; and
- enforcement authority and tools and civil and criminal penalties for unlawful actions.

The subsections with pointers for business, a discussion of trends, and list of resources are especially useful for those who need to navigate the chemical regulatory system in Canada.

CENTRAL AMERICA, MEXICO, AND SOUTH AMERICA

For those doing business involving chemicals in or with Mexico or the countries of Central America or South America, or for those advising businesses or otherwise engaging with the chemical management regulatory structures in these countries, the guiding principle is that there are few harmonizing threads among these jurisdictions. Accordingly, as chapters 4 and 5 make evident, there is no substitute for acquiring at least a basic familiarity with the regulatory regime in the country of interest, and it is unwise to go forward based simply on assumptions that regulatory approaches are similar. It is also important to know that in some jurisdictions, chemical substance control regulations are focused as much on impeding the manufacture of illegal drugs as on regulating industrial chemicals.

As noted in chapter 5, in South America, the presence of two trading blocs, each established more than 20 years ago, affects member countries of each by imposing across-the-board regulatory requirements that often do not otherwise exist within the individual jurisdictions. The two blocs are Mercosur (Mercado Común del Sur), made up of Argentina, Brazil, Paraguay, and Uruguay; and the Andean Community of Nations (Comunidad Andina), made up of Bolivia, Colombia, Ecuador, and Peru. Those doing business in or with any of these countries need to be aware of the potential for two sets of regulations, which are not necessarily coordinated.

Some similarities among Central and South American countries are in the absence of certain elements in their chemical regulatory structures. For example, few of these countries have adopted the Globally Harmonized System of Classification and Labelling of Chemicals (GHS), although Uruguay adopted it in 2009 and some others appear poised to do so in the near future. At the same time, Safety Data Sheets (SDS) are necessary in most countries. Each such country has its own requirements, though; usually these SDSs must be written in Spanish. Safety requirements associated with chemicals used in the workplace vary among countries, but they should not be overlooked by those responsible for compliance with hazard communication and other standards.

Some of the countries require registration and permitting of existing and/or new chemicals; examples include Costa Rica (both) and Panama (registration of a "hazardous substance" before its introduction into commerce). Importers of chemicals or chemical substances, as well as manufacturers, need to be aware of regulatory requirements specifically affecting them. For example, in Chile, importers of chemical substances are not subject to formal notification requirements, whereas companies seeking to import or manufacture defined "hazardous substances" must obtain prior government authorization. In Paraguay, however, notification of the import, manufacture, storage, sale, transport, or distribution of "toxic or dangerous substances" is required. As such, although pesticides must be registered, registration is not typically required for commercial and industrial chemicals.

Unique aspects in individual countries should not be overlooked. Venezuela, for instance, presents a variety of challenges, and is something of a moving target. It relinquished its membership in the Andean Community in favor of its anticipated entry into Mercosur, which was delayed. Venezuela's enforcement of chemical-related regulatory measures is notoriously lax, although some regulatory provisions are in place. It is a party to three global multilateral environmental agreements, apart from its own laws and any regional regulatory measures that will come into play when its entry into Mercosur is issued in final form. Overlaid on existing and anticipated regulatory requirements is the increasingly nationalistic approach of the government in recent years, with the effect that prospective new entrants into the marketplace may find themselves as joint venturers with government entities.

Mexico also presents challenges to those unfamiliar with its regulatory landscape, especially in determining the extent and applicability of the multiple sets of overlapping laws and regulations. Some six agencies are involved in administering and enforcing laws relating to chemicals, a regulatory regime addressed more specifically in chapter 4. Mexico has adopted a GHS standard, although compliance is still voluntary.

Europe

The European Union's (EU) Registration, Evaluation, Authorization, and Restriction of Chemicals (REACH) regulation is three decades younger than TSCA and more ambitious by virtue of its multinational coverage, extending to—and harmonizing chemical regulation in—all countries in the EU and the European Economic Area (EEA). REACH replaces an earlier scheme for chemical notification. Its objectives are to enhance safety and the quality and availability of information, encourage alternative means of risk assessment, and promote competition and commerce.

As detailed in chapter 6.1, REACH can be daunting for those inside and outside of Europe because of its relative newness. It is still a work in progress, looking toward a May 31, 2018, deadline, by which time all chemical substances manufactured in, or imported into, the EU with a volume exceeding one metric ton per annum must be registered. Registration dossiers must be supported by documentation submitted to the European Chemicals Agency (ECHA) for review; as discussed in the chapter, a surprisingly small percentage—currently only 5 percent—are subject to detailed review.

Chapter 6.1 illuminates features of REACH that may not be immediately apparent based on expectations formed through experience with TSCA, or that otherwise are novel or potentially confusing. For example, except where an exemption applies, "mixtures" are subject to registration. "Articles" (as defined under the regulation), however, are not. Activities of distributors, although covered under TSCA, are not subject to REACH, unless the distributor also is acting in a capacity, such as that of an importer, that REACH covers. Non-EU entities (if not to be considered downstream users) whose activities are subject to REACH registration either must create or work through an EU resident legal entity for registration purposes or, alternatively, may appoint an Only Representative (OR) to meet these obligations on its behalf. Whereas the TSCA Inventory is a major feature of the U.S. chemical regulatory landscape, no similar inventory is associated with REACH, although lists established under the previous regulatory scheme still exist. Chapter 6.1 explains these and many additional aspects of chemical regulation in Europe.

Separate from REACH is the EU legislation known as the Restriction of Hazardous Substances (RoHS), aimed at restricting the use of hazardous substances in electrical and electronic equipment, and the Waste from Electric and Electronic Equipment (WEEE), aimed at reducing waste from these types of equipment through collection and recycling, as well as more stringent controls on cross-border trading in such wastes. As described in chapter 6.2, the status of RoHS and WEEE as legislative directives means that they do not take immediate and uniform effect in all EU Member States but must be adopted individually by each. The resulting differences among Member States, although typically minor, mean that before placing electrical or electronic equipment or components on the EU market, it is worthwhile to become familiar with each Member State's legislation adopting RoHS and WEEE.

Another piece of EU legislation expected to become increasingly significant is the Biocidal Products Regulation (BPR), which took effect in 2013, superseding, though also building upon, an earlier Biocidal Products Directive (BPD), which it repealed. The BPR's objective is to harmonize the regulation of active substances and biocidal products on an EU-wide basis, rather than solely at the Member State level, and to control treated articles more stringently. Each of these terms of art—"active substance," "biocidal product," and, for the first time, "treated article"—are defined in the regulation. A variety of exemptions are provided for items regulated under other authorities.

As discussed in chapter 6.3, before they are made available on the market in the EU, all active substances must be approved, and all biocidal products must be authorized, as the BPR provides. "Making available on the market" is a term of art, defined in the regulation, and serves as a trigger. The onus is on the applicant to demonstrate, as a prerequisite to approval or authorization, respectively, that the active substance or biocidal product involved is not harmful when used as intended. Even with the EU-wide scope of the BPR, the Member State Competent Authorities take the lead in reviewing and decision making on applications for authorizations or approvals.

The BPR establishes responsibilities for manufacturers of active substances, producers of biocidal products, and importers of active substances and/or biocidal products. For non-EU-manufactured products, the importer is responsible for approval of the active substance; the importer also is responsible for securing authorization where an active substance is imported as a component of a biocidal product. The pathways through which a manufacturer or an importer, as appropriate, may support an application for approval to place an active substance on the market are (1) submitting a dossier to the ECHA, (2) supplying a letter of access to a dossier already approved, or (3) referring to data contained in a dossier for which the data protection period has expired. For biocidal product authorizations, these same requirements apply to all active substances contained in the product. Because the definition of "making available on the market" includes distribution, it is not anticipated that distributors will be involved actively with BPR regulatory responsibilities.

The approval process for active substances involves both scientific testing to determine hazard potential and efficacy testing to confirm effectiveness for the uses covered in the application. Active substances are approved for an initial period of ten years and can be renewed for up to an additional 15 years. Biocidal products are authorized for a period of up to ten years specified by the authorizing Member State Competent Authority and may be renewed for a term determined by the Competent Authority. Chapter 6.3 describes the many other key aspects of the BPR.

Asia

As a cost-effective locale for the manufacture of chemical substances, Asia is a draw for multinational companies and others seeking to import chemicals from abroad. With greater demand and ever-increasing activity in these markets, Asian governments have focused correspondingly on chemical control programs. In some instances, visible and unfortunate incidents have provided an additional incentive for updating and expanding regulatory programs and activities. Perhaps not surprisingly, aspects of some of these programs resemble those of the Asian nations' trading partners, most prominently the EU REACH model. In current approaches, the higher the volume of manufacture or import, the more rigorous the data required to support the registration of a chemical substance. In the context of a particular regulatory scheme, it is important to understand the extent to which the burden is on the manufacturer or importer to demonstrate that a chemical is safe enough to market, as opposed to the regulatory authorities undertaking this role.

Multinational corporations accustomed to Western regulatory systems typically anticipate a highly detailed written regulatory framework characterized by rigid rules, with compliance driven by the issuance or threat of penalties. In Asian nations, the regulatory framework often is markedly different from that in the West, as is the way business proceeds under that framework. Asian regulations typically provide an outline of applicable requirements rather than the far more comprehensive structure characteristic in Western systems, which include strict and frequently inflexible rules backed by penalties to be assessed on transgressors. Asian regulatory schemes rely for their implementation on gray areas—on what is not articulated—to provide flexibility in interpretation, as circumstances may warrant. This volume focuses on the examples of China and South Korea.

China

As discussed in chapter 7.1, chemical regulation in China proceeds primarily under two government decrees. The Ministry of Environmental Protection's (MEP) Decree No. 7, implementing largely through the MEP's Chemical Registration Center (CRC), addresses notification/registration of chemical substances, data submission and testing requirements, the compilation of an Inventory of Existing Chemical Substances in China (IECSC), use restrictions, and related functions. It is important for entities embarking on the notification/registration process to become familiar with data and study requirements. For example, Chinese laboratories must be used for data development to address certain (though not all) endpoints. Data sharing among registrants of the same active ingredient is not required.

The management of chemicals considered to be hazardous (defined as "highly toxic chemicals"), including safe transportation and handling and accident prevention, and the maintenance of a standardized Hazardous Chemicals Inventory, is implemented under the State Council's Decree No. 591. There is a separate registration process, as well as special licensing/permitting for hazardous chemicals. Decree No. 591 also is the primary vehicle in China for implementing the GHS for classification and labeling. The many layers of government agencies involved in carrying out the various missions under Decree No. 591 can make its navigation difficult.

China's approach to chemical regulation is different in many ways from what typically is found in the West, where the rules often are set out in considerable detail. In China, government-issued supporting documents, including guidance, can play an essential role in fleshing out the legislative framework, which may seem like an outline of what is required. The guidance does not necessarily provide cut-and-dried answers, however, and some guidance is left intentionally nonprescriptive. Additionally, experience has shown that in some instances, even unwritten guidance can play a role in the regulatory structure. All of these factors can be challenging for foreign manufacturers and importers of chemical who wish to participate in the marketplace in China. Still, the Chinese approach allows for more flexibility by the CRC and for cooperative problem-solving where uniformly brightline regulatory guidance would not.

South Korea

Of the several sets of regulations in South Korea that address chemicals, the two of greatest interest for purposes here are the Toxic Chemicals Control Act (TCCA), which focuses on managing industrial chemicals, and the Act on Registration and Evaluation, etc. of Chemical Substance, commonly known as Korea REACH or K-REACH, enacted in 2013 and scheduled to take effect January 1, 2015. As discussed in chapter 7.2, K-REACH is intended to be a broad regulatory measure, setting up a process for the registration, evaluation, and assessment of the risks and effects of chemical substances and products containing hazardous chemicals. Both new and existing chemicals will be subject to notification provisions.

The TCCA applies to both existing and new chemicals, but its primary objective is to assess the hazards that may be posed by a new chemical upon notification. As noted in chapter 7.2, its coverage (subject to exceptions for certain substances that fall outside its scope) is broad in terms of the activities involving toxic chemicals, and the persons engaging in those activities, to which it extends. While subject to various other TCCA requirements, existing chemicals are not subject to notification requirements; existing chemicals are those listed on the Korea Existing Chemicals Inventory (KCEI); those listed on recognized inventories maintained in other jurisdictions may be allowed to meet reduced notification requirements. New chemicals, unless exempt, are subject to notification to the Korea Chemicals Management Association (KCMA). In principle, the type of notification is determined by the type of substance and its production volume. Chapter 7.2 lists the core data requirements that apply to support a notification. The TCCA does not expressly provide for data sharing or set out the conditions under which it may occur; such arrangements are left to the affected parties. When K-REACH takes effect in 2015, it will not replace TCCA but will strengthen registration activities for both new and existing substances within its scope.

The chemical registration scheme under K-REACH is central to its operation. The registration requirement applies to all new chemical substances, to existing substances in quantities of one ton or more a year, and to those falling below this threshold quantity that may pose a serious risk to human health or the environment. K-REACH proceeds on the principle that the burden is on the manufacturer or importer to meet data requirements that demonstrate that a chemical substance to be marketed is safe for use. Where sufficient data are not available, a test plan must be submitted. K-REACH provisions allow for a joint submission in response to data requirements and for a Lead Registrant to be appointed. The data compensation period is 15 years, and sharing of vertebrate data is mandated to minimize animal testing. As under similar programs, K-REACH enables non-Korean entities to appoint an Only Representative to act as their agent.

Conclusion

Together, or individually, as needed, each of the focused chapters in this book helps to ground environmental professionals in the diverse regulatory structures that they may encounter in hands-on interactions, whether in the United States or abroad. Knowing what to expect, and how to prepare for it, are essential steps in successfully navigating these systems.

2
U.S. Chemical Control Laws

2.1

Toxic Substances Control Act (TSCA)

Lisa M. Campbell, Sheryl L. Dolan, and Kathleen M. Roberts

Executive Summary

Statutory Overview[1]

The Toxic Substances Control Act (TSCA) was enacted by Congress in 1976 to protect human health and the environment from potentially harmful chemical substances and mixtures.

---

1. Note that this chapter appropriately focuses on the legal aspects of TSCA compliance, but the authors do wish to acknowledge the important role that science plays in determining compliance with its legal requirements. *See*, for example, JOSEPH PLAMONDON, THE UNDERLYING FOUNDATION OF SCIENCE USED IN THE REGULATION OF INDUSTRIAL CHEMICALS (2009), which discusses the importance of such technical issues of chemical nomenclature and risk assessment under TSCA and its global regulatory counterparts.

Basic Provisions of TSCA

TSCA provisions include

- *Section 4*: Authority to promulgate rules requiring manufacturers, importers, and processors to test existing chemical substances or mixtures for their effects on human health and the environment.
- *Section 5*: Authority to regulate new chemical substances prior to their manufacture, import, processing, or distribution for commercial purposes, and to regulate existing chemical substances for significant new uses.
- *Sections 6 and 7*: Authority to regulate the manufacture, processing, distribution, use, or disposal of an existing chemical substance or mixture that the U.S. Environmental Protection Agency (EPA) determines poses an unreasonable risk to human health or the environment; and, for chemical substances or mixtures that the EPA determines will present an unreasonable risk of serious and widespread injury to health and the environment before a final TSCA section 6 rule could be published, the authority to seize that imminently hazardous chemical substance or mixture.
- *Section 8*: Authority to promulgate rules to require manufacturers and processors to collect, maintain, and submit data on certain chemical substances; maintain records of allegations of significant adverse reactions; submit health and safety data on certain chemical substances and mixtures; and report any information that a chemical substance or mixture presents a substantial risk of injury to health or the environment.
- *Sections 12(b) and 13*: Authority to require notification by persons intending to export certain chemical substances as well as promulgate rules regarding the importation of chemical substances.
- *Sections 11, 15, 16, and 17*: Authority to inspect facilities for TSCA compliance; issue civil and criminal penalties for TSCA violations; and seize any chemical substance or mixture manufactured, imported, processed, or distributed in commerce in violation of TSCA.

Scope of TSCA

Title I of TSCA was enacted by Congress in 1976 to protect human health and the environment from potentially harmful chemical substances and mixtures. Three amendments to TSCA have been enacted since that time: (1) Asbestos Hazard Emergency Response Act (AHERA), which is now Title II of TSCA, establishing asbestos abatement programs in schools, was enacted in 1986; (2) Title III of TSCA was enacted in 1988 and provides for indoor radon abatement; and (3) the Lead-Based Paint Exposure Reduction Act, which is now Title IV of TSCA, was enacted in 1992. This book focuses only on Title I.

Scope of Substances Subject to TSCA

TSCA regulates "chemical substances."[2] TSCA defines the term "chemical substance" as "any organic or inorganic substance of a particular molecular identity, including—(i) any combination of such substances occurring in whole or in part as a result of a chemical reaction or occurring in nature, and (ii) any element or uncombined radical."[3] EPA states: "TSCA defines 'chemical substance' broadly and in terms which cover microorganisms as well as traditional chemicals."[4] EPA has likewise stated that all substances, including nanoscale substances that meet the TSCA definition of chemical substance, are subject to TSCA.[5]

Substances Exempt from TSCA

Exclusions from the Definition of "Chemical Substance"

Mixtures: TSCA specifically excludes "mixtures" from the definition of "chemical substance." Mixture is defined as "any combination of

---
2. TSCA § 2(b), 15 U.S.C. § 2601(b) (policy to regulate "chemical substances").
3. TSCA § 3(2)(A), 15 U.S.C. § 2602(2)(A). *See also* 40 C.F.R. § 720.3(e).
4. Microbial Products of Biotechnology; Final Regulation under the Toxic Substances Control Act; Final Rule, 62 Fed. Reg. 17,910, 17,911 (Apr. 11, 1997) (to be codified at 40 C.F.R. pts. 700, 720, 721, 723, and 725) (promulgating final rule under TSCA section 5 to establish notification procedures for review of certain new microorganisms).
5. *See* Control of Nanoscale Materials under the Toxic Substances Control Act, http://www.epa.gov/oppt/nano/.

two or more chemical substances if the combination does not occur in nature and is not, in whole or in part, the result of a chemical reaction."[6] Also included in the definition of a "mixture" is any chemical substance that is the result of a chemical reaction, but that could have been manufactured for commercial purposes without a reaction.[7] Even though "mixtures" as defined under TSCA are excluded from the definition of a "chemical substance," they are still subject to certain TSCA requirements.[8] In addition, each component of a mixture is considered a chemical substance and subject to TSCA requirements.

Chemicals regulated under other statutes: TSCA excludes chemical substances regulated under other statutes, specifically pesticides under the Federal Insecticide, Fungicide, and Rodenticide Act (FIFRA); tobacco and tobacco products; certain materials regulated under the Atomic Energy Act; firearms and ammunition; and foods, food additives, drugs, cosmetics, and devices regulated under the Federal Food, Drug, and Cosmetic Act (FFDCA).[9]

A chemical substance can be a dual-use substance, in which some amount of the chemical is manufactured, used, and regulated under one statute, and some is subject to TSCA. For example, a chemical is subject to the FIFRA when used as a pesticide, but that same substance would be subject to TSCA when used as a general solvent.

Exemptions from Certain TSCA Provisions

In addition to those substances specifically excluded from TSCA, EPA has exempted other types of substances from certain TSCA requirements. Careful review and consideration of the criteria and the specific circumstances of use should be conducted before determining that reporting or recordkeeping requirements do not apply. The chemical substances listed below are not subject to section 5 new chemical notifications and, as such, do not have to be included on the TSCA Chemical Substance Inventory.[10] Although new chemical notifications are not required on chemicals meeting these exemptions, for certain business reasons, some companies decide to submit

---

6. TSCA § 3(8), 15 U.S.C. § 2602(8). *See also* 40 C.F.R. § 720.3(u).
7. TSCA § 3(8), 15 U.S.C. § 2602(8). *See also* 40 C.F.R. § 720.3(u).
8. *See, e.g.*, 40 C.F.R. pt. 799 (mixtures may be subject to TSCA section 4 testing requirements); *id.* at pt. 717 (TSCA section 8(c) recording of significant adverse reactions applies to mixtures).
9. TSCA § 3(2)(B), 15 U.S.C. § 2602(2)(B). *See also* 40 C.F.R. § 720.3(e).
10. TSCA § 8(b), 15 U.S.C. § 2607(b). *See also* 40 C.F.R. § 720.30.

formal notifications and add the chemicals to the Inventory. Chemicals on the Inventory that otherwise meet the exemption criteria are typically not subject to reporting under section 8 Chemical Data Reporting (CDR).

- *Impurity*: A chemical substance that is unintentionally present with another chemical substance;
- By-products that have no commercial purpose (i.e., a chemical substance produced without a separate commercial intent during the manufacture, processing, use, or disposal of another chemical substance or mixture that is not used for any commercial purpose);
- By-products whose only commercial purpose is use by public or private organizations that (1) burn them as fuel, (2) dispose of them as wastes, or (3) extract component chemical substances from them for commercial purposes;
- Any chemical substance resulting from a chemical reaction that occurs incidental to (1) exposure of another chemical substance, mixture, or article to environmental factors (e.g., air, moisture, sunlight); or (2) storage or disposal of another chemical substance, mixture, or article;
- Chemical substances that result from a chemical reaction that occurs either (1) upon end use of another chemical substance, mixture, or article that is not itself manufactured or imported for distribution in commerce or for use as an intermediate (e.g., adhesive, paint, fuel additive, battery); or (2) upon use during the manufacture of an article destined for the marketplace without further chemical change (e.g., inks, drying oils, metal finishing compounds);
- Chemical substances formed when (1) pH neutralizers and other specific reagents, or (2) other chemical substances "intended solely to impart a specific physiochemical characteristic" function as intended;
- Nonisolated intermediates (i.e., any intermediate that is not intentionally removed from the equipment in which it is manufactured, including the reaction vessel and ancillary equipment to the reaction vessel, but not including tanks or other vessels in which the chemical substance is stored);
- Chemical substances manufactured or imported solely for export (provided the chemical substance is labeled to indicate

that it is intended for export, and the chemical substance is distributed domestically only to persons who intend to export the chemical substance directly or process it solely for export);[11]
- *Articles:* An article is defined as a manufactured item (1) that is formed to a specific shape or design during manufacture, (2) that has end-use function(s) dependent in whole or in part upon its shape or design during end use, and (3) that has either no change of chemical composition during its end use or only those changes of composition that have no commercial purpose separate from that of the article and that may occur as described in 40 C.F.R. § 720.36(g)(5), except that fluids and particles are not considered articles regardless of shape or design;[12]
- Research and development (R&D) substances that are manufactured for noncommercial R&D work, or manufactured in small quantities solely for commercial R&D;[13] and
- Certain polymers that meet specific exemption criteria. Under EPA regulations, a "polymer" is

> a chemical substance consisting of molecules characterized by the sequence of one or more types of monomer units and comprising a simple weight majority of molecules containing at least 3 monomer units which are covalently bound to at least one other monomer unit or other reactant and which consists of less than a simple weight majority of molecules of the same molecular weight. Such molecules must be distributed over a range of molecular weights wherein differences in the molecular weight are primarily attributable to differences in the number of monomer units. In the context of this definition, sequence means that the monomer units under consideration are covalently bound to one another and form a continuous string within the molecule, uninterrupted by units other than monomer units.[14]

---

11. 40 C.F.R. § 720.30(e).
12. *See id.* § 720.3(c).
13. *Id.* § 720.36.
14. *Id.* § 723.250(b).

Another category of substances exempt from premanufacture notice (PMN) requirements are naturally occurring substances. For purposes of the TSCA Inventory, naturally occurring substances are a "combination that occurs in nature is a chemical substance and not a mixture."[15] Under EPA's regulations, certain naturally occurring chemical substances are automatically included on the TSCA Inventory. These include chemical substances that are naturally occurring and that are unprocessed or processed only by manual, mechanical, or gravitational means; by dissolution in water; by flotation; or by heating solely to remove water; or that are extracted from air by any means.[16] Because naturally occurring substances are automatically included on the TSCA Inventory, they are considered "existing" substances and are thus exempt from PMN requirements.

Scope of Activities Subject to TSCA

TSCA gives EPA broad authority to limit, restrict, or ban the manufacture, import, processing, distribution, use, or disposal of chemical substances that pose an unreasonable risk to human health or the environment. TSCA also provides EPA with authority to require chemical testing or submission of certain information.

Scope of Persons Subject to TSCA

TSCA may apply to any person who manufactures, processes, distributes in commerce, uses, or disposes of a chemical substance.[17] A "person" is defined broadly as "any natural or juridical person including any individual, corporation, partnership, or association, any State or political subdivision thereof, or any municipality, any interstate

---

15. EPA, TSCA Inventory Representation for Chemical Substances of Unknown or Variable Composition, Complex Reaction Products and Biological Materials: UVCB Substances at Section II.B, http://www.epa.gov/oppt/newchems/pubs/uvcb.txt.
16. 40 C.F.R. § 710.4(b).
17. *See generally* TSCA § 2(a)(2), 15 U.S.C. § 2601(a)(2) (Congress finding that there are some chemical substances and mixtures "whose manufacture, processing, distribution in commerce, use, or disposal may present an unreasonable risk of injury to health or the environment") and TSCA § 2(b), 15 U.S.C. § 2601(b) (Congressional policy statement that "adequate data should be developed with respect to the effect of chemical substances and mixtures on health and the environment and that the development of such data should be the responsibility of those who manufacture and those who process such chemical substances and mixtures").

body and any department, agency, or instrumentality of the Federal Government."[18]

TSCA imposes most of its requirements on manufacturers. "Manufacture" is defined in TSCA as meaning "to produce or manufacture in the United States or import into the customs territory of the United States."[19] This broad definition, therefore, includes importers of chemical substances.[20] Under the implementing regulations of TSCA sections 5 and 8, "manufacture" is defined to mean to manufacture for commercial purposes—"[t]o import, produce, or manufacture with the purpose of obtaining an immediate or eventual commercial advantage for the manufacturer."[21]

A person who contracts with a manufacturer "to manufacture or produce a chemical substance is also a manufacturer if (1) the manufacturer manufactures or produces the substance exclusively for that person, and (2) that person specifies the identity of the substance and controls the total amount produced and the basic technology for the plant process."[22]

Several TSCA provisions, including TSCA sections 4, 5, 6, and 8, apply to processors of chemical substances.[23] TSCA defines "process" as "the preparation of a chemical substance or mixture, after its manufacture, for distribution in commerce—(A) in the same form or physical state as, or in a different form or physical state from, that in which it was received by the person so preparing such substance or mixture, or (B) as part of an article containing the chemical substance or mixture."[24]

---

18. 40 C.F.R. § 710.2(s).
19. *See, e.g., id.* § 710.2(o).
20. TSCA § 3(7), 15 U.S.C. § 2602(7) ("'manufacture' means to import into the customs territory of the United States (as defined in general note 2 of the Harmonized Tariff Schedule of the United States), produce, or manufacture").
21. 40 C.F.R. §§ 704.3, 716.3, 717.3(e), 720.3(r).
22. *Id.* § 720.3(t).
23. *See id.* at pts. 712, 717, 721.
24. TSCA § 3(10), 15 U.S.C. § 2602(10); 40 C.F.R. § 720.3(aa). *See also* EPA, Question and Answer Summary: EPA Seminar on Industry Obligations under TSCA (June 10, 1986), at 1–2.

NOTIFCATION, APPROVAL REQUIREMENTS, AND COMPLIANCE

Summary of Authority Governing the Requirements for Notification

All chemicals used in commerce, not otherwise exempt, must be listed on the TSCA Inventory, per the authority under TSCA section 5(a), which states that no person may manufacture a chemical substance that is not on the TSCA Inventory as authorized under section 8(b).[25]

TSCA section 8(b) directs EPA to "compile, keep current, and publish a list of each chemical substance which is manufactured or processed in the United States."[26] This list is known as the TSCA Chemical Substance Inventory (TSCA Inventory).

TSCA Inventory and "New" and "Existing" Substances

TSCA defines chemical substances that are listed on the TSCA Inventory as "existing" chemical substances. Chemical substances that are not listed on the TSCA Inventory are considered "new" chemical substances.[27] In order for a "new" chemical to be added to the Inventory, it must go through a review by EPA under section 5, in which a PMN and subsequent notice of commencement (NOC) have been submitted.

The initial TSCA Inventory compiled by EPA covered chemical substances that had been in commerce since January 1975, based on information reported to EPA in 1978. The chemicals included in the initial Inventory are sometimes referred to as "grandfathered" chemicals. The additional chemicals that have been subject to review under section 5's new chemical notification are designated on the Inventory listing as "P," which stands for commenced PMN substance.

TSCA Inventory—Public and Confidential Lists

The TSCA Inventory has two components. The TSCA Public Inventory includes all existing chemical substances whose identity has not been claimed as confidential business information (CBI). The TSCA Confidential Inventory includes all existing chemical

---

25. TSCA § 5(a), 15 U.S.C. § 2604(a); 40 C.F.R § 720.22.
26. TSCA § 8(b)(1), 15 U.S.C. § 2607(b)(1).
27. 40 C.F.R. § 720.3(v).

substances whose identity has been claimed as CBI.[28] Chemical substances on the TSCA Public Inventory are listed by a specific chemical name and a Chemical Abstracts Service Registry Number (CASRN)—a unique number assigned by the Chemical Abstracts Service (CAS) to a specific chemical substance.[29] Chemical substances whose identities are claimed confidential, however, are listed in the TSCA Public Inventory by an accession number, assigned by EPA, and a generic chemical name, proposed by the submitter.[30]

Determining Inventory Status

Manufacturers and importers must determine the TSCA Inventory status of all chemical substances commercially manufactured or imported, including chemical substances that are components of a mixture and chemical substances contained in articles that are designed to be used or released. This determination of a chemical substance's TSCA Inventory status must be made for a chemical substance prior to manufacture or import for a nonexempt purpose.

- *Public Inventory:* The TSCA Public Inventory can be accessed through EPA's website at http://www.epa.gov/oppt/existingchemicals/pubs/tscainventory/howto.html. If a chemical substance is found to be listed on the TSCA Inventory, the manufacturer or importer must also ensure that other TSCA requirements are met before authorizing commercial manufacture or import, including existing TSCA section 5 Significant New Use Rule (SNUR) requirements and TSCA section 4 testing requirements.
- *Confidential Inventory:* If a search demonstrates that a chemical substance is not listed on the public TSCA Inventory, the manufacturer or importer should determine whether the chemical substance is on the confidential portion of the TSCA Inventory, which is only accessible to EPA. To do so, the manufacturer or importer can request that EPA search the TSCA Confidential Inventory by filing a Bona Fide Intent (BFI) request.[31] EPA is required to respond to the BFI request

---

28. Id. § 720.25(b)(1).
29. Id.
30. Id.
31. Id. § 720.25.

by searching the TSCA Confidential Inventory within 30 days of its receipt.[32] If the chemical substance is listed on the TSCA Confidential Inventory, the manufacturer or importer also should determine whether the chemical substance is subject to TSCA rules or orders before authorizing manufacture or import. If EPA's search of the TSCA Confidential Inventory reveals that the chemical substance is not listed, the manufacturer or importer must determine whether the chemical substance is subject to any exemptions from PMN requirements, or file a PMN, before commercial manufacture or import can commence.

Exemptions from New Chemical Notification Requirements

As noted in the "Scope of TSCA" section of this chapter, certain chemical substances are exempted from listing in the TSCA Inventory and thus do not require new chemical notification. As outlined below, some chemical substances that are not listed on the TSCA Inventory and do not meet the TSCA exemptions may still, however, be exempt from PMN and related TSCA section 5 requirements.

Exemptions for Chemical Substances Having No Separate Commercial Purpose and Other Related Exemptions

Substances that are not manufactured or imported for a separate commercial purpose are exempt from PMN requirements. A manufacturer or importer must determine if the chemical substance to be manufactured or imported is eligible for this exemption, which applies to the following categories of substances listed at 40 C.F.R. § 720.30(h):

- *Impurities*: See Scope of TSCA, Substances Exempt from TSCA, Exemptions from Certain TSCA Provisions
- *By-products*: See Scope of TSCA, Substances Exempt from TSCA, Exemptions from Certain TSCA Provisions
- *Chemical substances resulting from certain chemical reactions*: See Scope of TSCA, Substances Exempt from TSCA, Exemptions from Certain TSCA Provisions
- *Nonisolated intermediates*: See Scope of TSCA, Substances Exempt from TSCA, Exemptions from Certain TSCA Provisions

---

32. Id. § 720.25(b)(8).

Other exemptions set forth at 40 C.F.R. § 720.30 include

- *Mixtures and substances that are not defined as "chemical substances" under TSCA:* See Scope of TSCA, Substances Exempt from TSCA, Exclusions from the Definition of "Chemical Substance"
- *Chemical substances manufactured (including imported) solely for export:* See Scope of TSCA, Substances Exempt from TSCA, Exemptions from Certain TSCA Provisions

These are self-executing exemptions, which have no specific recordkeeping requirements. To demonstrate compliance with TSCA section 5, however, a manufacturer or importer may wish to retain documentation sufficient to demonstrate that the chemical substance is eligible for such a PMN exemption.

Polymer Exemption

Substances that meet polymer exemption criteria are exempt from PMN requirements.[33] A manufacturer or importer must determine if the chemical substance to be manufactured or imported is eligible for the polymer exemption. See Scope of TSCA, Substances Exempt from TSCA, Exemptions from Certain TSCA Provisions for definition of polymer.

EPA has issued guidance on how to identify a polymer and whether the polymer exemption can be applied.[34] Polymers that are excluded from the polymer exemption and thus must be listed on the TSCA Inventory if not otherwise exempt are (1) cationic polymers; (2) polymers that do not contain two or more of the atomic elements; (3) polymers that substantially degrade, decompose, or depolymerize; (4) certain water-absorbing polymers; or (5) polymers manufactured or imported from monomers, and/or polymers manufactured from other reactants that are not listed on the TSCA Inventory.[35]

The polymer exemption is also a self-executing exemption. If a chemical substance meets the definition of a polymer and at least one of the three polymer exemption criteria, the manufacturer or importer must maintain documentation sufficient to demonstrate that a chemical

---

33. *Id.* § 723.250(a); 60 Fed. Reg. 16,316 (Mar. 29, 1995).
34. *See* EPA, POLYMER EXEMPTION GUIDANCE MANUAL 18–24 (June 1997), http://www.epa.gov/oppt/newchems/pubs/polyguid.pdf.
35. 40 C.F.R. § 723.250(d). *See also* EPA, 1995 TSCA PMN Rule Amendments: Questions and Answers (Draft) (May 4, 1995), at 4–5; 60 Fed. Reg. at 16,318–20.

substance meets the polymer exemption's eligibility criteria and otherwise is in compliance with the exemption.[36] TSCA also requires for recordkeeping purposes that a company representative sign a certification statement stating that (1) the polymer meets the eligibility criteria of 40 C.F.R. § 723.250(e); (2) the polymer is manufactured or imported for a commercial purpose other than R&D; and (3) all information in the certificate is truthful.[37] In addition, manufacturers and importers must prepare a polymer exemption report, to be submitted to EPA by January 31 of each year, on polymers being manufactured or imported for the first time under the polymer exemption.[38]

R&D Exemption

Chemical substances that meet R&D criteria are exempt from PMN requirements[39] and are a self-executing exemption. To qualify for the R&D exemption, a chemical substance must:

- Satisfy the criteria for an R&D substance;
- Be used under the supervision of a technically qualified individual (TQI); and
- Comply with EPA requirements on volume, uses, labeling, handling and distribution, disposal, and recordkeeping.[40]

With regard to the first requirement, to qualify as an R&D substance, the chemical substance must be manufactured or imported only in "small quantities" for purposes of scientific experimentation or analysis or chemical research on, or analysis of, such substance or another substance, including such research or analysis for the development of a product.[41] The term "small quantities" is not defined quantitatively, but qualitatively, as quantities "that are not greater than reasonably necessary" for R&D purposes.[42] EPA states the following regarding what constitutes R&D activities:

---

36. 40 C.F.R. § 723.250(j); 60 Fed. Reg. at 16,324.
37. 40 C.F.R. § 723.250(h).
38. *Id.* § 723.250(f).
39. TSCA § 5(h)(3), 15 U.S.C. § 2604(h)(3); 40 C.F.R. § 720.36.
40. 40 C.F.R. § 720.36(a). *See also* EPA, New Chemical Information Bulletin: Exemptions for Research and Development and Test Marketing (Nov. 1986) [hereinafter R&D Exemption Bulletin].
41. 40 C.F.R. § 720.3(cc).
42. *Id. See also* R&D Exemption Bulletin, *supra* note 40, at 5.

R&D activities can be distinguished by their special purpose. Activities are considered R&D if they are intended solely as scientific experimentation, research, or analysis. . . . In the course of R&D activities, professional researchers using the substances must be engaged in collecting information about and monitoring tests of the chemical substances being studied or developed.[43]

Substances that satisfy the criteria for an R&D substance must be used by or under the supervision of a TQI, defined as a person:

(1) who, because of education, training, or experience, or a combination of these factors, is capable of understanding the health and environmental risks associated with the chemical substance which is used under his or her supervision, (2) who is responsible for enforcing appropriate methods of conducting scientific experimentation, analysis, or chemical research to minimize such risks, and (3) who is responsible for the safety assessments and clearances related to the procurement, storage, use, and disposal of the chemical substance as may be appropriate or required within the scope of conducting a research and development activity.[44]

The TQI thus is responsible for compliance with volume, prescribed uses,[45] hazard evaluation and notifications,[46] disposal,[47] and recordkeeping requirements.[48]

Exemptions Requiring EPA Approval

Exemptions requiring EPA approval are low volume exemptions (LVE), low release and low exposure exemptions (LoREX), and test marketing exemptions (TME). For these exemptions, a manufacturer or importer must submit, and EPA must approve, an exemption application before a company may commence the manufacture or import of the new chemical substance.

---

43. R&D Exemption Bulletin, *supra* note 40, at 2.
44. 40 C.F.R. § 720.3(ee).
45. Id. § 720.36(d), (e).
46. Id. § 720.36(a)(2), (b), (c).
47. Id. § 720.36(e).
48. Id. § 720.78(b).

LVE and LoREX

Substances that satisfy the criteria for an LVE or LoREX are exempt from PMN requirements.[49] Instead, EPA will conduct expedited 30-day reviews of chemical substances manufactured or imported.[50] To obtain a LVE, a manufacturer or importer must show that the chemical substance: (1) will be manufactured or imported in quantities of 10,000 kilograms or less per year; and (2) does not present an unreasonable risk of injury to health or the environment.[51] To obtain a LoREX, a manufacturer or importer must show (1) for consumers and the general population, no dermal or inhalation exposure and no exposure in drinking water greater than one milligram per year; (2) for workers, no dermal or inhalation exposure; (3) no releases to ambient surface water in concentrations above one part per billion; (4) no releases to the ambient air from incineration in excess of one microgram per cubic meter; (5) no releases to groundwater, land, or a landfill unless it is demonstrated there is negligible groundwater migration potential; and (6) no unreasonable risk of injury to health or the environment.[52]

For any chemical substance for which a company has obtained an LVE or LoREX, the company must prepare a label or written notice to any processor or user of that chemical substance: (1) notifying the processor or user of all applicable use, exposure, and related controls specified in the LVE or LoREX; and (2) stating that the chemical substance may be used only in accordance with the restrictions specified in the LVE or LoREX.[53]

Test Marketing Exemption

The TME—which is subject to an expedited 45-day review period—applies to chemical substances manufactured or imported for test marketing purposes that do not present an unreasonable risk of injury to health or the environment as a result of the test marketing.[54] To

---

49. *Id.* § 723.50(a); 60 Fed. Reg. 16,336 (Mar. 29, 1995).
50. 40 C.F.R. § 723.50(g).
51. *Id.* § 723.50(c)(1), (d).
52. *Id.* § 723.50(c)(2), (d). *See also* 60 Fed. Reg. at 16,337 ("[p]rospective submitters should be mindful that the principal focus of the LoREX exemption is on release and exposure, not toxicity").
53. 40 C.F.R. § 723.50(k).
54. TSCA § 5(h)(1), 15 U.S.C. § 2604(h)(1); 40 C.F.R. § 720.38(a).

satisfy the criteria for claiming the TME, a manufacturer or importer must distribute a predetermined amount of the chemical substance "to no more than a defined number of potential customers to explore market capability in a competitive situation during a predetermined testing period."[55]

If EPA notifies the company that the exemption is granted, or if the review period expires without notice from EPA, manufacture or import of the chemical substance may commence, consistent with the terms of the LVE, LoREX, or TME.[56]

Certain changes in use, exposure, releases, controls, production volume, or other exemption criteria require submission of a new exemption application.

Exemptions Requiring EPA Approval: Recordkeeping

Manufacturers and importers must maintain for each chemical substance manufactured or imported under an LVE, LoREX, or TME all documents regarding an LVE, LoREX, or TME for such chemical substance. For chemical substances qualifying for the LVE, manufacturers and importers must also maintain records of annual production or import volume concerning that chemical substance for five years after their preparation, or throughout the period the chemical substance is manufactured pursuant to the LVE or LoREX, whichever is longer.[57] For chemical substances qualifying for the TME, manufacturers and importers must also maintain documentation needed to demonstrate compliance with any restrictions imposed by the TME concerning that chemical substance for five years from the final date of manufacture or import under the exemption.[58]

Standard of Review/Burden of Proof

As stated in TSCA section 2(b)(2) and 2(b)(3), the standard for regulation under TSCA is "an unreasonable risk of injury to health or the environment," but that authority over substances should be exercised in a such a manner as not to impede unduly or create unnecessary

---

55. 40 C.F.R. § 720.3(gg). *See also* EPA, New Chemical Information Bulletin: Exemptions for Research and Development and Test Marketing, at 10–11 (explaining differences between scope of TME and R&D exemptions).
56. 40 C.F.R. § 723.50(g)(2).
57. *Id.* § 723.50(n).
58. *Id.* § 720.78(c).

economic barriers to technical innovation while fulfilling the primary purpose of the Act.[59] This standard can vary depending on the particular action at issue; for example, EPA can issue a SNUR under TSCA section 5(b)(4) based on a finding that a chemical "may present" an unreasonable risk of injury to health or the environment, while EPA can promulgate a section 6 rule upon a reasonable basis that a chemical substance "will present" an unreasonable risk of injury to health or the environment.

Inventories or Other Listing Procedures

TSCA section 5 governs the manufacture and import of new chemical substances in the United States, as well as the manufacture, import, and processing of existing chemical substances for new uses in the United States.[60] Chemical substances manufactured or imported that are not listed on the TSCA Inventory and that are not otherwise exempt from TSCA Inventory listing, are considered "new" chemical substances subject to TSCA section 5 PMN requirements.[61] Chemical substances listed on the TSCA Inventory—"existing" chemical substances—also may be subject to the TSCA section 5 SNUR requirements.[62]

PMN

Manufacturers and importers shall submit a PMN for any chemical substance to be manufactured or imported that is not listed on the public or confidential versions of the TSCA Inventory and that is not eligible for an exemption from PMN requirements. The PMN form seeks information on the submitter's identity, the chemical substance's identity, production volume, uses, exposures, and environmental fate.[63] TSCA does not require a submitter to test new chemical substances before submitting a PMN. Health and safety data relating to a new chemical substance's health or environmental effects that are in a submitter's possession or control, however, must be submitted with the PMN.[64] The submitter must provide this information to "the extent it is known

---

59. TSCA § 2(b), 15 U.S.C. § 2601(b).
60. TSCA § 5, 15 U.S.C. § 2604.
61. 40 C.F.R. § 720.3(v).
62. *See, e.g., id.* at pt. 721.
63. *Id.* at pt. 720, subpt. C; EPA Form 7710-25.
64. 40 C.F.R. § 720.50.

to or reasonably ascertainable by the submitter."[65] The review period for a PMN is 90 days, unless extended by EPA for up to an additional 90 days.[66] The PMN must be accompanied by a fee, which is currently $2,500.[67] This fee can be reduced under certain circumstances: (1) if a company meets the small business criteria, the fee is reduced to $100; (2) if a company submits a consolidated PMN for up to six substances with similar use, structure, and probable toxicology, the total fee for all PMNs is $2,500 (or for a small business, $100); and (3) if the PMN is for an intermediate substance that is submitted with a final product PMN, the fee is reduced to $1,000 for the intermediate substance.[68]

Prior to submitting a PMN, a manufacturer or importer may call one of EPA's prenotice coordinators and discuss EPA's likely decision on the PMN application (e.g., approval, TSCA section 5(e) Consent Order), considering the availability of toxicological data and type of chemical substance involved.

If EPA notifies the manufacturer or importer that the PMN is granted, or if the review period expires without notice from EPA,[69] manufacture or import of the chemical substance may commence as allowed under the conditions of the PMN.

The manufacturer or importer must compile and maintain records of production or import quantities for the first three years of manufacture or import for each PMN chemical substance manufactured or imported.[70] Certain changes in use, exposure, releases, controls, production volume, or other criteria require submission of a new PMN application only to the extent that such changes became known to the PMN submitter during the PMN review period.

NOC Preparation and Submission to EPA

Within 30 days of the first commercial manufacture (date of completion of production lot) or import (date new chemical substance clears U.S. Customs) of a chemical substance for which EPA has approved a PMN, the manufacturer or importer must prepare a NOC.[71] Upon receipt of the NOC, EPA places the PMN chemical substance on the TSCA Inventory.

---

65. Id. §§ 720.3(p), 720.40(d).
66. TSCA § 5(a), (c), 15 U.S.C. § 2604(a), (c); 40 C.F.R. § 720.75.
67. 40 C.F.R. § 700.45; EPA Form 7710-25.
68. 40 C.F.R. §§ 700.43, 700.45; 48 Fed. Reg. 21,722, 21,734–35 (May 13, 1983).
69. 40 C.F.R. § 720.75(d).
70. Id. § 720.78(a)(2).
71. Id. § 720.102(c).

Microbial Products of Biotechnology

In 1997, EPA issued comprehensive regulations on its notification process and reporting requirements for the review of certain "new" microorganisms under TSCA section 5.[72] EPA's regulations regarding microorganisms are codified at 40 C.F.R. part 725. A new microorganism, like a new chemical substance, is one that is not included on the TSCA Inventory.[73] New microorganisms for which manufacturers and importers are required to report under TSCA section 5 are those that are intergeneric, that is, those formed by the deliberate combination of genetic material originally isolated from organisms in different taxonomic genera.[74]

Persons who manufacture, import, or process new microorganisms for a commercial purpose must submit a microbial commercial activity notice (MCAN) instead of the traditional PMN form.[75] Review, notification, and listing procedures for microorganisms considered "new" substances are similar to the section 5 new chemical process.

Exemptions: Information regarding exemptions for R&D test marketing and general exemptions for new microorganisms is available under 40 C.F.R. 725, subparts E, F, and G.

EPA Review and Response for Chemicals Added to the Inventory

EPA has developed the PMN review process to estimate the risk attributable to a new chemical substance and to determine whether an unreasonable risk of injury to health or the environment may occur if the chemical substance is manufactured, imported, processed, or distributed in commerce. In most cases, EPA reviews PMN submissions and does not elect to control the manufacture, processing, distribution, use, or disposal of the new chemical substance. After expiration of the PMN period, therefore, the PMN submitter may commence commercial manufacture without restriction. In the event that EPA identifies potential risks associated with new chemical substances, however, EPA

---

72. Id. at pt. 721; Microbial Products of Biotechnology; Final Regulation under the Toxic Substances Control Act, 62 Fed. Reg. 17,910 (Apr. 11, 1997).
73. EPA provides for a procedure similar to a BFI request to determine if a microorganism is listed on the TSCA Confidential Inventory. See 40 C.F.R. § 725.15.
74. Id. § 725.3. See also id. § 725.8(c) (listing microorganisms that are not subject to 40 C.F.R. pt. 725).
75. Id. § 725.105 (describing those persons who must submit MCANs).

has authority, under TSCA section 5(e) and (f), to take actions to prevent or mitigate those risks.

TSCA Section 5(e) Consent Orders

TSCA section 5(e) grants to EPA the authority to issue administrative orders controlling new chemical substances where it finds (1) there is insufficient information reasonably to evaluate the risk, and (2) either the chemical may present an unreasonable risk to health or the environment, or it will be produced in substantial quantities that will enter the environment or to which there will be substantial or significant human exposure.[76] In its order, EPA can ban or limit the manufacture, processing, distribution, use, or disposal of the chemical. EPA must propose section 5(e) orders prior to the expiration of the 90-day PMN review period, and the orders become effective upon the expiration of that period unless the manufacturer files objections to the order.[77]

As a matter of practice, however, EPA typically enters into a consent order with manufacturers, rather than issuing a unilateral section 5(e) order. Under the consent order procedure, the manufacturer agrees to restrict the manufacture, processing, distribution, use, or disposal of a new chemical substance as provided in the order. The consent orders permit the manufacturer to distribute the chemical substances, subject to various restrictions, pending the development of data necessary to evaluate potential hazards.[78]

TSCA Section 5(f) Orders

EPA has authority under section 6 of TSCA to promulgate rules to protect the public against unreasonable risks presented by chemical substances. If EPA determines that a new chemical substance presents or will present an unreasonable risk *before* it can issue a section 6 rule, EPA must take action under TSCA section 5(f) to control the risk.[79] EPA can, at its election, (1) issue a proposed rule (which becomes effective immediately upon publication in the *Federal Register*) to limit or delay the manufacture, processing, distribution, use,

---

76. *See, e.g.*, Letter from Charles L. Elkins, EPA, to Geraldine V. Cox, Ph.D., CMA (Sept. 22, 1988) (discussing what EPA considers to be "substantial quantities" and "substantial or significant human exposure"); 58 Fed. Reg. 28,736 (May 14, 1993) (criteria for evaluating substantial or significant human exposure under TSCA section 4).
77. TSCA § 5(e)(1)(C), 15 U.S.C. § 2604(e)(1)(C).
78. 40 C.F.R. § 790.22. *See also* EPA, Enforceable Consent Agreements (ECAs), http://www.epa.gov/opptintr/chemtest/pubs/eca.html.
79. TSCA § 5(f)(1), 15 U.S.C. § 2604(f)(1).

or disposal of the chemical substance, or (2) issue a proposed order (which becomes effective upon expiration of the PMN review period) to prohibit the manufacture, processing, use, or distribution in commerce of the chemical substance, or apply for an injunction to prohibit such manufacture, processing, or distribution in commerce.[80]

Documents EPA requires manufacturers and importers to maintain to demonstrate compliance with TSCA section 5 include

- BFI requests;
- Any letter to the foreign manufacturer or supplier requesting that chemical identity information concerning the chemical substance be submitted to EPA;
- Any EPA response to the BFI request;
- PMN applications;
- Records of annual production or import volume for the first three years of production or import (including the date of commencement of manufacture or import);
- Communications from EPA, if any;
- Copies of data concerning that chemical substance in the manufacturer's or importer's possession as described by 40 C.F.R. § 720.50(b) (e.g., health and safety data in its possession or control, or descriptions of other health and safety data that are known to or are reasonably ascertainable by the manufacturer or importer); and
- Manufacturing records concerning that chemical substance.[81]

These records must be maintained for five years from the date of commencement of manufacture or import.[82]

REGULATION OF SUBSTANCES

Authority to Regulate Inventory/Listed Substances

For chemicals already listed on the Inventory, EPA has authority under TSCA section 5(a) to regulate significant new uses of existing chemicals, implement risk management measures under TSCA

---

80. TSCA § 5(f)(2)–(3), 15 U.S.C. § 2604(f)(2)–(3).
81. See 40 C.F.R. § 720.78.
82. Id.

section 6, and initiate certain actions to address imminent hazards under section 7.

Prioritization and Risk Assessment

TSCA does not require a prioritization of existing chemicals or specified process for conducting risk assessment. There is, however, no language in TSCA that prohibits EPA from taking those actions. Since TSCA was enacted, EPA has implemented several programs that have attempted to provide a more systematic review of existing chemicals to identify candidates for risk assessment and potential risk management. These programs include the High Production Volume Chemicals Program,[83] the Chemical Assessment and Management Program,[84] and TSCA Work Plan Chemicals.[85]

Use Restrictions

Section 5(a) requires manufacturers, importers, and processors of existing chemicals to provide notice to EPA of any use of a substance that EPA has determined is a "significant new use."[86] A determination that a use is significant and new must be made by rule, known as a SNUR. TSCA does not establish standards or criteria for establishing when a new use is deemed "significant," but it requires EPA to consider "all relevant factors" before promulgating a SNUR.[87]

When issuing a chemical-specific SNUR, EPA prescribes recordkeeping requirements from among those listed in its generic regulations.[88] EPA may require manufacturers, importers, and processors of chemicals to keep for five years records documenting dates and volume of manufacture and import; records documenting the names of customers; records documenting the establishment of hazard communication programs, or the implementation of worker safety requirements; and records documenting compliance with other applicable SNUR requirements.[89]

---

83. See EPA, High Production Volume Challenge, http://www.epa.gov/hpv/.
84. See EPA, Chemical Assessment and Management Program, http://www.epa.gov/champ/.
85. See EPA, TSCA Work Plan Chemicals, http://www.epa.gov/oppt/existingchemicals/pubs/workplans.html.
86. TSCA § 5(a)(1)(B), 15 U.S.C. § 2604(a)(1)(B).
87. TSCA § 5(a)(2), 15 U.S.C. § 2604(a)(2).
88. 40 C.F.R. § 721.125.
89. Id.

When EPA promulgates a SNUR designating a significant new use for a particular chemical substance, manufacturers, importers, and processors of that chemical substance must provide to EPA a significant new use notice (SNUN) at least 90 days before any manufacture, import, or processing for that use.[90] Although similar to PMN requirements, SNUR requirements apply to processors as well as manufacturers and importers.[91] Moreover, SNUR obligations are imposed on every person subject to the SNUR, not just the initial manufacturer/importer. Manufacturers, importers, and processors who themselves do not engage in the significant new use must nevertheless file a SNUN unless they can show that (1) they have notified in writing each person who purchases or otherwise receives the chemical from them of the regulatory requirement; (2) each recipient of the chemical has knowledge of the requirements; or (3) each recipient cannot undertake any significant new use.[92] SNUNs are submitted on the PMN form and contain virtually the same information as PMNs.

As with PMNs, a manufacturer can manufacture, import, or process the chemical for the significant new use at the end of the 90-day review period, absent EPA action to regulate the chemical. Substances currently subject to SNURs are listed at 40 C.F.R. part 721, subpart E.

TSCA section 6 authorizes EPA to restrict or ban the manufacture, processing, or distribution in commerce of chemical substances or mixtures upon showing that the activity "presents or will present an unreasonable risk of injury to health or the environment."[93] In exercising its section 6 authority, EPA can choose among a number of regulatory options. EPA can prohibit or limit the amount of manufacture, processing, or distribution of a substance; prohibit or limit the manufacture, processing, or distribution of specific uses; prescribe labeling or warning requirements; require recordkeeping; prohibit or otherwise regulate the chemical's disposal; or require that manufacturers notify purchasers or the general public about the risks involved with specific chemicals and replace or repurchase the chemical if requested. EPA must select the least burdensome of these options that is adequate to achieve the regulatory objectives.

Before taking any of these actions, however, EPA must initiate a rulemaking and make findings regarding (1) the effects of the

---

90. Id. § 721.25(a).
91. Id. § 721.1(a).
92. Id. § 721.5(a)(2).
93. TSCA § 6(a), 15 U.S.C. § 2605(a).

substance or mixture on health and the magnitude of human exposure; (2) the effects of such substance or mixture on the environment and the magnitude of environmental exposure; (3) the benefits of the substance or mixture for various uses and the availability of substitutes for such uses; and (4) the reasonably ascertainable economic consequences of the rule, after consideration of the effect on the national economy, small business, technological innovation, the environment, and public health.[94]

Chemical or Product Bans

Although section 6 provides EPA authority to ban chemical products, it has not yet successfully used that authority. EPA issued a final rule under section 6 in 1989 to ban the manufacture, import, and processing of almost all asbestos-containing products manufactured in the United States.[95] Most of the rule ultimately was set aside, however, by the U.S. Court of Appeals for the Fifth Circuit.[96]

EPA has successfully used its section 6 authority to restrict certain uses of chemical substances. EPA has prohibited the manufacture, processing, and distribution in commerce of fully halogenated chlorofluorocarbons (CFC) for aerosol propellant uses, except for certain limited uses.[97] EPA also has used its section 6 authority to issue rules to reduce the potential hazards arising from the use of metalworking fluids.[98] EPA additionally has promulgated regulations under section 6(a) to address exposure to airborne asbestos in school buildings[99] and the use of hexavalent chromium chemicals in certain heating, ventilation, air

---

94. TSCA § 6(c)(1), 15 U.S.C. § 2605(c)(1).
95. Asbestos; Manufacture, Importation, Processing, and Distribution in Commerce Prohibitions; Final Rule, 54 Fed. Reg. 29,460 (July 12, 1989) (codified at 40 C.F.R. pt. 763).
96. Corrosion Proof Fittings v. EPA, 947 F.2d 1201 (5th Cir. 1991).
97. Because CFCs are now regulated under the Clean Air Act (CAA), EPA revoked its TSCA regulations for CFCs. Chemical Substances; Deletion of Certain Chemical Regulations; Technical Amendments to the Code of Federal Regulations, 60 Fed. Reg. 31,917, 31,919 (June 19, 1995) (codified at 40 C.F.R. pts. 61, 704, 710, 712, 762, 763, 766, 790, 795, 796, 797, 798, and 799) (elimination of part 762 regulations).
98. 40 C.F.R. pt. 747.
99. Asbestos; Friable Asbestos-Containing Materials in Schools; Identification and Notification, 47 Fed. Reg. 23,360 (May 27, 1982) (codified at 40 C.F.R. pt. 763). When Congress enacted the AHERA to address asbestos hazards, however, EPA revoked its TSCA regulations and promulgated regulations implementing the new statute.

conditioning, and refrigeration systems based on its findings that such chemicals are human carcinogens.[100]

TSCA section 6(e) directs EPA to regulate the manufacture, importation, processing, distribution in commerce, use, disposal, and labeling of polychlorinated biphenyl (PCB) substances and items containing PCBs. Because of its concern about the health effects potentially associated with PCBs, Congress also banned, effective January 1, 1978, virtually all manufacture, processing, and distribution of PCBs in commerce.[101] PCBs still in use can remain in use provided they are used in a "totally enclosed manner," as determined by EPA.[102] EPA has promulgated additional regulations on the labeling, storage and disposal, spills, and tracking of PCB materials.[103]

TSCA section 7 permits EPA to initiate a district court action to seize any imminently hazardous chemical substance or mixture or any article containing such substance or mixture.[104] EPA also may seek other appropriate relief, including a requirement that manufacturers or others give notice to purchasers or the public of the risks associated with the substance, mixture, or article, as well as recall, replace, or repurchase the substance mixture or article.

Test Data Development

Authority to Require Testing/Testing Triggers

Section 4 of TSCA was enacted by Congress in response to the concern that the effects of chemical substances and mixtures on human health and the environment were insufficiently characterized or understood. This section gives EPA the authority to require the development of adequate test data on the health and environmental effects of such substances.[105]

Under TSCA section 4, EPA can require manufacturers, importers, and, in some cases, processors to conduct testing of any chemical

---

100. 40 C.F.R. pt. 749, subpt. D.
101. TSCA § 6(e)(2)(A), 15 U.S.C. § 2605(e)(2)(A).
102. In its PCB regulations, EPA has defined "totally enclosed manner" as "any manner that will ensure no exposure of human beings or the environment to any concentration of PCBs." 40 C.F.R. § 761.3.
103. Id. at pt. 761.
104. TSCA § 7(a), 15 U.S.C. § 2606(a).
105. TSCA § 4, 15 U.S.C. § 2603.

substance or mixture for which EPA makes certain findings. To require testing, EPA must first find that "there are insufficient data and experience upon which the effects of such manufacture, distribution in commerce, processing, use, or disposal of such substance or mixture or of any combination of such activities on health or the environment can reasonably be determined or predicted,"[106] and "testing of such substance or mixture with respect to such effects is necessary to develop such data."[107] EPA must also find that the chemical substance or mixture may present "an unreasonable risk" to human health or the environment,[108] or is produced in substantial quantities that could result in substantial or significant human exposure or environmental release.[109] To require testing of a mixture, EPA must find in addition that the effects caused by the mixture's manufacture, distribution, processing, use, or disposal may not be reasonably and more efficiently determined by testing the chemical substances that comprise the mixture.[110]

TSCA section 4 authorizes EPA to impose testing requirements on manufacturers (including importers), as well as processors. Manufacturers, however, are primarily responsible for testing. Under EPA's regulations, manufacturers are responsible for testing where testing is necessary to assess risk primarily from manufacturing; from both processing and manufacturing; or from distribution, use, and/or disposal activities.[111] Testing obligations will be imposed on processors where the testing is being conducted to evaluate the risks associated with processing of the chemical, or where no manufacturer submits a notice of intent to conduct testing.[112]

Chemical Substances/Products That Must Be Tested/Implementing Test Requirements

To determine which of the thousands of industrial chemicals should be tested by their manufacturers, TSCA established a committee of

---

106. TSCA § 4(a)(1)(A)(ii), 15 U.S.C. § 2603(a)(1)(A)(ii).
107. TSCA § 4(a)(1)(A)(iii), 15 U.S.C. § 2603(a)(1)(A)(iii).
108. TSCA § 4(a)(1)(A)(i), 15 U.S.C. § 2603(a)(1)(A)(i). See also Chem. Mfrs. Ass'n v. EPA, 859 F.2d 977 (D.C. Cir. 1988).
109. TSCA § 4(a)(1)(B)(i), 15 U.S.C. § 2603(a)(1)(B)(i). See also Chem. Mfrs. Ass'n v. EPA, 899 F.2d 360 (5th Cir. 1990); 58 Fed. Reg. at 28,736.
110. TSCA § 4(a)(2), 15 U.S.C. § 2603(a)(2).
111. 40 C.F.R. § 790.42.
112. Id. § 790.48.

government officials—the Interagency Testing Committee (ITC). The role of the ITC in recommending chemical substances for testing is outlined in TSCA section 4(e). Although many of EPA's test rules have been promulgated in response to ITC designations, EPA program offices also identify candidates for testing.

Although TSCA section 4 speaks of testing "by rule," EPA has developed regulations governing the procedures under which it may decide to enter into enforceable consent agreements (ECA) with manufacturers to conduct chemical testing.[113] EPA often prefers such agreements because they avoid the costs and delays associated with notice-and-comment rulemaking. Manufacturers often favor ECAs because EPA regulations permit them to become involved at an early phase and potentially influence EPA's preliminary testing determinations. EPA does not, however, enter into a consent agreement unless EPA and the manufacturers can reach consensus on the testing requirements and timetable.

EPA has promulgated detailed regulations governing the development of test rules.[114]

Data Sharing and Compensation

To comply with a test rule, a manufacturer must submit, for each test required, a letter of intent to conduct the testing or an application for an exemption from test requirements.[115] If a manufacturer opts to apply for an exemption from the test rule, it is not relieved of the obligation to bear some of the testing costs. Persons who are granted an exemption are required to provide "fair and equitable" reimbursement to other parties who do conduct the testing. If the parties cannot agree on allocation of testing costs, EPA must determine the amount of reimbursement applicable to each liable party. Any approval of an exemption is expressly conditioned on the test sponsor's successful completion of testing.[116]

---

113. Id. § 790.22.
114. Id. at pt. 790.
115. Id. § 790.45(a).
116. Id. § 790.87(c).

Reporting and Recordkeeping Requirements

Reports to Be Submitted to EPA

TSCA section 8 imposes reporting and recordkeeping requirements on manufacturers, importers, processors, and distributors of chemical substances and mixtures.

TSCA Section 8(a)—Preliminary Assessment Information Rule (PAIR)

TSCA section 8(a) authorizes EPA to issue rules requiring manufacturers and importers of chemical substances and mixtures to maintain records and submit reports to EPA.[117] The PAIR, issued under TSCA section 8(a), requires manufacturers and importers of certain chemical substances listed on the TSCA Inventory to submit a one-time report by the date established in the *Federal Register* announcing the PAIR on (1) the quantities of chemical substances manufactured, imported, used as a reactant, used in industry and consumer products, or lost to the environment; and (2) worker exposure.[118] Manufacturers and importers must submit such information to the extent it is "readily obtainable by management and supervisory employees responsible for manufacturing, processing, distributing, technical services, and marketing. Extensive file searches are not required."[119]

The list of chemical substances and mixtures that have been subject to the PAIR is codified at 40 C.F.R. § 712.30.

Manufacturers who produce 500 kilograms (1,100 pounds) or more of a listed chemical substance or mixture for commercial purposes at a single site, and persons who import 500 kilograms (1,100 pounds) or more of a listed chemical substance in "bulk form" at a single site, must complete and submit PAIR forms to EPA.[120] Separate forms must be completed for each site where a chemical substance or mixture listed in the PAIR is manufactured or imported above specified quantities, unless an exemption applies, as discussed below.[121] The information included in the PAIR response must cover the last

---

117. TSCA § 8(a), 15 U.S.C. § 2607(a).
118. *See* 40 C.F.R. pt. 712.
119. *Id.* § 712.7.
120. *Id.* § 712.20.
121. *Id.* § 712.28(b).

complete fiscal year as of the effective date of the chemical substance's listing.[122]

Exemptions: Persons who manufacture or import a chemical substance or mixture listed in the PAIR are exempt from PAIR if, during the applicable reporting period—the last complete fiscal year as of the effective date of the chemical substance's listing—the chemical substance or mixture was manufactured or imported only as a byproduct that has no commercial purpose, an impurity, a nonisolated intermediate, an R&D substance or an article. (*See* Scope of TSCA, Substances Exempt from TSCA, Exclusions from the Definition of "Chemical Substance.") In addition, small manufacturers and importers are not subject to PAIR reporting.[123]

TSCA Section 8(a)—CDR

TSCA section 8(a) authorizes EPA to issue rules requiring manufacturers and importers of chemical substances and mixtures to maintain records and submit reports to EPA.[124] EPA issues its CDR under TSCA section 8(a). The CDR (previously known as Inventory Update Reporting (IUR)) requires certain manufacturers and importers of certain chemical substances listed on the TSCA Inventory to report, every four years, data on production volume, plant site, physical forms of the chemical substance, number of potentially exposed workers, and downstream process and use information.[125] Manufacturers and importers are required to submit information to the extent it is "known to or reasonably ascertainable by" the manufacturer or importer.[126] EPA has numerous training and guidance materials available.[127]

Chemical substances subject to the CDR are those that are listed on the TSCA Inventory and that are not exempt from CDR reporting.[128] Excluded from CDR reporting are the following:

---

122. *Id.* § 712.30(a)(2).
123. *Id.* § 712.25(c).
124. TSCA § 8(a), 15 U.S.C. § 2607(a); 40 C.F.R. pt. 711.
125. *See* EPA, Chemical Data Reporting, Basic Information and Resources, http://www.epa.gov/cdr/index.html.
126. 40 C.F.R. § 704.3.
127. *See* EPA, Training and Workshops, and Guidance Documents, http://www.epa.gov/cdr/tools/index.html.
128. 40 C.F.R. § 710.25.

- Chemical substances manufactured or produced in quantities less than 25,000 pounds;[129]
- Polymers, microorganisms, and naturally occurring chemical substances that are not subject to: a final or proposed rule under TSCA sections 4, 5(a)(2), 5(b)(4), or 6; an order issued under TSCA sections 5(e) or 5(f); or relief granted pursuant to a civil action under TSCA sections 5 or 7;
- Chemical substances manufactured or imported only in small quantities solely for R&D purposes;
- Chemical substances imported only as part of an article; or
- Chemical substances manufactured as impurities, noncommercial by-products, nonisolated intermediates, or in a manner incidental to another operation or upon end use of another substance or mixture, as described under 40 C.F.R. § 720.30(g), (h).[130]

All manufacturers and importers must maintain production volume records for all chemical substances manufactured or imported in quantities of less than 25,000 pounds to document and support the company's decision not to report pursuant to the CDR rules.[131]

All domestic manufacturers and importers are subject to reporting, although there is an exemption for small manufacturers.[132] CDR reports are filed every four years.[133]

The CDR amendments implemented in August 2011[134] were to be phased in over two reporting periods, 2012 and 2016. Starting

---

129. Prior to the 2006 IUR reporting cycle, the reporting threshold was 10,000 lbs. See EPA, Changes to the IUR Since 2002 Reporting, http://www.epa.gov/cdr/pubs/guidance/changes.html.
130. 40 C.F.R. §§ 710.4(d), 710.26, 710.30.
131. Id. § 710.37.
132. Small manufacturers, as defined in 40 C.F.R. § 704.3, are exempt from CDR requirements unless they manufacture (including import) 25,000 lbs or more of a chemical substance that is the subject of a rule proposed or promulgated under sections 4, 5(b)(4), or 6 of TSCA, or is the subject of an order in effect under section 5(e) of TSCA, or is the subject of relief that has been granted under a civil action under sections 5 or 7 of TSCA, 15 U.S.C. § 2607(a)(3)(A)(ii); 40 C.F.R. § 711.9.
133. The IUR reporting cycle had been every four years since the IUR regulations were promulgated in 1986. In 2005, EPA amended the IUR reporting cycle to every five years, beginning in 2006. In 2011, EPA amended the IUR again, changing the name to the CDR rule and returning the reporting frequency to once every four years.
134. See EPA, TSCA Inventory Update Reporting Modifications; Chemical Data Reporting, available at http://www.regulations.gov/#!documentDetail;D=EPA-HQ-OPPT-2009-0187-0393 (posted Aug. 16, 2011).

with the 2016 reporting cycle, full reporting, including information on process and use, is required on any nonexempt chemical substance manufactured or imported at 25,000 pounds or more at a single site in any year between reporting cycles. If a report is required, production volume information must be reported for all intervening years, whether the threshold was met during that particular year or not. In addition, if a facility manufactures or imports 2,500 pounds or more of a chemical substance subject to the TSCA actions listed below[135] in a calendar year between reporting cycles, it must report during the subsequent reporting cycle:

- Rule *proposed or promulgated* under TSCA section 5(a)(2), 5(b)(4), or 6;
- Order issued under TSCA section 5(e) or 5(f); or
- Relief that has been granted under a civil action under TSCA section 5 or 7.

For importers, a "site" for reporting purposes is the operating unit within the company that is directly responsible for importing the chemical substance and controlling the transaction.[136] EPA acknowledges that the importing site may be the corporate headquarters or other nonmanufacturing site and that the imported chemical substance may never physically be at said importing site.

TSCA Section 8(c)—Allegations of Significant Adverse Health or Environmental Reactions

TSCA section 8(c) requires manufacturers, importers, processors, and distributors to maintain records of allegations of significant adverse health or environmental reactions.[137] Examples of "significant adverse reactions" that must be reported are provided at 40 C.F.R. § 717.12.[138]

An allegation may be made orally or in writing. Any employee who receives an oral allegation must inform the alleger that the allegation

---

135. A list of chemicals subject to the actions listed above is available in Appendix B, Chemical Substances That are the Subject of Certain TSCA Orders, Proposed or Final TSCA Actions or Relief Granted under Civil Actions (as of July 2011), *in* EPA's Instructions for the 2012 TSCA Chemical Data Reporting (linked at http://www.epa.gov/cdr). Note that the list is for information purposes only and subject to change.
136. 40 C.F.R. § 711.3.
137. TSCA § 8(c), 15 U.S.C. § 2607(c).
138. 40 C.F.R § 717.12.

may be subject to TSCA section 8(c) recordability and must request that the alleger submit a signed written statement to the company or have the company transcribe the allegation.[139] TSCA section 8(c) allegations shall be kept at the company's headquarters or at any other appropriate location central to the company's chemical operations.[140]

Each allegation should be reviewed for TSCA section 8(c) recordability. A TSCA section 8(c) allegation form should be prepared for those cases determined to be recordable. EPA has not created a specific form to use to record TSCA section 8(c) allegations.[141] Instead, companies may develop their own formats for a TSCA section 8(c) allegation form, provided each recordable allegation contains the following information:

- The name and address of the site that received the allegation;
- The date the allegation was received;
- The identity of the chemical substance or mixture;
- A description of the person making the allegation (e.g., employee, customer, plant neighbor);
- A description of the alleged reaction(s); and
- The gender and birth year of any person alleged to have experienced a health effect.[142]

The original allegation must be attached to the TSCA section 8(c) allegation form.[143] If applicable, attached to the TSCA section 8(c) allegation form should be the results of any self-initiated investigation with respect to the allegation and copies of any other reports concerning the allegation, such as Occupational Safety and Health Administration (OSHA) reports.[144]

---

139. Id. § 717.10(b)(1). See also Records and Reports of Allegations That Chemical Substances Cause Significant Adverse Reactions to Health or the Environment; Recordkeeping and Reporting Procedures, 48 Fed. Reg. 38,178 (Aug. 22, 1983) (codified at 40 C.F.R. pt. 717) (discussing written and oral allegations).
140. 40 C.F.R. § 717.15(a).
141. 48 Fed. Reg. at 38,178.
142. 40 C.F.R. § 717.15(b).
143. Id. § 717.15(b)(1).
144. Id. § 717.15(b)(3)–(4).

EPA does not require a company to keep records of allegations that do not meet TSCA section 8(c) recording requirements.[145]

EPA may issue a letter or *Federal Register* notice requiring the submission of TSCA section 8(c) allegation records to EPA.[146] The *Federal Register* notice will specify which records or portion of records must be submitted, and whether allegations relating to mixtures must be reported.[147]

Exemptions: Allegations that are exempt from TSCA section 8(c) recording include the following:

- Allegations of known human effects as described in a Safety Data Sheet (SDS), the public scientific literature, or product labeling;
- Allegations of environmental harm where the alleged cause is attributable to an incident of environmental contamination that has already been reported to the federal government; and
- Unsigned written allegations.[148]

TSCA Section 8(d)—Health and Safety Studies Reporting

Under TSCA section 8(d), manufacturers, importers, and processors who fall within the North American Industry Classification System (NAICS) Subsector 325 (chemical manufacturing and allied products) or Industry Group 32411 (petroleum refineries)[149] must submit, for certain chemical substances and mixtures, unpublished health and safety studies that are in their possession, lists of all unpublished health and safety studies on those chemical substances and mixtures known to them but not in their possession, and lists of all health and safety studies being conducted or initiated on those chemical substances and mixtures.[150] TSCA defines a "health and safety study" as "any study of any

---

145. *See* EPA, Questions and Answers Concerning the TSCA Section 8(c) Rule (July 1984), at 1; 48 Fed. Reg. at 38,178.
146. 40 C.F.R. § 717.17(b); TSCA § 8(c), 15 U.S.C. § 2607(c) ("[u]pon request . . . , each person who is required to maintain records under this subsection shall permit the inspection of such records and shall submit copies of such records.").
147. 40 C.F.R. § 717.17(b).
148. *Id.* §§ 717.10(b), 717.12(b), (d). *See also id.* § 717.3(c) (definition of "known human effects").
149. *Id.* § 716.5(a).
150. TSCA § 8(d), 15 U.S.C. § 2607(d); 40 C.F.R. § 716.1(a).

effect of a chemical substance or mixture on health or the environment or on both, including underlying data and epidemiological studies, studies of occupational exposure to a chemical substance or mixture, toxicological, clinical, and ecological or other studies of a chemical substance or mixture, and any test performed under TSCA."[151]

Compliance Procedure Pursuant to regulations that EPA issued April 1, 1998, and that were effective June 30, 1998, TSCA section 8(d) requires manufacturers, importers, and processors to review their files and submit, for a chemical substance or mixture newly listed under TSCA section 8(d), unpublished health and safety studies that are in their possession, unpublished lists of health and safety studies known to them but not in their possession, and lists of all health and safety studies being conducted by or initiated by them during a one-time 60-day period, starting on the effective date of the chemical substance or mixture's listing.[152] In certain circumstances, however, EPA can require the submission of health and safety studies on a chemical substance or mixture for longer than 60 days, but not longer than two years.[153]

File Search for TSCA Section 8(d) Health and Safety Studies Notices of TSCA section 8(d) reporting requirements are issued in the *Federal Register*. A company is subject to the TSCA section 8(d) reporting rule if it manufactures, imports, or processes a listed chemical substance or mixture and if it is in possession of health and safety studies on such chemical substance or mixture. A manufacturer, importer, or processor subject to the TSCA section 8(d) rule must coordinate and/or conduct a file search to identify all health and safety studies in the possession of the company, or of which company staff are aware, that might be reportable under TSCA section 8(d).[154] The scope of the search may be limited to records where such information normally is kept, and to the records of employees who have the responsibility for advising the company on the health and safety of chemical substances and mixtures.[155] File searches are required only for reportable information dated on or after January 1, 1977, unless a rule specifies otherwise.[156] An importer need search only files located

---

151. 40 C.F.R. § 716.3.
152. Id. §§ 716.30, 716.35, 716.60.
153. Id. § 716.65.
154. Id. § 716.25.
155. Id.
156. Id. *See also* 63 Fed. Reg. at 15,769.

in the United States—it is not required to obtain health and safety studies in the possession of a foreign subsidiary or parent.[157]

Submission of Copies of Health and Safety Studies Manufacturers, importers, and processors of chemical substances and mixtures listed under TSCA section 8(d) must submit copies of unpublished health and safety studies on the chemical substance or mixture that are in their possession.[158] Affected manufacturers, importers, and processors are responsible for submitting copies of health and safety studies on only the listed chemical substances or mixtures that they have manufactured, imported, or processed, or have proposed to manufacture, import, or process (1) within the ten years preceding the effective date for reporting on the listed chemical substances or mixtures; (2) on the effective date for reporting; or (3) following the effective date for reporting.[159]

Submission of Lists of Health and Safety Studies Manufacturers, importers, and processors of chemical substances and mixtures listed under TSCA section 8(d) must submit lists of ongoing health and safety studies relating to those chemical substances and mixtures being conducted by or for their companies, and unpublished health and safety studies known to them, but of which they do not have copies.[160] In the case of health and safety studies not in a company's possession, the company must provide the name and address of the person known to be in possession of the health and safety study.[161] Companies that list studies as ongoing or initiated must submit those studies when they are completed.[162]

TSCA Section 8(d) Report Preparation and Submission to EPA A company must prepare a TSCA section 8(d) report containing copies of health and safety studies and lists of health and safety studies determined to be reportable to EPA under TSCA section 8(d). Copies of studies must identify the CAS number of the applicable

---

157. EPA, General Questions and Answers about Reporting under the TSCA § 8(d) Health and Safety Study Reporting Rule (Feb. 16, 1989), at 6.
158. 40 C.F.R. § 716.30.
159. Id. §§ 716.5(a), 716.30(a)(1).
160. Id. § 716.35.
161. Id.
162. Id. § 716.30(a)(1).

chemical substance or mixture on the study's cover page or otherwise.[163] A cover letter accompanying the TSCA section 8(d) report must include the following information:

- The name, job title, address, and telephone number of the person submitting the information;
- The name and address of the company;
- The identity of any impurity or additive known to have been present in the chemical substance or mixture as studied, if such information is not noted in the study itself; and
- If the study is submitted by a trade association, a statement that the study was submitted to satisfy TSCA section 8(d) reporting requirements.[164]

Final and, if applicable, sanitized copies of the TSCA section 8(d) report of health and safety studies and lists of health and safety studies must be sent by certified mail, within 60 days of the effective date of the chemical substance or mixture's addition to the TSCA section 8(d) list.

Exemptions Studies not subject to TSCA section 8(d) reporting are listed at 40 C.F.R. § 716.20, unless specifically included per the 8(d) rulemaking.

Recordkeeping

TSCA Section 8(a)—Preliminary Assessment Information Rule (PAIR)

Documents that the manufacturer or importer may wish to maintain to demonstrate compliance with the PAIR include copies of all submitted PAIR forms (hard copy or electronic submission), documentation used to complete all PAIR submissions, and documentation demonstrating that a chemical substance or mixture is exempt from PAIR reporting requirements. TSCA section 8(a) and its PAIR regulations do not specify how long manufacturers and importers should maintain documents demonstrating compliance with TSCA.

---

163. Id. § 716.30(b).
164. Id.

Generally, however, TSCA regulations provide that documents must be maintained for five years.[165]

TSCA Section 8(a)—CDR

Under CDR regulations, EPA requires manufacturers and importers to "maintain records that document any information reported to EPA."[166] In accordance with EPA requirements, for each chemical substance manufactured or imported that is subject to the IUR, documents that must be maintained include

- Copies of all electronically submitted Form Us;
- Documentation used to complete all Form Us (e.g., documents used to determine production volume, plant size, and site-limited status of chemical substances manufactured or imported);
- Documentation demonstrating that the chemical substance is exempt from CDR reporting requirements; and
- Production volume records for all chemical substances manufactured or imported in quantities less than 25,000 pounds.

EPA states: "As long as the records are maintained in a manner consistent with normal business practice, [the company] may determine their exact format."[167] Parties subject to reporting under the CDR rule must retain records that document any information reported to EPA.

Records relevant to reporting during a submission period must be retained for a period of five years beginning on the last day of the submission period.[168] EPA encourages submitters to retain records longer than five years so that they are available as a reference when new submissions are being generated.

---

165. *See, e.g., id.* § 720.78 (TSCA section 5 PMN and related records must be kept for five years); *id.* § 717.15 (TSCA section 8(c) records of significant adverse reactions of nonemployees must be kept for five years); *id.* §§ 721.40, 721.125 (SNUR records must be kept for five years). *See also* 3M Co. v. Browner, 17 F.3d 1453 (D.C. Cir. 1994) (applying five-year statute of limitations to TSCA administrative civil penalty actions).
166. 40 C.F.R. § 710.37.
167. EPA, Instructions for Reporting for the 1998 Partial Updating of the TSCA Chemical Inventory Data Base, at 25.
168. 40 C.F.R. § 711.25.

TSCA Section 8(c)—Allegations of Significant Adverse Health or Environmental Reactions

For section 8(c), in accordance with EPA requirements, for each chemical substance or mixture manufactured, imported, or processed for which a significant adverse reaction has been alleged, the following records must be maintained:

- The written allegation or transcription of an oral allegation concerning that chemical substance or mixture;
- TSCA section 8(c) allegation forms concerning that chemical substance or mixture;
- Any pertinent Worker Compensation report or OSHA report or litigation documents concerning that chemical substance or mixture;
- The results of any investigation into an allegation concerning that chemical substance or mixture; and
- TSCA section 8(c) reports or other documents submitted to EPA concerning that chemical substance or mixture.[169]

The allegations and related documents, in accordance with EPA requirements, must be filed so that they are retrievable under the following categories, based on the alleged source of the adverse reaction: (1) chemical substances; (2) mixtures; (3) articles; (4) processes or operations; and (5) site discharges and effluents.[170]

In accordance with EPA requirements, each TSCA section 8(c) allegation and accompanying documents involving health effects in employees must be retained for 30 years from the date the significant adverse reaction was first reported or known to the company.[171] In addition, in accordance with EPA requirements, TSCA section 8(c) allegations and accompanying documents involving other health effects must be maintained for five years from the date the significant adverse reaction was first reported or known to the company.[172]

---

169. Id. § 717.15.
170. Id. § 717.15(c).
171. TSCA § 8(c), 15 U.S.C. § 2607(c); 40 C.F.R. § 717.15(d).
172. TSCA § 8(c), 15 U.S.C. § 2607(c); 40 C.F.R. § 717.15(d).

TSCA Section 8(d)—Health and Safety Studies Reporting

Documents that the manufacturer, importer, or processor may wish to maintain to demonstrate compliance with TSCA section 8(d) include copies of all submitted TSCA section 8(d) reports and a copy of each health and safety study and list of health and safety studies provided to EPA. TSCA section 8(d) does not specify how long manufacturers, importers, and processors should maintain documents demonstrating compliance with TSCA. Generally, however, TSCA regulations provide that documents must be maintained for five years.[173]

Risk Notification

Pursuant to TSCA section 8(e), any person who manufactures, imports, processes, or distributes a chemical substance or mixture and who obtains information that reasonably supports the conclusion that the chemical substance or mixture poses a substantial risk of injury to human beings or the environment must provide the information to EPA immediately.[174] EPA's TSCA section 8(e) guidance provides that "immediately" means within 30 calendar days after person obtained such information.[175] In the event a company becomes aware of an emergency incident involving a chemical substance or mixture known to be a serious human or environmental toxicant, the information must be reported "as soon as reasonably possible" after the company obtains such information.[176] EPA states that a company is regarded as having obtained information "at the time any officer or employee capable of appreciating the significance of such information obtains it."[177] "Known" information includes that information about which a prudent person of similar training, job function, etc., could be reasonably expected to know.[178] Failure to report substantial risk information can give rise to civil and criminal penalties.

---

173. *See, e.g.*, 40 C.F.R. § 720.78 (TSCA section 5 PMN and related records must be kept for five years); *id.* § 717.15 (TSCA section 8(c) records of significant adverse reactions of nonemployees must be kept for five years); *id.* §§ 721.40, 721.125 (SNUR records must be kept for five years). *See also* 3M Co. v. Browner, 17 F.3d 1453 (D.C. Cir. 1994) (applying five-year statute of limitations to TSCA administrative civil penalty actions).
174. TSCA § 8(e), 15 U.S.C. § 2607(e).
175. *See* EPA, TSCA 8(e) Notices, Basic Information, http://www.epa.gov/oppt/tsca8e/.
176. EPA, TSCA Section 8(e) Reporting Guide 11 (June 1991), http://www.epa.gov/opptintr/tsca8e/pubs/1991guidance.pdf [hereinafter Section 8(e) Reporting Guide].
177. 43 Fed. Reg. at 11,111.
178. Section 8(e) Reporting Guide, *supra* note 176, at 6.

Determining Whether Information Is Substantial Risk Information

All employees should be trained to recognize substantial risk information that may be reportable under TSCA section 8(e).[179] A procedure should be established to (1) specify what information is substantial risk information; (2) determine whether an incident is reportable as an emergency incident; (3) explain how information will be prepared and submitted; (4) note the federal penalties for failing to report substantial risk information; and (5) notify officers and employees, in writing, of the company's disposition of the information, explaining what decision was made on whether to report the information, and, if the decision was made not to report, explaining why and inform the officers and employees of their right to report the information individually to EPA under TSCA section 23.[180]

EPA has emphasized that "substantial risk" information "need not and most typically does not establish conclusively that a substantial risk exists."[181] EPA has also said that in deciding whether information is "substantial risk" information, one must consider "1) the seriousness of the adverse effect, and 2) the fact or probability of the effect's occurrence."[182] According to EPA, the two criteria should be weighed differently, depending on the seriousness of the effect or the extent of the exposure (the more serious the effect, the less heavily one should weigh exposure, and vice versa).[183]

Sources of substantial risk information include (1) designed controlled studies such as in vitro or in vivo tests, epidemiological studies, environmental monitoring studies, medical and health surveys, clinical studies, and news reports; (2) TSCA section 8(c) allegations, where a pattern appears or a single instance of a serious human health effect is strongly associated with one or a few chemicals; and (3) oral communications to an employee.[184] If the information was provided orally, the person who received the information must prepare a summary of the information.

Recent enforcement cases highlight how broadly EPA can interpret the scope of section 8(e) reporting and potential consequences when EPA determines information that was not reported is reportable.

---

179. 43 Fed. Reg. at 11,111 (it is "incumbent upon business organizations to establish procedures for expeditiously processing pertinent information" under TSCA section 8(e)).
180. *Id.*; SECTION 8(E) REPORTING GUIDE, *supra* note 176, at 4.
181. SECTION 8(E) REPORTING GUIDE, *supra* note 176, at 2.
182. 43 Fed. Reg. at 11,111; SECTION 8(E) REPORTING GUIDE, *supra* note 176, at 2.
183. 43 Fed. Reg. at 11,111.
184. *Id.* at 11,112; SECTION 8(E) REPORTING GUIDE, *supra* note 176, at 7.

A November 2013 administrative decision ordered Elementis Chromium, Inc., "one of the largest manufacturers of chromium chemicals in the world," to pay a penalty of $2,571,800 for failing to disclose information about substantial risk of injury to human health from exposure to hexavalent chromium.[185] EPA alleged TSCA violations for failing to report the results of an industry-commissioned study that documented significant occupational impacts to workers in modern chemical plants. According to EPA, the study filled a gap in scientific literature regarding the relationship between hexavalent chromium exposure and respiratory cancer in modern chromium production facilities.

In addition, on December 21, 2005, the Environmental Appeals Board (EAB) approved a settlement with DuPont involving $10.25 million in fines and resolving alleged violations related to perfluorooctanoic acid (PFOA).[186] Seven of the eight counts involved violations of TSCA section 8(e). The consent agreement and final order settles EPA's allegations that DuPont violated TSCA section 8(e) by failing to report immediately a single blood test result that DuPont had obtained in 1981, which showed the presence of PFOA in umbilical cord blood and "confirm[ed] the transplacental movement of PFOA."[187] A 2006 settlement imposed $1.5 million in fines to resolve alleged TSCA reporting violations with regard to PFOA data that 3M voluntarily disclosed to EPA under the terms of a TSCA corporate-wide audit agreement.[188]

Reporting Emergency Incidents of Environmental Contamination

Emergency incidents that must be reported immediately—"as soon as reasonably possible"—under TSCA section 8(e) can include accidental or intentional environmental contamination involving any chemical substance or mixture known to be a serious human or environmental toxicant, where the extent, pattern, and amount of the contamination either (1) seriously threatens humans with cancer, birth defects,

---

185. *In re* Elementis Chromium, Inc., Docket No. TSCA-HQ-2010-5022 (Nov. 12, 2013), available at http://www.epa.gov/oalj/orders/2013/TSCA-HQ-2010-5022_Elementis Chromium_13-11-12_ID_Biro.pdf.
186. Consent Agreement and Final Order at 8, *In re* E.I. du Pont de Nemours & Co., TSCA-HQ-2004-0016, RCRA-HQ-2004-0016, TSCA-HQ-2005-5001 (filed Dec. 14, 2005), available at http://www2.epa.gov/enforcement/ei-dupont-de-nemours-and-company-settlement.
187. Complaint at 11–12, *In re* E.I. du Pont de Nemours & Co., SCA-HQ-2004-0016, TSCA-HQ-2005-5001, RCRA-HQ-2004-0016 (July 8, 2004).
188. Consent Agreement and Final Order, *In re* 3M Co., TSCA-HQ-2006-5004 (Apr. 25, 2006), available at http://www2.epa.gov/sites/production/files/documents/3m-consentagr.pdf.

mutation, death, or serious injury; or (2) seriously threatens nontarget organisms with large-scale population destruction.[189] EPA allows companies to report emergency incidents of environmental contamination to EPA or the National Response Center.

When a company becomes aware of an incident that is determined to be a reportable emergency incident, it must immediately call the appropriate EPA regional office or the National Response Center and provide the following information:

- The time and location of the incident;
- The name and address of the facility where the incident occurred;
- The name of the chemical substance or mixture and CAS number (if known);
- A summary of the nature and extent of the contamination and risks and adverse effects involved;
- A statement that the information is being provided under TSCA section 8(e); and
- The name, address, job title, and telephone number of the person providing the information.[190]

EPA has provided additional guidance on what it considers to be substantial risk information for "non-emergency situations involving environmental contamination; environmental effects."[191]

EPA notes that, for purposes of determining whether there is a non-emergency situation for environmental contamination, information about contamination found below established levels that are "presumed to present no risk to human health or the environment" would not trigger TSCA section 8(e) reporting.[192] If information about contamination is found at or above benchmark values that trigger regulatory requirements (e.g., Resource Conservation and Recovery Act (RCRA) Toxicity Characteristic Limits), EPA states that TSCA section 8(e) reporting obligations must be considered.

---

189. 43 Fed. Reg. at 11,113.
190. Id.
191. 68 Fed. Reg. at 33,138.
192. Id. EPA provides the following example: "If a person found groundwater contaminated with a chemical at a level that did not exceed the RfD for that substance, the person could assume that a substantial risk does not exist." Id. at 33,132–33.

A written follow-up TSCA section 8(e) report for emergency incidents of environmental contamination is not required.[193]

TSCA Section 8(e)—Report Preparation and Submission to EPA

If a decision is made that substantial risk information must be reported, a TSCA section 8(e) report must be prepared. EPA has not created a specific form to use to record TSCA section 8(e) information.[194] Instead, companies can develop a TSCA section 8(e) report, which must include the following information:

- A statement that the report is being submitted under TSCA section 8(e);
- The name of the chemical substance or mixture and CAS number (if known);
- The name, job title, address, telephone number, and signature of the person reporting;
- The name and address of the site with which the person reporting is affiliated;
- A summary of the adverse effects being reported; and
- Identification of the specific source of information and supporting technical data.[195]

EPA notes that it considers information submitted under TSCA section 8(e) to be "health and safety information" that, under TSCA section 14, may be released to the public unless a particular exemption applies.[196] EPA states that if a submitter wants to claim CBI to, for example, protect information regarding manufacturing processes, the submitter must provide, at the time the section 8(e) report is submitted to EPA, a CBI claim substantiation.[197]

---

193. Id. at 33,138.
194. 48 Fed. Reg. at 38,178.
195. 43 Fed. Reg. at 11,113; SECTION 8(E) REPORTING GUIDE, *supra* note 176, at 13.
196. 68 Fed. Reg. at 33,140.
197. The list of substantiation questions that must be answered to substantiate a CBI claim can be found at EPA, Confidential Business Information, http://www.epa.gov/oppt/tsca8e/pubs/confidentialbusinessinformation.html.

The final, and, if applicable, sanitized copy of the TSCA section 8(e) report must be sent by certified or registered mail, or other method permitting receipt to be verified.[198]

Exemptions: Categories of Information That Do Not Require TSCA Section 8(e) Reporting

Section 8(e) does not provide exemptions for small businesses, small production, or small importation volumes. Nor are chemicals exempted for noncommercial activities, such as manufacture for export only or R&D chemicals.[199]

Section 8(e) does not require a subject person to submit information about a chemical substance or mixture that the person does not manufacture, import, process, or distribute commercially. Furthermore, a person who obtains substantial risk information about a chemical or mixture that the person did at one time, but does not any longer, manufacture, import, process, or distribute in commerce, is not required to submit the information under section 8(e).[200]

Information otherwise subject to the reporting requirements of TSCA section 8(e) need not be submitted if the information

- Is contained in an EPA study or report;
- Is published in the open scientific literature or major U.S. news publication;
- Has been reported to EPA under mandatory reporting requirements of TSCA or other authority administered by EPA;
- Is contained in a formal publication, report, or statement made available to the public by another federal agency;
- Is corroborative of a well-established adverse effect (and does not newly identify any serious adverse effects or confirm a previously suspected serious adverse effect);[201] or
- Is information for which EPA has waived compliance in accordance with TSCA section 22.[202]

---

198. Reporting and Recordkeeping Requirements; Technical Amendment; Final Rule, 52 Fed. Reg. 20,083 (May 29, 1987) (codified at 40 C.F.R. pts. 704, 710, 712, 716, and 717); SECTION 8(E) REPORTING GUIDE, *supra* note 176, at 12.
199. *See* EPA, Who Must Report to EPA under TSCA 8(e)?, http://www.epa.gov/oppt/tsca8e/pubs/reportingrequirements.html.
200. *Id.*
201. 68 Fed. Reg. at 33,139.
202. 43 Fed. Reg. at 11,112; SECTION 8(E) REPORTING GUIDE, *supra* note 176, at 8–10.

In addition, EPA has added exemptions to "relieve persons who are potentially subject to reporting under section 8(e) from the burden of considering information from secondary sources when the secondary source does not provide sufficient information for a person to judge whether the information should be reported."[203]

EPA also has exemptions "to reduce the potential for duplicative submission under TSCA section 8(e) authorities by allowing an exemption to reporting under section 8(e) for all information which is required to be reported under other EPA statutes including where implementation had been delegated to the States, and where such reporting was required to be submitted within 90 days of being obtained."[204]

EPA states that these exemptions do not apply to information reported under state or local authorities, "other than those reporting requirements that originate in laws administered by EPA in which the United States Congress has provided for delegation to the States, and such delegation has occurred."[205] EPA also notes that TSCA section 8(e) reporting will continue to be required for chemical product contamination that could be required to be submitted to the Consumer Product Safety Commission (CPSC).[206]

Lastly, EPA provides additional exemptions to address situations when potential TSCA section 8(e) information regarding "non-emergency situations of chemical contamination involving humans and/or the environment" or "emergency incidents of environmental contamination" is obtained at a site undergoing remediation, at a site not owned or operated by the person obtaining such information, or at a site outside the United States.[207]

Recordkeeping

Documents that the manufacturer or importer may wish to maintain to demonstrate compliance with TSCA section 8(e) include copies of documents containing substantial risk information (or a summary of information provided orally), TSCA section 8(e) reports, and the

---

203. 68 Fed. Reg. at 33,133.
204. Id. at 33,134.
205. Id.
206. Id. at 33,135 (reporting of chemical product contamination will continue because "EPA, uniquely among Federal agencies, has the authority to address all potential health and environmental risk aspects of a chemical's life cycle").
207. Id. at 33,139–40.

procedures established to review and report under TSCA section 8(e). TSCA section 8(e) does not specify how long manufacturers, importers, processors, or distributors should maintain documents demonstrating compliance with TSCA. Generally, however, TSCA regulations provide that documents must be maintained for five years.[208]

IMPORTS AND EXPORTS

Authority to Regulate Imports and Exports

Export Requirements and Exemptions

TSCA section 12(b) requires exporters to notify EPA, in writing, if they export, or intend to export, chemical substances or mixtures that are subject to certain TSCA rules or orders. Failure to comply with TSCA section 12(b) can give rise to civil penalties, enforcement actions, and/or seizures under TSCA sections 16 and 17.[209]

Identifying Chemical Substances and Mixtures Subject to TSCA Section 12(b)

Chemical substances and mixtures subject to TSCA section 12(b) export notification requirements are those for which EPA has taken one or more of the following actions:

- Data have been required under TSCA sections 4 or 5(b);[210]
- An order has been issued under TSCA section 5(e);

---

208. *See, e.g.*, 40 C.F.R. § 720.78 (TSCA section 5 PMN and related records must be kept for five years); *id.* § 717.15 (TSCA section 8(c) records of significant adverse reactions of nonemployees must be kept for five years); *id.* §§ 721.40, 721.125 (SNUR records must be kept for five years). *See also* 3M Co. v. Browner, 17 F.3d 1453 (D.C. Cir. 1994) (applying five-year statute of limitations to TSCA administrative civil penalty actions).
209. 40 C.F.R. § 707.60(e).
210. TSCA regulations clarify that TSCA section 12(b) requirements are triggered only by final TSCA section 4 test rules or final ECAs and not proposed test rules or ECAs. *See* Procedures Governing Testing Consent Agreements and Test Rules under the Toxic Substances Control Act, 51 Fed. Reg. 23,706, 23,710 (June 30, 1986) (codified at 40 C.F.R. pts. 790 and 799) ("For the purposes of TSCA export notification requirements, the Agency considers a final testing consent agreement to be the equivalent of a final section 4 test rule. It is an equivalent data gathering requirement under section 4 of TSCA because it represents the Agency's commitment to proceed with data collection with respect to specific substances. Therefore, the Agency has determined that consent agreements will trigger TSCA section 12(b) export notification requirements. Persons who export or intend to export a substance which is the subject of a final consent agreement will be subject to section 12(b)[.]").

2.1. Toxic Substances Control Act (TSCA)

- A rule has been proposed or promulgated under TSCA sections 5 (e.g., SNUR) or 6; or
- An action is pending or relief has been granted under TSCA sections 5(f) or 7.[211]

TSCA regulations clarify that for a new chemical substance—a chemical substance not listed in the TSCA Inventory—TSCA section 12(b) export notification is not triggered by compliance with TSCA section 5(a) PMN requirements. TSCA section 5 rules and orders trigger TSCA section 12(b) export notification only when the new chemical substance is:

- Included on the TSCA section 5(b)(4) list;
- Subject to a TSCA section 5(e) or 5(f) order; or
- Subject to a proposed or final SNUR.[212]

In addition, export notification is only required for chemical substances and mixtures known to be in exported material.[213] That is, a company is not required to test any material to be exported to determine if it contains chemical substances or mixtures subject to export notification requirements.[214]

Furthermore, TSCA section 12(b) requirements are not applicable after the TSCA rules or orders triggering TSCA section 12(b) requirements expire. EPA maintains a list of chemicals subject to section 12(b), cautioning, however, the list "is intended merely as an information resource to facilitate compliance with TSCA Section 12(b). While EPA has attempted to be as accurate as possible in compiling this list, exporters subject to the requirements of Section 12(b) should be aware that this list may contain unintended errors or omissions and may not be completely up to date, so the absence of a chemical from this list does not necessarily mean that export of the chemical is exempt from Section 12(b) reporting obligations."[215]

---

211. TSCA § 12(b)(1), (2), 15 U.S.C. § 2611(b)(1), (2).
212. 45 Fed. Reg. at 82,849.
213. Id. at 82,845.
214. Id.
215. EPA, Chemical Substances Subject to Section 12(b), http://www.epa.gov/oppt/import-export/pubs/12blist.html.

TSCA Section 12(b) Export Notice Preparation and Submission to EPA

Information on the content and timing of a TSCA section 12(b) export notification is available on EPA's Section 12(b) Export Notification web page.[216]

Confidential Business Information

A TSCA section 12(b) export notice can include information on more than one export (e.g., it can include several chemical substances and mixtures and/or different importing countries). The exporter may assert a CBI claim for any information contained in the TSCA section 12(b) export notice.

Exemptions

Generally, articles are exempt from TSCA section 12(b) export notification. Articles that are not exempt include (1) PCB articles being exported for purposes other than disposal and (2) those articles addressed in individual TSCA sections 5, 6, or 7 actions.[217]

Notification of export is generally not required for de minimis concentrations of less than 1 percent by weight or volume, unless[218]

- The chemical substance or mixture is a known or potential human carcinogen, in which case the de minimis concentration is 0.1 percent (by weight or volume). A chemical is considered to be a known or potential human carcinogen, for purposes of TSCA section 12(b) export notification, if that chemical is
  - A chemical substance or mixture listed as a "known to be human carcinogen" or "reasonably anticipated to be human carcinogen" in the Report on Carcinogens (latest edition) issued by the U.S. Department of Health and Human Services, Public Health Service, National Toxicology Program,
  - A chemical substance or mixture classified as "carcinogenic to humans" (Group 1), "probably carcinogenic to humans" (Group 2A), or "probably carcinogenic to humans" (Group 2B) in the Monographs and Supplements on the Evalua-

---

216. EPA, Section 12(b) Export Notification, http://www.epa.gov/oppt/import-export/pubs/sec12.html#exporterscontent.
217. 40 C.F.R. § 707.60(b).
218. *Id.* § 707.60(c).

tion of Carcinogenic Risks to Humans issued by the World Health Organization International Agency for Research on Cancer (IARC), Lyons, France (latest editions), or
- Alpha-naphthylamine (CASRN 134-32-7) or 4-nitrobiphenyl (CASRN 92-93-3).
- The exported chemical is polychlorinated biphenyl chemicals (PCBs) (see definition in 40 C.F.R § 761.3), where such chemical substances are present in a concentration of less than or equal to 50 ppm (by weight or volume).

Recordkeeping

Documents that the exporter may wish to maintain to demonstrate compliance with TSCA section 12(b) include all TSCA section 12(b) export notices and any correspondence with EPA regarding a chemical substance or mixture's compliance with TSCA section 12(b). TSCA section 12(b) and its regulations do not specify how long exporters should maintain documents demonstrating compliance with TSCA. Generally, however, TSCA regulations provide that documents must be maintained for five years.[219]

Import Requirements and Exemptions

Under TSCA section 13, the secretary of the treasury (U.S. Customs Service or Customs) must refuse entry into the United States any chemical substance—including chemical substances that are components of mixtures and chemical substances or mixtures contained in articles that are designed to be used or released from the articles—that fails to comply with TSCA.[220] Under rules adopted by the U.S. Customs Service,[221] importers must certify at the point of entry, for each shipment, that

---

219. *See, e.g., id.* § 720.78 (TSCA section 5 PMN and related records must be kept for five years); *id.* § 717.15 (TSCA section 8(c) records of significant adverse reactions of nonemployees must be kept for five years); *id.* §§ 721.40, 721.125 (SNUR records must be kept for five years). *See also* 3M Co. v. Browner, 17 F.3d 1453 (D.C. Cir. 1994) (applying five-year statute of limitations to TSCA administrative civil penalty actions).
220. TSCA § 13(a)(1), 15 U.S.C. § 2612(a)(1).
221. 19 C.F.R. §§ 12.118–12.127, 127.28(i).

- Chemical substances, mixtures, and articles being imported comply with applicable rules and orders under TSCA and are not offered for entry in violation of TSCA (positive certification); or
- Chemical substances, mixtures, and articles being imported are not subject to TSCA (negative certification).[222]

TSCA Section 13 Import Requirements

TSCA section 13 import requirements apply to importers of chemical substances, mixtures, and articles. For each import, an importer must (1) establish that the chemical substance or mixture is in compliance with TSCA sections 5, 6, and 7, or is not subject to TSCA, and (2) prepare an import certificate.

Establishing Compliance with TSCA

An importer must establish the identity of each chemical substance, including those contained in a mixture or article, that is intended to be imported to establish that the chemical substance, including those contained in a mixture or article, in an import shipment complies with TSCA sections 5, 6, and 7 requirements.[223]

Where an importer does not know the identity of a chemical substance, including those contained in a mixture or article, it wishes to import, it must make a good-faith effort to determine whether that unidentified chemical substance complies with TSCA, including asking the foreign supplier for the information necessary to confirm the TSCA status of each chemical substance, chemical substance component of a mixture, and chemical substance or mixture contained in an article that is designed to be used or released (e.g., SDS, product label, CAS number).[224]

If the supplier claims the chemical substance's identity as trade secret and certifies that the chemical substance is listed on the confidential portion of the TSCA Chemical Substance Inventory, the importer may either (1) submit, and obtain EPA's response to, a BFI request to confirm that the chemical substance is listed on the TSCA

---

222. Id. § 12.121.
223. TSCA § 13(a)(1), 15 U.S.C. § 2612(a)(1).
224. 40 C.F.R. § 707.20(c)(1)(iii).

Confidential Inventory before importing that chemical substance;[225] or (2) rely upon a supplier's certification letter.[226] A supplier's certification letter is a certification from a supplier of a chemical substance's compliance with TSCA.

The second option is not risk-free. If a company imports a noncomplying shipment based on an incorrect statement in a supplier's certification letter, U.S. Customs or EPA can detain the shipment and/or assess civil penalties. TSCA regulations state, however: "[T]he greater the effort an importer makes to learn the identities of the imported substances and their compliance with TSCA, the smaller his chance of committing a violation by importing a noncomplying shipment. If a shipment is ultimately determined to have violated TSCA, the good faith efforts of the importer to verify compliance, as evidenced by documents contained in his files, may obviate or mitigate the assessment of a civil penalty."[227]

Import Certificate Preparation

In accordance with EPA requirements, import certificates—certification statements regarding a chemical substance's compliance with TSCA—must accompany shipments of chemical substances from the foreign supplier to an importer upon presentation for entry by U.S. Customs.

There are three types of import certificates: positive, negative, and blanket. One of these types of import certificates must accompany each shipment containing chemical substances, including those contained in mixtures, or articles that are subject to the TSCA importation requirements. Import certificates must be preprinted, typed, or stamped on the invoice for each shipment or attached to such invoice or other shipping document, and presented to U.S. Customs Service upon entry.[228]

Positive Import Certificates

If the chemical substances in a shipment to be imported comply with TSCA, the importer shall prepare a positive import certificate.

---

225. *See id.* § 720.25 for BFI request requirements.
226. *Id.* § 707.20(c)(1)(iii) ("the importer should attempt to discover the chemical constituents of the shipment by contacting another party to the transaction").
227. *Id.*
228. *Id.* § 707.20(c).

Categories of substances that can receive a TSCA positive certification include

- "Chemical substances" as defined under TSCA section 3 that are listed on the TSCA Inventory or otherwise exempt from TSCA Inventory listing and otherwise in compliance with TSCA;
- Chemical substances that are components of a mixture that are listed on the TSCA Inventory or otherwise exempt from TSCA Inventory listing and otherwise in compliance with TSCA;
- An article containing a chemical substance or mixture designated to be used or released (e.g., ink in pens) that is listed on the TSCA Inventory or otherwise exempt from TSCA Inventory listing and otherwise in compliance with TSCA;
- TSCA R&D substances and/or samples eligible for PMN exemption and otherwise in compliance with TSCA; and
- TSCA test marketing substances and/or samples eligible for PMN exemption and otherwise in compliance with TSCA.

Each positive import certificate must contain the following language:

> I certify that all chemical substances in this shipment comply with all applicable rules or orders under TSCA and that I am not offering a chemical substance for entry in violation of TSCA or any applicable rule or order under TSCA.[229]

Negative Import Certificates

If the chemical substances in a shipment are exempt from TSCA, the importer shall prepare a negative import certificate, unless the chemical substance, mixture, or article is exempt from TSCA import certification requirements, as discussed below. Categories of chemical substances subject to TSCA negative certification include

- Pesticides not accompanied by EPA Form 3540-1 (the entry form required by EPA for pesticides offered for import into the United States);

---

229. Id. § 707.20(b)(2)(i).

- Any food, food additive, drug, drug intermediate, cosmetic, cosmetic intermediate, or medical device subject to the FFDCA not accompanied by the U.S. Food and Drug Administration (FDA) Form FD701 (the entry form required by FDA for food, food additives, drugs, drug intermediates, cosmetics, cosmetic intermediates, and medical devices offered for import into the United States);
- Nuclear source materials, special nuclear materials, or nuclear byproduct materials; and
- Firearms and ammunition.[230]

Each negative import certificate must contain the following language:

> I certify that all chemicals in this shipment are not subject to TSCA.[231]

Blanket Import Certificates

A blanket import certificate is a single import certificate that is used to certify TSCA compliance for multiple shipments of the same chemical substances, including those contained in mixtures, or articles over a one-year period on a calendar year basis.[232] Blanket import certificates must be approved by the appropriate district director of the U.S. Customs Service and filed with each U.S. Customs district (not at every port of entry within that district).[233] If a positive blanket import certificate has been filed, the importer must reference on commercial invoices or entry documents for covered shipments that a positive blanket import certificate has been filed with the following statement:

> Importation of the products described above are subject to certificate on file with the District Director in respect of compliance with TSCA executed by [IMPORTER] on *month day, year*, the terms of which, including the fact of its execution are incorporated herein by this reference.[234]

---

230. EPA, TOXIC SUBSTANCES CONTROL ACT: A GUIDE FOR CHEMICAL IMPORTERS/EXPORTERS 3–4, 25 (Apr. 1991) [hereinafter GUIDE FOR CHEMICAL IMPORTERS/EXPORTERS]. This document can be found by using the search engine at http://nepis.epa.gov.
231. 40 C.F.R. § 707.20(b)(2)(ii).
232. GUIDE FOR CHEMICAL IMPORTERS/EXPORTERS, *supra* note 230, at 20–21.
233. Id.
234. Id. at 24.

If a negative blanket import certificate has been filed, the importer must reference on commercial invoices or entry documents for covered shipments that a negative blanket import certificate has been filed with the following statement:

> Importation of the products described above are subject to certificate on file with the District Director indicating that they are not subject to TSCA executed by [IMPORTER] on *month day, year*, the terms of which, including the fact of its execution are incorporated herein by this reference.[235]

Shipments without Import Certificates

Customs Shipments Procedure U.S. Customs detains any shipment containing chemical substances, mixtures, or articles that is not accompanied by an import certificate at the port of entry or when Customs has reasonable grounds to believe that the shipment is not in compliance with TSCA.[236] EPA can also notify Customs to detain a shipment that EPA has reasonable grounds to believe is not in compliance with TSCA.[237]

Non-Customs Courier Shipments Procedure Compliance with FedEx, DHL, and other shipping companies' import clearance procedures does not guarantee that shipments comply with TSCA importation requirements. Foreign suppliers must be directed, therefore, to ensure that all shipments are in compliance with TSCA. Postimportation import certificates—certificates prepared by an importer after a shipment containing chemical substances, including those contained in mixtures or articles, is sent in non-Customs courier shipments to the importer without an import certificate—may be filed after receipt of the shipment directly with EPA to the TSCA Document Control Office.[238]

Exemptions

TSCA exempts the following shipments from any TSCA importation certification requirements:

---

235. Id.
236. 19 C.F.R. § 12.122(b).
237. Id.
238. GUIDE FOR CHEMICAL IMPORTERS/EXPORTERS, *supra* note 230, at 32.

- Pesticides and FDA-regulated substances accompanied by appropriate FDA and EPA import forms (e.g., EPA Form 3540-1, FDA Form FD701);
- Tobacco or any tobacco products, unless EPA specifies otherwise; and
- Articles containing chemical substances or mixtures that are not intended to be used or released and have no separate commercial purpose.[239]

Recordkeeping

Documents that the importer may wish to maintain to demonstrate compliance with TSCA section 13 include

- All import certificates;
- Any foreign supplier's invoice concerning that chemical substance, mixture, or article;
- Documents (e.g., supplier's certification letters) used in good faith to determine a chemical substance, mixture, or article's compliance with TSCA section 5, 6, or 7; and
- If applicable, any correspondence with EPA regarding a chemical substance or mixture's compliance with TSCA section 13.

TSCA section 13 and its regulations do not specify how long documents demonstrating compliance with TSCA should be maintained. Generally, however, TSCA regulations provide that documents must be maintained for five years.[240]

---

239. *Id.* at 3–5, 25.
240. *See, e.g.*, 40 C.F.R. § 720.78 (TSCA section 5 PMN and related records must be kept for five years); *id.* § 717.15 (TSCA section 8(c) records of significant adverse reactions of nonemployees must be kept for five years); *id.* §§ 721.40, 721.125 (SNUR records must be kept for five years). *See also* 3M Co. v. Browner, 17 F.3d 1453 (D.C. Cir. 1994) (applying five-year statute of limitations to TSCA administrative civil penalty actions).

CONFIDENTIAL AND TRADE SECRET INFORMATION

Authority

TSCA section 14(a) prohibits EPA, except in certain limited circumstances, from disclosing to the public trade secrets and commercial or financial information that is privileged or confidential.[241] TSCA section 14(a) provides that EPA shall disclose CBI to federal officers or employees in connection with their official duties under laws for the protection of health or the environment, or for specific law enforcement purposes, and to contractors with the United States, if deemed necessary for the satisfactory performance of their duties in connection with TSCA.[242] Section 14(a) also mandates disclosure of otherwise CBI when EPA determines that disclosure is necessary to protect against an "unreasonable risk of injury to health or the environment."[243] Finally, section 14(a) permits disclosure of such information when relevant in proceedings under TSCA, but provides that disclosure is to be made "in such manner as to preserve confidentiality to the extent practicable without impairing the proceeding."[244]

TSCA section 14(b) provides that the prohibitions on disclosure set forth in section 14(a) do not prohibit the disclosure of a health and safety study and its underlying data, and that a Freedom of Information Act (FOIA) request for such information shall not be denied on the basis of FOIA Exemption 4 (trade secrets). Section 14(b) also specifies, however, that EPA shall not release data that disclose either processes used in manufacturing or processing a chemical substance or mixture. Section 14(b) also specifies that EPA shall not release data that discloses the portion (i.e., percentage) of a mixture comprised by any chemical substances in a mixture.

TSCA Provisions Relating to Confidential Business Information

EPA has promulgated regulations establishing procedures that manufacturers must follow to invoke and substantiate a claim that information is confidential.[245] These regulations apply to claims of confidential-

---

241. TSCA § 14(a), 15 U.S.C. § 2613(a).
242. TSCA § 14(a)(1), (2), 15 U.S.C. § 2613(a)(1), (2).
243. TSCA § 14(a)(3), 15 U.S.C. § 2613(a)(3).
244. TSCA § 14(a)(4), 15 U.S.C. § 2613(a)(4).
245. 40 C.F.R. §§ 2.201–2.215.

ity under all the statutes that EPA administers. In addition, EPA has promulgated separate regulations that modify the general CBI provisions for various statutes it administers, including TSCA.[246]

A person making a claim of confidentiality must assert such claim at the time the information is submitted to EPA "by placing on (or attaching to) the information . . . a cover sheet, stamped or typed legend, . . . employing language such as *trade secret, proprietary*, or *company confidential*."[247] For most CBI claims, EPA does not require substantiation of the claim of confidentiality at the time of submission and typically does not evaluate a claim of CBI until it receives a request for the information under the FOIA. When it determines that information claimed as confidential is subject to disclosure, through receipt of a FOIA request, for example, EPA must notify in writing and by certified mail the manufacturer, processor, or distributor who submitted the information. EPA is prohibited from releasing such information until the expiration of 30 days after the manufacturer, processor, or distributor has received the required notice.[248] EPA's notice requests the submitter to provide substantiation of the CBI claim.

If EPA upholds the claim of confidentiality, it notifies the claimant that EPA honors the confidentiality of the materials. If EPA rejects the claim of confidentiality, EPA advises the claimant that its claim has been denied and that the determination is a final agency action subject to judicial review.[249] EPA will disclose the information unless the claimant acts, within 30 days, to commence an action in federal district court seeking injunctive relief to prevent EPA from disclosing the information.

In addition to its generic regulations governing information obtained by EPA under TSCA generally, EPA has established by regulation requirements applicable to the confidentiality of specific kinds of information under particular TSCA sections. For this reason, a person submitting records to EPA and wishing to claim confidentiality should not rely entirely on EPA's regulations governing CBI, even as modified for claims made generally under TSCA. Rather, submitters should examine the specific TSCA section governing the reports or other submissions at issue.

---

246. *Id.* § 2.306.
247. *Id.* § 2.203(b).
248. TSCA § 14(c), 15 U.S.C. § 2613(c)(2)(A).
249. *Id.* § 2.205(f)(1), (2).

Enforcement and Penalties

Unlawful Activities

Under TSCA section 15, it is unlawful to (1) fail or refuse to comply with a section 4 test rule or order, or any rule, order, or requirement issued under sections 5 or 6; (2) use for commercial purpose a chemical substance or mixture that the person knew or had reason to know was manufactured, processed, or distributed in violation of sections 5 or 6, or a rule or order issued under sections 5, 6, or 7; (3) fail or refuse to establish or maintain records, submit reports, notices, or other information, or permit access to or copying of records required under TSCA; or (4) fail or refuse to permit entry or inspection as required by section 11.[250] Like most environmental laws, TSCA is a strict liability statute, and there is no requirement that a violator's conduct be knowing or willful before civil penalties can be imposed under section 15.

Civil Penalties

Under section 16(a) of TSCA, as updated by regulation, EPA can impose civil penalties of up to $37,500 for each violation of section 15, with each day constituting a separate violation.[251] In determining the amount of any civil penalty, EPA must consider (1) the nature, circumstances, extent, and gravity of the violation; (2) the violator's ability to pay, including its ability to continue to do business; (3) the history of prior violations; (4) the violator's culpability; and (5) "such other matters as justice may require."[252] Before assessing a penalty, EPA must provide to the violator written notice of the proposed penalty and must provide 15 days from the date of notice for the violator to request a hearing.

TSCA Section 16 Settlements

Most TSCA civil penalty proceedings result in negotiated settlements. EPA's procedures for the administration of civil penalty proceedings encourage settlement, where consistent with the goals of

---

250. TSCA § 15, 15 U.S.C. § 2614.
251. Civil Monetary Penalty Inflation Adjustment Rule, 73 Fed. Reg. 75,340, 75,345 (Dec. 11, 2008).
252. TSCA § 16(a)(2)(B), 15 U.S.C. § 2615(a)(2)(B).

TSCA.[253] A respondent may request a settlement conference at any time during the course of an administrative proceeding.

Criminal Penalties

TSCA section 16(b) authorizes EPA to seek criminal penalties against any person who "knowingly or willfully" violates any provision of TSCA section 15. EPA can seek criminal fines of up to $37,500 for each day the violation continues and/or imprisonment for up to one year. EPA can seek criminal penalties in lieu of, or in addition to, civil penalties.

Specific Enforcement

TSCA section 11 grants to EPA broad authority to conduct inspections to enforce the Act. Under TSCA section 11(a), EPA may inspect any establishment, facility, or other premises in which chemical substances or mixtures are manufactured, processed, stored, or held before or after their distribution in commerce, and any conveyance being used to transport chemical substances in connection with their distribution in commerce.[254] Failure or refusal to permit entry or inspection as required under section 11 constitutes an unlawful act under TSCA section 15, giving rise to civil penalties.

TSCA inspections may extend to all things within the facility or conveyance inspected (including records, files, papers, processes, controls, and facilities) bearing on compliance with TSCA. Inspections cannot, however, extend to financial data, sales data, pricing data, or research data (other than data required to be retained under TSCA), unless the nature and extent of such data are described with reasonable specificity in a written notice of inspection.

TSCA section 11 permits EPA to designate persons to conduct TSCA inspections, and EPA frequently has delegated its TSCA inspection authority to state employees. EPA's authority to make such designations has been affirmed in both administrative and judicial

---

253. 40 C.F.R. § 22.18.
254. TSCA § 11(a), 15 U.S.C. § 2610(a).

fora.[255] Under TSCA section 11(a), an inspection "may only be made upon the presentation of appropriate credentials and of a written notice to the owner, operator, or agent in charge of the premises or conveyance to be inspected."[256]

EPA is authorized to seek a search warrant to conduct inspections under TSCA and can do so if it is denied entry, or in exceptional circumstances when it considers a surprise inspection crucial to its enforcement goals.[257] In most cases, however, a TSCA inspection takes place only after the facility has received a written notice of inspection. At the beginning of the inspection, the inspector presents his or her credentials and seeks the consent of the owner or operator to the inspection. Each inspection must be "commenced and completed with reasonable promptness" and must be conducted at reasonable times, within reasonable limits, and in a reasonable manner.[258]

Subpoena Power

TSCA section 11(c) authorizes EPA to issue administrative subpoenas to require (1) the attendance and testimony of witnesses; (2) the production of reports, papers, documents, and answers to questions; and (3) other information that the EPA administrator "deems necessary."[259]

A recipient of an EPA subpoena may refrain from complying with it, without penalty, until directed otherwise by a federal court order; but the EPA administrator is authorized to petition a federal district court to order compliance.[260]

Specific Enforcement and Seizure

TSCA section 17(a) grants the U.S. district courts the authority to restrain any violation of TSCA section 15; compel the taking of any action required under TSCA; or direct any manufacturer or processor of a chemical substance, mixture, or product in violation of TSCA section

---

255. *See In re* Lazarus, Docket No. TSCA-V-C-32-93, 1995 EPA ALJ LEXIS 17 (May 25, 1995) (TSCA section 11 permits EPA to designate state officials to conduct TSCA inspections); Aluminum Co. of Am. v. DuBois, No. C80-1178V, slip op. at 7 (W.D. Wash. May 29, 1981) (EPA authorized to designate private contractors as inspectors under TSCA section 11).
256. TSCA § 11(a), 15 U.S.C. § 2610(a).
257. *See* Boliden Metech, Inc. v. EPA, 695 F. Supp. 77 (D.R.I. 1988) (EPA is authorized to obtain an ex parte warrant to fulfill its inspection authority under TSCA section 11).
258. TSCA § 11(a), 15 U.S.C. § 2610(a).
259. TSCA § 11(c), 15 U.S.C. § 2610(c).
260. TSCA § 11(c), 15 U.S.C. § 2610(c).

5 or 6, or Subchapter IV (relating to lead exposure reduction), or a rule or order under section 5 or 6, or Subchapter IV, and distributed in commerce, (1) to give notice of such fact to distributors in commerce of each such substance, and to the extent reasonably ascertainable, to other persons in possession of such substance; (2) to give public notice of such risk of injury; and (3) to either replace or repurchase such substance, whichever the person to which the requirement is directed elects.[261]

Under TSCA section 17(b), district courts may seize and condemn any chemical substance, mixture, or product manufactured, processed, or distributed in violation of TSCA.[262] Section 7 of TSCA gives EPA similar authority to institute an action in the district court to seize any "imminently hazardous chemical substance or mixture or any article containing such a substance or mixture."[263]

CITIZENS' PETITIONS

Under TSCA section 21, any person may petition the EPA administrator to "initiate a proceeding for the issuance, amendment, or repeal of a rule" under TSCA sections 4, 6, or 8, or an order issued under TSCA sections 5(e) or 6(b)(2).[264] A TSCA section 21 petition must set forth facts that the petitioner believes establish the need for the action requested.

EPA is required to grant or deny the petition within 90 days of its receipt.[265] The administrator may hold a public hearing or may conduct such investigation or proceeding as deemed appropriate to determine whether the petition should be granted.[266] If EPA grants the petition, it must promptly commence an appropriate proceeding.[267] If EPA denies the petition, it must publish its reasons for the denial in the *Federal Register*.[268] Within 60 days of denial or no action, petitioners may commence a civil action in a U.S. district court to compel initiation of the requested rulemaking.[269] When reviewing a petition for

---

261. TSCA § 17(a), 15 U.S.C. § 2616(a).
262. TSCA § 17(b), 15 U.S.C. § 2616(b).
263. TSCA § 7(a)(1)(A), 15 U.S.C. § 2606(a)(1)(A).
264. TSCA § 21(a), 15 U.S.C. § 2620(a).
265. TSCA § 21(b)(3), 15 U.S.C. § 2620(b)(3).
266. TSCA § 21(b)(2), 15 U.S.C. § 2620(b)(2).
267. TSCA § 21(b)(3), 15 U.S.C. § 2620(b)(3).
268. TSCA § 21(b)(3), 15 U.S.C. § 2620(b)(3).
269. TSCA § 21(b)(4)(A), 15 U.S.C. § 2620(b)(4)(A).

a new rule, the court must provide an opportunity for de novo review of the petition.[270] After hearing the evidence, the court can order EPA to initiate the requested action.

Key Business Issues

Practical Tips for Commonly Encountered Issues

Recognizing Applicability of TSCA and Confirmation of TSCA Listing

Many entities do not recognize the materials that they import or manufacture as "chemicals" and thus, do not appreciate that TSCA is directly applicable to their business. It is critical that TSCA Inventory status, including chemical substances that are components of a mixture and chemical substances contained in articles that are designed to be used or released, is confirmed and that this confirmation occurs before manufacture or importation. Likewise, companies should be fully aware that changes in supplied raw materials, formulation processing, and other modifications can affect the manufactured substance or mixture, which may require new chemical notification.

Problems in confirming TSCA Inventory listing can be encountered when foreign suppliers of raw materials are not willing to provide sufficient information to confirm whether a substance is listed. See discussion on options to address this issue in the "Establishing Compliance with TSCA" section of this chapter.

Companies engaged in contract and/or toll manufacturing should establish clear contractual language regarding which entity is responsible for reporting and recordkeeping obligations for TSCA regulations.

As a matter of good management practices, companies should maintain complete records of chemical inventories and the TSCA status of those chemical substances, components, and products.

TSCA Import Certifications

Although U.S. Customs can refuse entry of chemicals that are not in compliance with TSCA, companies should not assume that inaction by Customs on their chemical products indicates TSCA compliance.

---

270. TSCA § 21(b)(4)(B), 15 U.S.C. § 2620(b)(4)(B).

2.1. Toxic Substances Control Act (TSCA)

TSCA Awareness

Companies should develop and implement standard operating procedures (SOPs) for all application aspects of TSCA on its businesses. SOPs should be reviewed and updated on an annual basis. Companies should also engage in TSCA awareness training for all applicable employees, including marketing personnel, production supervisors, occupational health workers, and others who may have a role in TSCA compliance.

EPA makes TSCA notices and rulemakings public through publication in the *Federal Register*. EPA does not contact companies to inform them of reporting or testing obligations. As such, companies should have a process in place for daily review of the *Federal Register* to track ongoing rules and policies. Likewise, companies should sign up for automatic e-mail notifications from EPA's Office of Pollution Prevention and Toxics, in which EPA provides prepublication notice of certain activities and offers webinar training and other information.

Another approach to maintain TSCA awareness is to engage in opportunities to network within consortium groups.

Trends

TSCA Reform and EPA Actions

There have been recent efforts to reform TSCA, in part but not limited to EPA's inability to use its TSCA section 6 authority to ban successfully a chemical product. Although congressional efforts on TSCA reform continue, the administration will likely ramp up its efforts to address perceived TSCA deficiencies within the confines of current TSCA rules.

EPA Efforts to Address New Technologies under TSCA

As noted in the "Scope of TSCA" section of this chapter, EPA states: "TSCA defines 'chemical substance' broadly and in terms which cover microorganisms as well as traditional chemicals."[271] EPA has likewise stated that all substances, including nanoscale substances that meet the TSCA definition of chemical substance, are subject to TSCA.[272]

---

271. Microbial Products of Biotechnology; Final Regulation under the Toxic Substances Control Act; Final Rule, 62 Fed. Reg. 17,910 (Apr. 11, 1997) (to be codified at 40 C.F.R. pts. 700, 720, 721, 723, and 725) (promulgating final rule under TSCA section 5 to establish notification procedures for review of certain new microorganisms).
272. *See* EPA, Control of Nanoscale Materials under the Toxic Substances Control Act, http://www.epa.gov/oppt/nano/.

The rapidly evolving industry of biobased chemical products has catalyzed further debate as to how existing TSCA regulations and policies should be applied to biobased chemicals.[273]

EPA Focus on Chemical Nomenclature, Policy "Reinterpretations"

EPA recently enacted several significant enforcement procedures on companies relying on listing of chemicals on the Inventory, which EPA claims are not accurate or appropriate for the manufactured or imported chemicals. These enforcement actions seem to be focused on Class 2 substances (chemical substance that cannot be fully represented by a complete, specific chemical structure diagram).[274]

In a similar trend, EPA has indicated its intent to "clarify" how certain statutory mixtures should be reported. The statutory mixtures at issue are Cement, Portland, Chemicals (CASRN 65997-15-1); Cement, Alumina, Chemicals (CASRN 65997-16-2); Glass, Oxide Chemicals (CASRN 65997-17-3); Frits, Chemicals (CASRN 65997-18-4); Steel Manufacture, Chemicals (CASRN 65997-19-5); and Ceramic Materials and Wares, Chemicals (CASRN 66402-68-4). The "clarification" would negate companies' abilities to report the mixture under the TSCA Inventory listed CAS number. Many people view this "clarification" as a reversal of long-standing guidance. The EPA action has caused confusion among regulated entities on reporting obligations.[275]

EPA's apparent desire to modify nomenclature approaches relied on by industry since the inception of the TSCA Inventory through enforcement cases or "clarification" notices, rather than through public review and comment, is troubling.

EPA Expanding TSCA Risk Management Authorities

In what could be viewed as an effort to circumvent the hurdles associated with risk management actions under section 6, EPA has begun using its section 5 SNUR authority to achieve restrictions

---

273. *See* Lynn L. Bergeson, Charles M. Auer & R. David Peveler, *TSCA and the Regulation of Renewable Chemicals*, 8(5) INDUS. BIOTECHNOL. 262–71 (Oct. 2012), doi:10.1089/ind.2012.1539.
274. *See* Lynn L. Bergeson, *$1.4 Million Civil Penalty for TSCA Violations*, POLLUT. ENG'G, Apr. 20, 2012.
275. *See* Lisa R. Burchi, Charles M. Auer, Kathleen M. Roberts & Lynn L. Bergeson, *Are TSCA Section 8(b)(2) Statutory Mixture Categories Subject to Reporting under the Chemical Data Reporting Rule?*, Daily Env't Rep. (Bloomberg BNA) No. 16, Jan. 26, 2012.

on certain chemicals. For example, in 2012, EPA attempted to combine SNUR requirements with a section 4 test rule for certain polybrominated diphenyl ethers (PBDE), thus forcing industry to either abide by the use restrictions in the SNUR or agree to significant testing costs to maintain the commercial market.

Other EPA proposals demonstrate a trend in which the existing article exemption is waived. Examples include a proposed section 8(d) reporting requirement on cadmium and cadmium compounds (later withdrawn based on concerns raised by industry), a final section 8(d) rule for lead, and SNURs for nanomaterials and certain brominated chemicals.

EPA also has announced several actions to enhance EPA's chemical management program, including development of action plans on ten chemicals and chemical groups, efforts to prioritize existing chemicals and initiate focused risk assessment programs, and increasing transparency. In early 2013, EPA issued five draft risk assessments developed under its TSCA Work Plan Chemicals program, and it has published a list of 83 other chemicals that may be subject to future assessment activities.[276]

EPA Actions to Increase Transparency—May Lessen the Ability to Claim CBI

Recent EPA policy to increase the transparency of health and safety information means that companies' ability to claim CBI for certain information will be reduced or will have increased burden. In an updated policy on CBI, EPA states that TSCA section 14(b) "does not extend confidential treatment to health and safety studies, or data from health and safety studies, which, if made public, would not disclose processes used in the manufacturing or processing of a chemical substance or mixture or, in the case of a mixture, the release of data disclosing the portion of the mixture comprised by any of the chemical substances in the mixture."[277] In a January 21, 2010, *Federal Register* notice, EPA announced that "[w]here a health and safety study submitted under section 8(e) of TSCA involves a chemical identity that is already listed on the public portion of the TSCA Chemical Substance Inventory, EPA expects to find that the chemical identity clearly is not entitled to

---

276. More details on these efforts are *available at* EPA, Enhancing EPA's Chemical Management Program, http://www.epa.gov/opptintr/existingchemicals/pubs/enhanchems.html.
277. 75 Fed. Reg. 29,754 (May 27, 2010).

confidential treatment."[278] Under the CDR, EPA expanded the information elements that require upfront substantiation for CBI claims, including processing and use information reported.[279]

EPA Requires Electronic Reporting

On April 17, 2012, EPA proposed a rule to require electronic reporting[280] for most of the reporting obligations under TSCA. Rules for mandatory electronic reporting already exist for reporting under the CDR rule under section 8(a) and for PMN under section 5. To adhere to the required electronic reporting, any entity that is expected to report under TSCA needs to register under EPA's Central Data Exchange (CDX).[281] The CDX system includes provisions for submitting CBI electronically.

Resources

EPA's Office of Pollution Prevention and Toxics (OPPT) manages programs under TSCA. The various programs and regulatory requirements are included on the OPPT website, at http://www.epa.gov/oppt. Specific areas of interest include

- Summary of the Toxic Substances Control Act at http://www.epa.gov/lawsregs/laws/tsca.html; includes an overview of the various sections of TSCA, information on compliance and enforcement, a summary of EPA history in implementing TSCA, and a link to the TSCA legislation text.
- EPA's New Chemicals Program at http://www.epa.gov/oppt/newchems/; includes information, guidance, and other resources for submitting PMNs, polymer, LVE, LoREX, and R&D exemption, SNURs, and EPA's Sustainable Futures Initiative, which provides chemical developers the same risk-screening models that EPA uses to evaluate new chemicals before they enter the market.
- TSCA 8(e) notices information at http://www.epa.gov/oppt/tsca8e/; provides reporting requirements, guidance, frequently asked questions, and links to the most recent 8(e) and FYI submissions.
- TSCA Importing and Exporting Requirements at http://www.epa.gov/oppt/import-export/; includes basic information on reporting obligations for chemical exports and imports.

---

278. 75 Fed. Reg. 3462 (Jan. 21, 2010).
279. 40 C.F.R. § 711.30(d)(1).
280. Electronic Reporting under the Toxic Substances Control Act; Proposed Rule, 77 Fed. Reg. 22,707 (Apr. 17, 2012).
281. *See* EPA, Central Data Exchange, http://www.epa.gov/cdx/.

2.2

Federal Insecticide, Fungicide, and Rodenticide Act (FIFRA)

Lisa M. Campbell, Lisa R. Burchi, Sheryl L. Dolan, and Henry M. Jacoby, MS

Executive Summary

FIFRA Overview

Pesticides are regulated by the United States Environmental Protection Agency (EPA) under the Federal Insecticide, Fungicide, and Rodenticide Act (FIFRA)[1] and the Federal Food, Drug, and Cosmetic Act (FFDCA),[2] as amended in 1996 by the Food Quality Protection Act (FQPA).

Basic Provisions

FIFRA is a dynamic law with flexibility to meet emerging scientific issues and the development of new pesticidal products. The following

---

1. 7 U.S.C. § 136–136y.
2. 21 U.S.C. §§ 301–397.

list identifies those sections of FIFRA that affect the registration, sale, distribution, and use of pesticide products and devices:

- *FIFRA section 3:* Requires all pesticides to be registered; defines the conditions for registration and limitations of distribution and sale.
- *FIFRA section 4:* Requires reregistration of all pesticides and review of tolerances.
- *FIFRA section 5:* Authorizes experimental use permits.
- *FIFRA section 6:* Addresses suspensions, cancellation procedures, and existing stock provisions.
- *FIFRA section 7:* Requires registration of pesticide-producing establishments.
- *FIFRA section 11:* Addresses federal certification of applicators for restricted use pesticides (RUP).
- *FIFRA section 12:* Defines unlawful acts.
- *FIFRA section 22:* Authorizes the collection of fees.

Scope of FIFRA

Scope of Substances/Products Subject to and/or Exempt from FIFRA

FIFRA regulates pesticides. FIFRA defines a pesticide as "any substance or mixture of substances intended for preventing, destroying, repelling, or mitigating any pest."[3] EPA regulations further define the term "pesticide" as "any substance (or mixture of substances) intended for a pesticidal purpose, i.e., use for the purpose of preventing, destroying, repelling, or mitigating any pest."[4] FIFRA defines the term "pest" broadly to mean "(1) any insect, rodent, nematode, fungus, weed, or (2) any other form of terrestrial or aquatic plant or animal life or virus, bacteria, or other micro-organism (except viruses, bacteria, or other micro-organisms on or in living man or other animals)."[5]

---

3. FIFRA § 2(u), 7 U.S.C. § 136(u). FIFRA also defines the term "active ingredient," in pertinent part, to mean "an ingredient which will prevent, destroy, repel, or mitigate any pest." FIFRA § 2(a)(1), 7 U.S.C. § 136(a)(1).
4. 40 C.F.R. § 152.15.
5. FIFRA § 2(t), 7 U.S.C. § 136(t); *see also* 40 C.F.R. § 152.5.

In addition to a chemical substance, as that term is commonly understood, a pesticide substance also may be a microbial agent[6] or a plant-incorporated protectant.[7]

Active Ingredient

EPA considers a substance to be intended for a pesticidal purpose, and thus to be a pesticide requiring registration, if

- (a) The person who distributes or sells the substance claims, states, or implies (by labeling or otherwise):
  - (1) That the substance (either by itself or in combination with any other substance) can or should be used as a pesticide; or
  - (2) That the substance consists of or contains an active ingredient and that it can be used to manufacture a pesticide; or
- (b) The substance consists of or contains one or more active ingredients and has no significant commercially valuable use as distributed or sold other than (1) use for pesticidal purpose (by itself or in combination with any other substance), (2) use for manufacture of a pesticide; or
- (c) The person who distributes or sells the substance has actual or constructive knowledge that the substance will be used, or is intended to be used, for a pesticidal purpose.[8]

This regulation thus provides two alternative tests for identifying pesticides. Under the first test, established by subpart (a), EPA considers whether pesticidal claims are made for the substance. Under the second test, established by subparts (b) and (c), EPA considers whether the substance will have, or is intended to have, a pesticidal use. The substance that meets one or both of these tests is also known as the active ingredient.[9]

Exemptions

There are several provisions in EPA's regulations that provide exemptions from FIFRA regulation. These include the following:

---

6. 40 C.F.R. § 158.2100(b).
7. Id. § 174.3.
8. Id. § 152.15(a)–(c).
9. Id. § 152.3.

- Certain substances or articles are not intended for use against "pests" and thus are not pesticides. Examples include substances or articles intended to control bacteria and fungi in or on living man or animals;[10] and certain products intended only to aid the growth of desirable plants.[11]
- Certain products are not intended to prevent, destroy, mitigate, or repel pests and thus are not pesticides. Examples include deodorizers, certain bleaches, and cleaning agents;[12] and products intended to exclude pests only by providing a physical barrier against pest access and that contain no toxicants.[13]
- Certain pesticides are regulated by other federal agencies and are exempt from all FIFRA requirements. Examples include certain biological control agents and human drugs.
- Certain pesticides or classes of pesticides are of a character not requiring FIFRA registration. Examples include treated articles or substances[14] and enumerated minimum risk pesticides.[15]
- If prescribed criteria are met, certain otherwise regulated pesticides may be transferred, sold, or distributed without registration. Examples include pesticides transferred between registered establishments, either operated by the same producer[16] or not;[17] pesticides distributed or sold under an experimental use permit (EUP);[18] and pesticides transferred solely for export.[19]

---

10. Id. § 152.6(c)–(d); see also FIFRA § 2(t), 7 U.S.C. § 136(t).
11. See id. § 152.6(g).
12. See id. § 152.10(a).
13. See id. § 152.10(c).
14. See id. § 152.25(a). Certain treated articles are exempt from FIFRA and need not be registered as pesticides. First, the pesticide used to treat the article must be registered for that use, and second, the claims must be limited to the protection of the article itself. Paint treated with a pesticide to protect the paint coating, or wood products treated to protect the wood against insect or fungus infestation may be considered treated articles. See also Pesticide Registration (PR) Notice 2000-1, Applicability of the Treated Articles Exemption to Antimicrobial Pesticides (Mar. 6, 2000), http://www.epa.gov/PR_Notices/pr2000-1.pdf, addressing permissible claims for treated article products that are exempt from registration requirements.
15. See id. § 152.25(f).
16. See id. § 152.30(a).
17. See id. § 152.30(b).
18. See id. § 152.30(c).
19. See id. § 152.30(d).

Devices are not subject to registration under FIFRA section 3. EPA has declared, however, through policies published in the *Federal Register*, that certain devices are subject to establishment registration and recordkeeping requirements, labeling requirements, import and export requirements, and other enumerated requirements.[20] Devices that EPA explicitly states to be subject to FIFRA include certain ultraviolet light systems, ozone generators, water and air filters, ultrasonic devices, high-frequency sound generators, carbide cannons, foils, rotating devices, black-light traps, fly traps, electronic and heat screens, fly ribbons, flypaper, and mole thumpers. EPA exempts from regulation devices that depend for their effectiveness more on the performance of the person using the device than on the performance of the device itself and those that operate to entrap vertebrate animals. Products falling within these two categories include rat and mouse traps, flyswatters, tillage equipment for weed control, and fish traps.

In certain cases, a pesticide can be used for research and development to accumulate information needed to register the pesticide or a new use of the pesticide without the need to register the pesticide.[21] There is, however, a permit (EUP application (Form 8570-17)) that must be granted by EPA before a pesticide may be used for nonexempt experimental purposes. The EUP application must address the purpose of the proposed testing and include required data and the names and addresses of the participants in the experimental testing. All pesticides shipped or used under an EUP must bear labels containing the statements "For Experimental Use Only" and "Not for Sale to Any Person Other Than a Participant or Cooperator of the EPA-Approved Experimental Use Program," the permit number, directions for use, warning or caution statements, and other information. Data gathered under the EUP must be submitted to EPA following the testing, along with information regarding the disposition of the pesticide containers and unused pesticides. If the experimental pesticide is to be used on a food crop and has no residue tolerance, the EUP applicant must request EPA to establish a tolerance or temporary tolerance or must certify that the crop or any food/feed resulting from the testing program will be destroyed or fed to experimental animals.

---

20. *See* FIFRA §§ 2(h), 2(p), 2(q), 7, 8, 17, 25(c)(3), 25(c)(4), 7 U.S.C. §§ 136(h), 136(p), 136(q), 136e, 136f, 136o, 136w(c)(3), 136w(c)(4); 40 C.F.R. § 152.500.
21. *See* FIFRA § 5, 7 U.S.C. § 136c; 40 C.F.R. § 152.30(c), pt. 172.

Experimental uses of pesticides are presumed not to need an EUP (and thus are exempt from the need to obtain an EUP) when the experimental use of the pesticide is limited to (1) laboratory or greenhouse tests; (2) limited replicated field trials meeting specified requirements and involving no more than ten acres of land or one surface acre of water per pest (except when testing for more than one pest at the same time and location); or (3) certain other animal treatment tests whose purpose is only to assess the pesticide's potential efficacy, toxicity, or other properties, when the person conducting the test does not expect to receive any benefit in pest control from the pesticide's use. For other experimental uses of a pesticide, an EUP is presumed to be needed "absent a specific determination by EPA to the contrary."[22]

Inert or Other Ingredients

FIFRA defines an "inert ingredient" as any ingredient other than an "active ingredient."[23] In registering any pesticide, a registrant must provide the complete product formula, including such inert ingredients.[24] The only inert ingredients approved for use in pesticide products applied to food are those for which EPA has published either a tolerance or tolerance exemption[25] or those where no residues are found in food. EPA has developed lists of inert ingredients for food and nonfood use, as well as lists of acceptable perfumes and proprietary formulations (trade name chemicals). EPA regulations currently do not require as a general matter identification on product labeling of all inert ingredients, although EPA may require that inert ingredients be listed in the ingredient statement if it is determined that such ingredients may pose a hazard to humans or the environment.[26] EPA encourages pesticide registrants to substitute the term "other ingredients" for the term "inert ingredients" because EPA believes that the term "inert ingredients" may be misconstrued by consumers to be water or otherwise harmless ingredients.[27]

---

22. 40 C.F.R. § 172.3(d).
23. FIFRA § 2(m), 7 U.S.C. § 136(m).
24. FIFRA § 3(c)(1)(D), 7 U.S.C. § 136a(c)(1)(D).
25. 40 C.F.R. pt. 180, especially §§ 180.910–.960.
26. Id. § 156.10(g)(7).
27. PR Notice 97-6, Use of Term "Inert" in the Label Ingredients Statement (Sept. 11, 1997), http://www.epa.gov/PR_Notices/pr97-6.html.

Scope of Activities Subject to FIFRA

FIFRA affects all activities associated with pesticides or products containing pesticides (such as paints/coatings or other products that contain material preservatives, also known as treated articles).

Any production of a pesticide must occur in a registered establishment. "Produce" is interpreted broadly to mean "to manufacture, prepare, compound, propagate, or process any pesticide," including any pesticide produced pursuant to FIFRA section 5, any active ingredient or device, or to package, repackage, label, relabel, or otherwise change the container of any pesticide or device.[28] Whether pesticide production is taking place is important in determining whether establishment registration and other FIFRA requirements apply.

FIFRA section 12 makes unlawful the sale or distribution of an unregistered pesticide. "To distribute or sell" is defined as "to distribute, sell, offer for sale, hold for distribution, hold for sale, hold for shipment, ship, deliver for shipment, release for shipment, or receive and (having so received) deliver or offer to deliver."[29] Whether distribution or sale of a pesticide occurs is a key factor in determining the applicability of many FIFRA requirements and in determining, in many instances, whether FIFRA violations have taken place.[30] For example, a pesticide that is distributed and sold must be registered and must bear the label approved by EPA for that pesticide. Its composition must also be, at the time of its distribution or sale, as described in the Confidential Statement of Formula (CSF) submitted with the application for its registration. In addition, anyone who sells or offers for sale, or delivers or offers for delivery a pesticide must maintain records showing the movement of the pesticide for two years and must make the records available to EPA inspectors. Pesticide producers must report annually to EPA the amount of pesticides they distributed or sold during the previous year.

As part of the registration process, pesticide manufacturers must submit and adhere to process and formulation information; all aspects of the production process, including packaging and labeling, are subject to FIFRA requirements.

Imported pesticide products or treated articles must be registered in the United States for the intended use(s). Imports of pesticide products

---

28. FIFRA § 2(w), 7 U.S.C. § 136(w). *See also* 40 C.F.R. § 167.3.
29. FIFRA § 2(gg), 7 U.S.C. §§ 136(gg).
30. *See* FIFRA §§ 2(gg), 8(b), 12(a)(1), 12(a)(2)(F), 7 U.S.C. §§ 136(gg), 136f(b), 136j(a)(1), 136j(a)(2)(F); 40 C.F.R. §§ 167.85, 169.2(c), 169.2(d).

trigger a reporting requirement (notice of arrival), as noted in the "Imports and Exports" section of this chapter, that must be completed prior to the shipment's arrival in the United States. Pesticide residues on imported agricultural products must meet the applicable tolerance or tolerance exemption requirements.

Research efforts related to pesticides are subject to EUP requirements, unless specific exemptions are met. Data developed on pesticides are subject to adverse effects reporting requirements under FIFRA section 6(a)(2).

Finally, use of a pesticide in a manner inconsistent with the labeling requirements is unlawful under FIFRA section 12(a)(2)(G). Accordingly, applicators and handlers must use the pesticide according to the use directions. The label addresses a range of additional actions, including selection and use of personal protective equipment, and product storage and disposal.

Scope of Persons Subject to FIFRA

FIFRA affects the actions of anyone engaged in an activity that involves pesticides.

Pesticide product manufacturers must meet the applicable requirements to register their products and properly maintain these registrations once approved, including, for example, ongoing data development as required by EPA. Manufacturers must comply with requirements related to registered establishments, product development (including, for example, the transfer of unregistered pesticides and EUPs), supplemental distribution as applicable, advertising and claims, recordkeeping and reporting, and storage and disposal.

Product producers and distributors further along the supply chain must comply with registered establishment requirements if any pesticide production occurs. Supplemental distributors,[31] along with product registrants, must conform with the stringent requirements

---

31. Supplemental distribution is permitted upon notification to EPA if the pesticide is produced, packaged, and labeled by a registered establishment, if the pesticide is not repackaged, and if the pesticide's label is identical to that of the originally registered product, except for minor changes such as the addition of the distributor's name, address, and brand name. See FIFRA § 3(e), 7 U.S.C. § 136a(e); 40 C.F.R. § 152.132. The registration of a supplementally distributed product belongs to the registrant, not the supplemental distributor, and the registrant remains responsible for the product in all circumstances. The distributor is considered an agent of the registrant, however, and both the registrant and distributor may be held liable for violations pertaining to the distributor product.

that apply to supplementally distributed products. Toll manufacturing agreements with the product registrants are often warranted.

Because only approved inert ingredients are permitted in registered formulated pesticides, raw material suppliers of these ingredients may wish to confirm and/or support approval of their chemicals as inert ingredients to enable these markets.

Importers of pesticide products or products containing pesticides (treated articles) must confirm that these products are registered in the United States for the intended use(s). Importers of pesticide products and devices must complete and submit a notice of arrival, as noted in the "Imports and Exports" section of this chapter, prior to the shipment's arrival in the United States. Importers of agricultural products that contain pesticide residues must confirm that these residues comply with the applicable tolerance or tolerance exemption requirements.

EPA regulates pesticide applicators and end users primarily through pesticide labels and labeling. The labels and labeling provide information about the terms and conditions of use, including application methods and use rates, number of permissible applications, worker safety precautions and requirements, toxicity and hazard information, protective clothing and equipment requirements, geographic and other use limitations, field reentry limitations, and disposal and storage requirements, among other things. Current FIFRA labeling requirements are found in 40 C.F.R. part 156. Subpart K of 40 C.F.R. part 156 contains label requirements added by EPA's Worker Protection Standard (WPS) regulations.[32] For further information on labeling requirements, *see* Regulation of Substances/Products, Use Restrictions, Label Language.

The WPS also requires certain workplace practices designed to reduce or eliminate exposure to pesticides and establishes procedures for responding to exposure-related emergencies.[33] These requirements apply to workers and handlers involved in the production of agricultural plants on farms or in nurseries, greenhouses, and forests.

---

32. *See* Worker Protection Standard, 57 Fed. Reg. 38,102 (Aug. 21, 1992) (to be codified at 40 C.F.R. pts. 156 and 170); PR Notice 93-7, Labeling Revisions Required by the Worker Protection Standard (WPS) (Apr. 20, 1993); PR Notice 93-11, Supplemental Guidance for PR Notice 93-7—Labeling Revisions Required by the Worker Protection Standard (WPS) (Aug. 13, 1993); Pub. L. No. 103-231 (Apr. 6, 1994); Pesticide Worker Protection Standard; Grace Period for Providing Worker Safety Training, 60 Fed. Reg. 21,944 (May 3, 1995) (to be codified at 40 C.F.R. pt. 170); Exception to Worker Protection Standard Early Entry Restrictions for Limited Contact Activities, 60 Fed. Reg. 21,955 (May 3, 1995) (administrative exception decision to 40 C.F.R. pt. 170).
33. 40 C.F.R. pt. 170.

Employers are required to ensure that workers follow these requirements. Handlers are required to use the protective clothing and equipment specified on the product labeling.[34] Use of a pesticide in a manner inconsistent with the labeling requirements is unlawful under FIFRA section 12(a)(2)(G). Additional use restrictions are placed on products that are classified for "restricted use."[35] A RUP may only be sold to and applied by certified applicators or persons under their direct supervision and only for the uses covered by the certified applicator's certification.[36]

NOTIFICATION/REGISTRATION/APPROVAL REQUIREMENTS AND COMPLIANCE

Summary of Authority Governing the Requirements for Notification/Registration/Approval

No person may distribute or sell any pesticide that is not registered pursuant to FIFRA.[37] Under FIFRA, several types of pesticide registrations are established, including (1) unconditional registrations; (2) conditional registrations; (3) supplemental registrations; (4) state special local needs registrations; (5) RUPs; and (6) emergency exemptions from registration.

Separately, EPA regulates pesticide residues in or on food under FFDCA section 408, as amended in 1996 by the FQPA.[38] Maximum permissible pesticide residue limits are known as "tolerances." EPA does not register a pesticide under FIFRA unless necessary tolerances have been established under the FFDCA because pesticide residues in or on food render it adulterated. Tolerances "are usually expressed in terms of parts of the pesticide residue per million parts of the food (ppm), by weight."[39] EPA is authorized to set tolerances in both raw

---

34. *Id.* § 170.240.
35. *See* FIFRA § 3(d)(1)(C), 7 U.S.C. § 136a(d)(1)(C).
36. *See* 40 C.F.R. § 156.10(j)(2)(i)(B); EPA, OPP, LABEL REVIEW MANUAL, ch. 6: Use Classification.
37. FIFRA § 3(a), 7 U.S.C. § 136a(a).
38. The FDA has the authority to enforce the tolerances set by EPA. Legal and Policy Interpretation of the Jurisdiction under the Federal Food, Drug, and Cosmetic Act of the Food and Drug Administration and the Environmental Protection Agency over the Use of Certain Antimicrobial Substances, 63 Fed. Reg. 54,532, 54,534 (Oct. 9, 1998).
39. 40 C.F.R. § 177.3.

2.2. Federal Insecticide, Fungicide, and Rodenticide Act (FIFRA)

agricultural commodities (RAC) and processed foods.[40] A separate tolerance for a processed food is not necessary as long as the residues in the processed food do not exceed the tolerance for the corresponding RAC. If the residues in the processed food concentrate above the tolerance for the RAC, EPA must set a separate tolerance under section 408 for the processed food.

Standard of Review/Burden of Proof

Registrations

Pursuant to FIFRA section 3(c)(5), EPA shall register a pesticide if it determines that (1) its composition is such as to warrant the proposed claims for it; (2) its labeling and other material required to be submitted comply with the requirements of FIFRA; (3) it will perform its intended function without unreasonable adverse effects on the environment; and (4) when used in accordance with widespread and commonly recognized practice, it will not generally cause unreasonable adverse effects on the environment.[41] In such circumstances, EPA issues to an applicant an unconditional registration for a pesticide product.

Under FIFRA section 3(c)(7), EPA may issue a "conditional registration." For a pesticide product containing an existing pesticide active ingredient, EPA may issue a conditional registration if it determines that "(i) the pesticide and proposed use are identical or substantially similar to any currently registered pesticide and use thereof, or differ only in ways that would not significantly increase the risk of unreasonable adverse effects on the environment, and (ii) approving the registration or amendment in the manner proposed by the applicant would not significantly increase the risk of any unreasonable adverse effect on the environment."[42] An applicant seeking to obtain a conditional registration must commit typically to submit data that are required to support the registration.[43] In addition, EPA may conditionally register a product containing an active ingredient not presently registered for a period reasonably sufficient for the generation of required data, on the condition that by the end of such period the registrant submits the

---

40. Under the new law, pesticide residues are excluded from the definition of "food additive" in FFDCA section 201(s) and therefore are no longer subject to the section 409 Delaney Clause. 21 U.S.C. § 321(s).
41. FIFRA § 3(c)(5), 7 U.S.C. § 136a(c)(5). *See also* 40 C.F.R. § 152.112.
42. FIFRA § 3(c)(7)(A), 7 U.S.C. § 136a(c)(7)(A). *See also* 40 C.F.R. § 152.113.
43. FIFRA § 3(c)(7)(A), 7 U.S.C. § 136a(c)(7)(A).

required data and the data do not meet or exceed applicable risk criteria and such other conditions as EPA may prescribe.[44] In granting a conditional registration, EPA must determine that use of the pesticide during that time period will not cause any unreasonable adverse effect on the environment, and that use of the pesticide is in the public interest.[45] It is not uncommon for pesticide products to be initially registered on a conditional basis, subject to requirements for additional data.

Under EPA regulations, a pesticide registration may allow for "supplemental distribution" of a registered pesticide product under another person's name and address, instead of, or in addition to, the name and address of the registrant.[46] The supplemental distributor is considered an agent of the registrant for all purposes under FIFRA, and both the registrant and the distributor may be held liable for violations pertaining to the distributor product.[47]

In addition to EPA-issued registrations under FIFRA section 3, states may regulate the sale or use of any federally registered pesticide or device, except that such state regulation may not permit any sale or use prohibited under FIFRA.[48] States may provide, however, for additional uses of federally registered pesticides or devices to meet "special local needs." Such authorization, subject to possible disapproval by EPA, allows distribution and use only within the affected state.[49] State authorization is not effective for more than 90 days if disapproved subsequently by EPA.[50] EPA may not, however, disapprove a state "special local needs" authorization based on grounds that the use of the pesticide is not essential, unless the composition and use patterns of the pesticide are "substantially similar" to those of a federally registered pesticide.[51]

Under FIFRA section 3(d), EPA may classify a pesticide for general use or restricted use or may distinguish among uses of the products in determining its classification.[52] EPA classifies a pesticide for "restricted use" if EPA determines that a pesticide, when applied in accordance with its directions for use, warnings, and precautionary statements for

---

44. FIFRA § 3(c)(7)(C), 7 U.S.C. § 136a(c)(7)(C). *See also* 40 C.F.R. §§ 152.114–.115.
45. FIFRA § 3(c)(7)(C), 7 U.S.C. § 136a(c)(7)(C).
46. 40 C.F.R. § 152.132.
47. Id.
48. FIFRA § 24(a), 7 U.S.C. § 136v(a).
49. FIFRA § 24(c)(1), 7 U.S.C. § 136v(c)(1).
50. FIFRA § 24(c)(2), 7 U.S.C. § 136v(c)(2).
51. FIFRA § 24(c)(2), 7 U.S.C. § 136v(c)(2).
52. FIFRA § 3(d), 7 U.S.C. § 136a(d).

the uses for which it is registered, or when applied in accordance with widespread and commonly recognized practice, may generally cause, without additional regulatory restrictions, unreasonable adverse effects on the environment, including injury to the applicator.[53]

Under FIFRA section 18, EPA may exempt any federal or state agency from any provision of FIFRA.[54] In response to requests submitted by state government authorities, this provision is used most commonly to permit the use of a pesticide product, without federal registration under FIFRA section 3, to control pests on crops for which the product is not registered, when necessary to address emergency conditions. In issuing a FIFRA section 18 emergency exemption from the requirement of registration, EPA requires that there be established a tolerance that defines the maximum permissible level of the pesticide residue on a food or feed crop.[55]

Tolerances

Under FFDCA section 408(b)(2)(A), the standard for establishing a tolerance is whether there is "a reasonable certainty that no harm will result from aggregate exposure to the pesticide chemical residue, including all anticipated dietary exposures and all other exposures for which there is reliable information." Aggregate exposure refers to dietary exposure under all tolerances established for the pesticide, as well as exposure from all nonoccupational sources (for example, drinking water) and the cumulative effects of the pesticide residues "and other substances that have a common mechanism of toxicity."[56] EPA also must make a specific determination that there is a reasonable certainty that no harm will result to infants and children from aggregate exposure to the pesticide. FQPA also directs EPA to develop a screening program for estrogenic effects and require the testing of all pesticide chemicals under this program.

FQPA authorizes EPA to apply this standard differently for threshold and nonthreshold effects.[57] For a threshold effect, the standard is met if the aggregate exposure is lower than the no observed adverse effect level (NOAEL) by an ample margin of safety. Congress directed EPA to apply a hundredfold safety factor to the no-effect level observed

---

53. FIFRA § 3(d)(1)(C), 7 U.S.C. § 136a(d)(1)(C).
54. FIFRA § 18, 7 U.S.C. § 136p.
55. FFDCA § 408, 21 U.S.C. § 346a.
56. FFDCA § 408(b)(2)(D)(v)–(vi), 21 U.S.C. § 346a(b)(2)(D)(v)–(vi).
57. *See* H.R. Rep. No. 104-669, pt. 2, at 41 (1996).

in animal data.[58] An additional tenfold margin of safety must be applied to take into account "potential pre- and post-natal developmental toxicity and completeness of the data" with regard to infants and children.[59] EPA can dispense with the extra tenfold margin of safety, in whole or part, if EPA determines "on the basis of reliable data" that the lesser safety factor will be safe for infants and children.[60]

For pesticides that operate by a nonthreshold mechanism, the safety standard is satisfied if the increased lifetime risk is "negligible," defined as no greater than a one-in-a-million lifetime risk.

Inventories or Other Listing Procedures

FIFRA defines in broad terms the procedures for applying to register a pesticide.[61] In addition, EPA has issued extensive additional requirements and guidance through regulations, guidelines, and other less formal guidance.

Each applicant for a pesticide registration must submit to EPA, among other information, (1) a complete copy of the labeling proposed for the product, including a statement of all claims made for it, and any directions for its use; (2) the complete formula of the pesticide (i.e., CSF);[62] and (3) a full description of any tests made and the results thereof upon which the claims are based, or alternatively, citation to data either in the public literature or previously submitted to EPA.[63] An application for a new registration of a pesticide product must be approved before any product may legally be distributed or sold.[64] The contents of an application for registration are specified in detail by EPA regulations, including extensive requirements for data on which EPA can make an evaluation of the environmental effects,

---

58. This safety factor (also called an "uncertainty" factor or a "safety margin") accounts for potential interspecies sensitivity and a potential range of sensitivities among different individuals.
59. EPA, 1996 FOOD QUALITY PROTECTION ACT—IMPLEMENTATION PLAN 10 (Mar. 1997), http://www.epa.gov/opp00001/regulating/laws/fqpa/impplan.pdf.
60. Id.
61. See FIFRA § 3, 7 U.S.C. § 136a.
62. The CSF must list each ingredient in the pesticide, its concentration, its purpose, any associated impurities, and, for some ingredients, the certified limits. The CSF defines a registered pesticide and its composition. A pesticide that does not have the composition listed in its CSF may be considered unregistered. The CSF is deemed confidential business information. See FIFRA § 3(c)(1)(D), 7 U.S.C. § 136a(c)(1)(D); 40 C.F.R. § 158.320.
63. FIFRA § 3(c)(1), 7 U.S.C. § 136a(c)(1).
64. 40 C.F.R. § 152.42.

2.2. Federal Insecticide, Fungicide, and Rodenticide Act (FIFRA)

health effects, and safety of the product.[65] These data requirements address product chemistry, mammalian and environmental toxicology, environmental fate, residue chemistry, reentry protection and spray drift, and efficacy. The precise data requirements for a particular product registration vary greatly depending on the particular active ingredient and its intended uses.

EPA distinguishes between a manufacturing use product (MUP)[66] and an end-use product (EP).[67] An MUP typically contains a high concentration, relatively speaking, of the pesticide active ingredient and is sold to other registrants to formulate an EP. Because most of EPA's data requirements attach to the active ingredient, the larger data development costs usually are associated with MUP registrations. If an EP registrant purchases a registered MUP, then the EP registrant may take the benefit of the "formulator's exemption" and may rely on the data submitted by the MUP registrant in support of the active ingredient; the EP registrant must still meet data requirements applicable to its formulated product. See "Test Data Development," on page 112, for additional information.

Registrants must distribute pesticide products that conform to the product's composition and the labeling approved by EPA in connection with the application for registration.[68] A registrant thus must amend its registration prior to changing the composition or labeling for a pesticide, although certain minor amendments may be accomplished merely by notification to EPA (e.g., adding or deleting a pest that does not pose a public health threat, adding alternate brand names, adding or deleting advisory statements that are not mandatory, and making certain changes to warranty statements).[69] The application to amend a registration must include the information concerning

---

65. *Id.* § 152.50, pt. 158.
66. "*Manufacturing use product* means any pesticide product that is not an end-use product." 40 C.F.R. § 152.3.
67. "*End use product* means a pesticide product whose labeling (1) Includes directions for use of the product (as distributed or sold, or after combination by the user with other substances) for controlling pests or defoliating, desiccating, or regulating the growth of plants, and (2) Does not state that the product may be used to manufacture or formulate other pesticide products." 40 C.F.R. § 152.3.
68. FIFRA § 12(a)(1)(B), (C), (E), (F), 7 U.S.C. § 136j(a)(1)(B), (C), (E), (F).
69. 40 C.F.R. §§ 152.44, 152.46; PR Notice 98-10, Notifications, Non-Notifications and Minor Formulation Amendments (Oct. 22, 1998), http://www.epa.gov/PR_Notices/pr98-10.pdf; PR Notice 95-2, Notifications, Non-Notifications and Minor Formulation Amendments (May 31, 1995), at 3, http://www.epa.gov/PR_Notices/pr95-2.html.

the amendment of the registration to the same extent as though such an amendment were submitted to obtain a new registration.[70]

An inert ingredient is any ingredient in a formulated pesticide product other than an active ingredient. EPA now usually requires submission of data in support of inert ingredients not approved previously by EPA. The required data set generally would include physical/chemical properties, mammalian toxicity (including acute and chronic data), environmental fate and effects, and ecotoxicity data. The data set to support a food-use inert ingredient typically would be more robust than for a nonfood use, and would also need to include a discussion of anticipated dietary (from food and drinking water), residential (dermal, inhalation, and incidental oral) and occupational exposures from the proposed and existing pesticidal uses of the chemical.

Separately, for both active and inert ingredients intended for formulation into products with food uses, EPA sets a maximum permissible pesticide residue level, or tolerance, in the specified treated commodity or grants an exemption from the requirement of a tolerance. EPA sets a tolerance or tolerance exemption in response to a petition. The petitioner must address several information requirements, including data showing the recommended amount, frequency, method, and time of application of the pesticide chemical; full reports of tests and investigations made with respect to the safety of the pesticide chemical; full reports of tests and investigations made with respect to the nature and amount of the pesticide chemical residue that is likely to remain in or on the food; a practical method for detecting and measuring the levels of the pesticide chemical residue in or on the food, or for exemptions, a statement why such a method is not needed; and other reasonable grounds in support of the petition.[71]

EPA Review and Response

The EPA Office of Pesticide Programs (OPP) is organized into nine divisions, including three that review applications for pesticide registrations and amendments; those three are the Registration Division (RD), the Antimicrobials Division (AD), and the Biopesticides and Pollution Prevention Division (BPPD). Each division is organized

---

70. 40 C.F.R. § 152.44.
71. Id. § 180.7.

2.2. Federal Insecticide, Fungicide, and Rodenticide Act (FIFRA) 99

into product teams, by type of pesticide, to review applications for registration. The Health Effects and Environmental Fate and Effects Divisions provide science support and analysis for RD; AD and BPPD have scientists housed within their divisions. As noted in the "Standard of Review/Burden of Proof" section of this chapter, EPA may register a pesticide only if it determines that the product will not cause any "unreasonable adverse effects on the environment."[72]

EPA is authorized to charge an applicant a fee for registration and related actions (e.g., EUPs, study protocol reviews, registration amendments) under the Pesticide Registration Improvement Act of 2003 (PRIA)[73] and its subsequent reauthorizations, most recently the Pesticide Registration Improvement Extension Act of 2012 (PRIA 3).[74] PRIA 3 organizes the universe of pesticide-related submissions that are subject to PRIA 3 into 189 categories. Each PRIA 3 category has its own fee and a decision time by which EPA must reach a decision about the application. Depending on the action, PRIA 3 fees may range from the thousands to the hundreds of thousands of dollars, with decision periods from four to 24 months. There are also several types of actions that are not subject to fees, including minor registration amendments and EPA-initiated amendments.

Upon receipt, a registration or amendment application first must clear a series of administrative reviews that confirm the submission conforms with applicable format and content requirements and that the applicable fee has been paid. If the submitter has requested a fee reduction based on its status as a qualifying small business or other basis, eligibility for the fee reduction is assessed. The submission is then delivered to the product manager in the appropriate branch of the appropriate division (RD, AD, or BPPD), who divides the submission and shepherds its parts to the appropriate reviewers.

Consistent with an overall approach to increase process transparency, starting October 1, 2009, EPA has implemented procedural practices intended to allow the public to review and comment on risk assessments and proposed registration decisions for pesticides. EPA has applied these practices to all new pesticide active ingredients and first food uses, first outdoor uses, first residential uses, and "other actions of

---

72. FIFRA § 3(a), 7 U.S.C. § 136a(a).
73. Pesticide Registration Improvement Act of 2003, S. 1664.
74. Pesticide Registration Improvement Extension Act of 2012, S. 3552.

significant interest."[75] For these registration actions, EPA adds its risk assessments and proposed decisions to a public online docket and makes them available for a 30-day public comment period. Following the comment period, EPA publishes its decision and a response-to-comment document. EPA also publishes notices of filing and decisions related to EPA's review of petitions for tolerances and tolerance exemptions.

If once it has begun its review EPA determines that an application is incomplete, it informs the applicant of the deficiencies by letter. The applicant then has 75 days from the date of EPA's letter to submit additional information to complete the application, inform EPA of the date by which additional information will be submitted to complete the application, or withdraw the application.[76] If the applicant does not respond within the 75 days, or if the applicant fails to complete the application by the deadline stated in the applicant's response, then EPA considers the application to be withdrawn.

Separately, PRIA 3 sets deadlines for EPA action for different types of registration and amendment applications. If EPA determines that an application is incomplete and there is insufficient time for the applicant to submit additional information and for EPA to review that information, the applicant may work with EPA to set a new deadline or to withdraw the application. If the applicant does not respond with regard to the PRIA 3 deadline, EPA may issue a determination not to grant the application. By doing so, EPA discharges its obligation under FIFRA section 33 to make a determination by a specified date. This determination is neither a denial of the application under FIFRA section 3(c)(6) nor a withdrawal of the application. EPA states that in this case, it will continue to work on such an application diligently as long as the applicant responds to the 75-day letter discussed above. The applicant loses the benefit of a decision deadline, however, and other PRIA 3 submissions are given priority for review resources.

The standards of review that EPA must meet to register a product are discussed above. See "Standard of Review/Burden of Proof." In its regulations, EPA further elaborates on the determinations it must make to register a product either conditionally[77] or unconditionally.[78] EPA

---

75. EPA, Public Participation Process for Registration Actions (Mar. 31, 2010), http://www.epa.gov/pesticides/regulating/public-participation-process.html.
76. 40 C.F.R. § 152.105.
77. Id. §§ 152.113–152.115.
78. Id. § 152.112.

notifies an applicant of the approval of the submitted application by a notice of registration for a new registration or by a letter for an amended existing registration.[79] For a new registration or a label amendment, this written communication is accompanied by an EPA stamped-approved product label.

If EPA determines that the pesticide product does not meet the criteria for registration, then EPA may deny the application.[80] If EPA determines that an application should be denied, it notifies the applicant by a certified letter that states the reason(s) and factual basis for the denial, as well as the conditions that must be met for registration approval. If the applicant elects to do so, the applicant has 30 days from receipt of the certified letter to take specific corrective action to remedy the pending application. If the applicant fails to correct the deficiencies within the 30-day period, EPA may issue a notice of denial, which sets forth the reasons and the factual basis for the denial. Within 30 days after publication of the notice of denial, the applicant may request a hearing in accordance with FIFRA section 6(b).

REGULATION OF SUBSTANCES/PRODUCTS

Authority to Regulate Inventory/Listed Substances or Products

As discussed in the preceding section, EPA has the authority to approve pesticide registrations under FIFRA section 3. EPA also has the authority to set tolerances or grant exemptions from the requirement of a tolerance under FFDCA section 408, as amended by the FQPA. EPA additionally has authority to restrict a product through approved labeling as well as by classifying the product for restricted use. EPA also has authority to restrict or cancel a product registration under FIFRA section 6. Finally, as discussed in the "Enforcement and Penalties" section of this chapter, EPA is authorized under FIFRA section 12 to prohibit many acts, including the distribution and sale of adulterated or misbranded pesticide products.

---

79. *Id.* § 152.117.
80. *Id.* § 152.118.

Prioritization and Risk Assessments

Risk Assessment/Management

To approve a pesticide registration, EPA must conclude that the pesticide performs its intended function without unreasonable adverse effects on the environment and that when used in accordance with widespread and commonly recognized practices, it does not generally cause unreasonable adverse effects on the environment.[81] FIFRA defines "unreasonable adverse effects on the environment" to include "(1) any unreasonable risk to man or the environment, taking into account the economic, social, and environmental costs and benefits of the use of any pesticide, or (2) a human dietary risk from residues that result from a use of a pesticide in or on any food inconsistent with the standard" under FFDCA section 408.[82] EPA uses risk assessment to characterize the nature and magnitude of the health risk to humans (e.g., mixers/loaders/applicators, residents, recreational visitors) and ecological receptors (e.g., birds, fish, other wildlife) from pesticides and their decomposition products.

During its evaluation of a pesticide, whether for a new registration, a registration amendment, or EPA's registration review process, EPA assesses a wide variety of potential human health and environmental effects associated with use of the pesticide. To support this analysis, EPA requires registrants to address data requirements for many endpoints, as discussed in the "Test Data Development" section of this chapter. Registrants typically conduct a wide range of laboratory and field studies related to mammalian and environmental toxicity, environmental fate, physical and chemical characteristics, residues, spray drift, and exposure, usually according to EPA-established guidelines. With these data, EPA evaluates whether a pesticide has the potential to cause adverse effects on humans and terrestrial and aquatic wildlife, including endangered species and nontarget organisms, as well as possible contamination of surface water or groundwater from leaching, runoff, and spray drift. EPA also must consider in its assessment several additional factors, including the pesticide's composition, its intended use and use sites, its application rates, the opportunity for exposure, its intrinsic hazards, and its storage and disposal.

Fundamentally, risk is a product of hazard and exposure. An EPA risk assessor must evaluate the frequency and magnitude of human

---

81. FIFRA § 3(c)(5), 7 U.S.C. 136a(c)(5).
82. FIFRA § 2(bb), 7 U.S.C. 136(bb).

and ecological exposures that may occur as a consequence of contact with the residues of the pesticide and other ingredients caused by its registered or proposed uses. This evaluation of exposure is then combined with information on the inherent toxicity of the chemical (that is, the expected response to a given level of exposure) to predict the probability, nature, and magnitude of the adverse health effects that may occur—the final risk characterization.

Optimally, EPA's risk assessment is predicated on robust and complete data. Because there may be data gaps in the database for a pesticide, EPA risk assessors often have to make estimates and use judgment when performing risk calculations. Accordingly, an important part of EPA's risk assessment process is transparency and a fair and open presentation of the uncertainties in the risk calculations. The risk assessment process also typically is iterative. New data and other information and revised assessment techniques may support a refined risk assessment. FIFRA authorizes EPA to obtain any additional data deemed necessary to maintain "in effect an existing registration."[83]

EPA risk managers use the information from the risk assessment as an aid to decide how to protect humans and the environment. Risk management decisions typically result in use or other label restrictions, as discussed in the next section.

Reduced Risk Prioritization

FIFRA directs EPA to develop expedited registration procedures for pesticides that may reasonably be expected to accomplish one or more of the following goals: reduce the risks of pesticides to human health; reduce the risk of pesticides to nontarget organisms; reduce the potential for contamination of groundwater, surface water, or other valued environmental resources; or broaden the adoption of integrated pest management strategies, or make such strategies more available or more effective.[84] EPA has published reduced risk eligibility criteria in Pesticide Registration (PR) Notice 97-3, to address these goals.[85] Under PRIA 3, conventional pesticide products that meet these criteria are eligible for reduced application review periods.

---

83. FIFRA § 3(c)(2)(B), 7 U.S.C. § 136a(c)(2)(B).
84. FIFRA § 3(c)(10), 7 U.S.C. § 136a(c)(10).
85. PR Notice 97-3, Guidelines for Expedited Review of Conventional Pesticides under the Reduced-Risk Initiative and for Biological Pesticides (Sept. 4, 1997), http://www.epa.gov/PR_Notices/pr97-3.html.

As stated in PR Notice 97-3, "EPA believes that biological pesticides generally pose less risk than most conventional pesticides. Therefore, EPA established [Biopesticides and Pollution Prevention Division] BPPD to provide an expedited review to all biological pesticide products."[86] EPA has since published separate data requirements for biopesticides, which when compared with conventional pesticides, often constitute a reduced data set. Under PRIA, biopesticides generally have shorter review periods than corresponding applications for conventional pesticides.

Reregistration/Registration Review

EPA periodically reviews pesticides to assess whether they meet current scientific and regulatory standards. In 2008, in a process called reregistration, EPA completed an assessment of pesticides registered prior to November 1984. EPA efforts to implement risk mitigation measures as a result of reregistration are ongoing. Meanwhile, EPA began its registration review process in fall 2006. Under registration review, EPA reviews all registered pesticides every 15 years to confirm that the products continue to meet registration standards.

Use Restrictions

Label Language

Perhaps the most common pesticide maxim is "The label is the law." The outcome of the majority of EPA's risk assessment, risk characterization, and risk management decisions is given effect through the approved label language. The label specifies approved uses and use directions, approved crops or use sites, engineering controls, personal protective equipment, restricted entry intervals, and other requirements and limitations applicable to the product's use. It is a violation of FIFRA "to use any registered pesticide in a manner inconsistent with its labeling."[87] In its definition of this phrase, FIFRA also clarifies which activities are not inconsistent with pesticide labeling, including applying a pesticide at a lower dosage, concentration or frequency than specified on the labeling unless specifically prohibited; employing any method of application not prohibited by the labeling unless the labeling specifically requires certain application method(s); and mixing a pesticide with a fertilizer

---

86. *Id.* § XI.B.
87. FIFRA § 12(a)(2)(G), 7 U.S.C. § 136j(a)(2)(G).

when such a mixture is not expressly prohibited by the labeling.[88] EPA currently uses its own risk characterization metrics and associated precautionary language but is increasingly engaged in the global discussion of harmonized hazard communication.[89]

A registrant may distribute or sell a registered product with the composition, packaging, and labeling currently approved by EPA. If the product labeling is amended on the initiative of the registrant, the registrant usually may distribute or sell under the previously approved labeling for a period of 18 months after approval of the revision.[90] If EPA takes an action directing product label revisions, EPA typically specifies the period of time that previously approved labeling may be used by the registrant. If it elects to do so, EPA also may establish a date after which no product may be distributed or sold by any person (e.g., downstream in commerce from the registrant) unless it bears revised labeling. EPA states that in setting such a date, it provides sufficient time for product in channels of trade to be distributed or sold to users or otherwise disposed of.[91]

Use Classification

As stated above (see "Standard of Review/Burden of Proof"), under FIFRA section 3(d), EPA may classify a pesticide for general use or restricted use or may distinguish among uses of the products in determining its classification.[92] EPA, however, typically does not classify products for general use; products that are not restricted remain unclassified.[93]

EPA classifies a pesticide for "restricted use" if EPA determines that a pesticide, when applied in accordance with its directions for use, warnings, and precautionary statements for the uses for which it is registered, or when applied in accordance with widespread and commonly recognized practice, may generally cause, without additional regulatory restrictions, unreasonable adverse effects on the environment,

---

88. FIFRA § 2(ee), 7 U.S.C. § 136(ee).
89. *See, e.g.,* PR Notice 2012-1, Material Safety Data Sheets as Pesticide Labeling (Apr. 20, 2012), http://www.epa.gov/PR_Notices/pr2012-1.pdf, in which EPA provides guidance to registrants on how to manage conflicts that may arise between statements on EPA-approved pesticide labels and statements on Safety Data Sheets, which are subject to the Occupational Safety and Health Administration's Hazard Communication Standard but are considered by EPA as labeling when accompanying a pesticide product.
90. 40 C.F.R. § 152.130.
91. *Id.*
92. FIFRA § 3(d), 7 U.S.C. § 136a(d).
93. 40 C.F.R. § 152.160(a).

including injury to the applicator.[94] There are general criteria, as well as specific toxicity criteria, some of which differ for different types of use patterns, such as residential, or different product types, such as a granular product.[95] EPA states that it typically reviews and classifies a pesticide as restricted use under only the following circumstances:

- During review of an application to register a product containing a new active ingredient.
- As part of the review of an application for a new use of a product, if existing uses of that product have previously been classified for restricted use. The purpose of the review is to determine whether the new use should also be restricted. EPA generally does not conduct initial classification review for existing uses of individual products in conjunction with an application for amended registration.
- As part of the periodic chemical review process of developing or amending a registration standard, EPA often conducts a classification review of all uses of a currently registered pesticide.
- As part of a special review of a pesticide in accordance with the procedure in 40 C.F.R. part 154.[96]

When EPA determines that a product meets one or more of the criteria for restricted use and should be considered a candidate for restricted use classification, EPA considers whether additional label restriction may mitigate the identified hazard without a restricted use classification.[97]

If EPA classifies a pesticide as an RUP because its acute dermal or inhalation toxicity poses a hazard to the applicator or other persons, the pesticide may be applied only by a "certified applicator" licensed by EPA or by appropriate state or other authorities.[98] If EPA determines that a pesticide should be classified for restricted use because it may, without additional regulatory restriction, cause unreasonable adverse effects on the environment, EPA must require that the pesticide be applied only by, or under the direct supervision of, a certified applicator, or subject to

---

94. FIFRA § 3(d)(1)(C), 7 U.S.C. § 136a(d)(1)(C).
95. 40 C.F.R. § 152.170.
96. *Id.* § 152.164(b).
97. *Id.* § 152.170(e)(1).
98. FIFRA § 3(d)(1)(C)(i), 7 U.S.C. § 136a(d)(1)(C)(i).

such other restrictions as EPA may determine by regulation.[99] The term "under the direct supervision of a certified applicator" means that the "application of a pesticide is made by a competent person acting under the instructions and control of a certified applicator who is responsible for the actions of that person and who is available if and when needed, even though such certified applicator is not physically present at the time and place the pesticide is applied."[100]

Applicators are certified by the states (or Indian reservation governing body) in accordance with certain minimal standards required by EPA's regulations at 40 C.F.R. part 171. There are two general categories of certified applicators: "private applicators" and "commercial applicators."[101] A "private applicator" is a certified applicator who uses or supervises the use of a RUP "for purposes of producing any agricultural commodity on property owned or rented by him or his employer."[102] A "commercial applicator" is a certified applicator who uses or supervises the use of a RUP "for any purpose or on any property other than as provided by the definition of 'private applicator.'"[103]

Restricted use products are identified as such on the EPA-approved label, along with any other terms of restricted use imposed by EPA. It is a violation of FIFRA, subject to criminal and civil penalties, to distribute or sell a RUP in a manner prohibited by the restrictions on its use or to use a RUP in a manner inconsistent with its labeling.[104] Advertising of a restricted use product must identify the product as a restricted use product.[105]

EPA publishes a list of active ingredients and use patterns that are subject to restricted use in 40 C.F.R. § 152.175.

Chemical or Product Bans

FIFRA provides EPA with authority under both FIFRA sections 6(b) and 6(e) to cancel a pesticide registration. A registrant also may voluntarily cancel its pesticide registration under FIFRA section 6(f).

---

99. FIFRA § 3(d)(1)(C)(ii), 7 U.S.C. § 136a(d)(1)(C)(ii).
100. 40 C.F.R. § 171.2(a)(28).
101. FIFRA § 2(e), 7 U.S.C. § 136(e).
102. 40 C.F.R. § 171.2(a)(20).
103. *Id.* § 171.2(a)(9).
104. FIFRA § 12(a)(2)(F)–(G), 7 U.S.C. § 136j(a)(2)(F)–(G).
105. 40 C.F.R. § 152.168.

EPA-Initiated Cancellation: Section 6(b)

FIFRA section 6(b) authorizes EPA to issue a notice of intent to cancel the registration of a pesticide, or modify the terms and conditions of registration, if "it appears . . . that a pesticide or its labeling . . . when used in accordance with widespread and commonly recognized practice, generally causes unreasonable adverse effects on the environment." The term "unreasonable adverse effects on the environment" is defined by FIFRA section 2(bb) as "any unreasonable risk to man or the environment, taking into account the economic, social, and environmental costs and benefits of the use of any pesticide." EPA can also issue a cancellation notice where a pesticide, its labeling, or other materials required to be submitted to EPA in connection with a registration do not comply with FIFRA.

The U.S. Court of Appeals for the Fifth Circuit, in *Ciba-Geigy Corp. v. EPA*, held that the term "generally causes unreasonable adverse effects" means that the pesticide "creates unreasonable risks, though not necessarily actual adverse consequences, with considerable frequency."[106] An "unreasonable risk" must be "significant" or "substantial" to support a cancellation action.[107] The Eighth Circuit has ruled that a violation of the Endangered Species Act may be a basis for cancellation.[108]

Before issuing a notice of intent to cancel, EPA must issue a proposed notice of intent to cancel. FIFRA section 6(b) requires EPA to submit the proposed notice to the secretary of agriculture, with an analysis of the impact of the proposed action on the agricultural economy, at least 60 days prior to publication of the proposed notice, or notification of the registrant of the proposal, whichever comes first. In the same time frame, EPA must submit the proposed notice to the Scientific Advisory Panel (SAP) for comment on the effect of the proposed action on health and the environment.[109] EPA must publish the comments, evaluations, and recommendations of the secretary of agriculture and the SAP, and EPA's responses, in the final notice of intent to cancel.

In determining whether to issue a notice of intent to cancel, EPA must take into account the impact of the cancellation action on the

---

106. Ciba-Geigy Corp. v. EPA, 874 F.2d 277, 280 (5th Cir. 1989).
107. See EDF v. Ruckelshaus, 465 F.2d 528, 535 (D.C. Cir. 1972); EDF v. Ruckelshaus, 489 F.2d 1247, 1252 (D.C. Cir. 1973).
108. Defenders of Wildlife v. Adm'r, EPA, 882 F.2d 1294 (8th Cir. 1989).
109. FIFRA § 25(d), 7 U.S.C. § 136w(d).

production and prices of agricultural commodities, retail food prices, and otherwise on the agricultural economy.[110] An analysis of these impacts must be published in the final notice of intent to cancel. The statute also directs EPA to consider restricted use classification as an alternative to cancellation.

A cancellation notice becomes final 30 days after publication in the *Federal Register* or notification of the registrant, whichever occurs later, unless a hearing is requested, or, where appropriate, the registrant submits an amendment to the terms and conditions of registration. A cancellation order states the length of time that existing stocks of the canceled pesticide may be sold or distributed by the manufacturer and others in the chain of commerce.[111]

A hearing request must identify the specific uses for which a hearing is sought and set forth the objections and basis for objections to EPA's actions. Copies of labels of the products must be included with the hearing request.[112] If a hearing is requested, the cancellation action is stayed pending the outcome of the hearing proceeding. The rules of practice governing cancellation hearings are found in 40 C.F.R. part 164.

The standard of proof applicable to a cancellation decision is the "preponderance of the evidence."[113] On appeal, the standard is "substantial evidence when considered on the record as a whole."[114] EPA's hearing regulations state that the registrant has the ultimate burden of persuasion for establishing that the benefits exceed the risks in a cancellation proceeding.[115] EPA has the "burden of going forward to present an affirmative case."[116]

---

110. FIFRA § 6(b), 7 U.S.C. § 136d(b).
111. *See* Existing Stocks of Pesticide Products, Statement of Policy, 56 Fed. Reg. 29,362 (June 26, 1991); Existing Stocks of Pesticide Products, Amendment to Statement of Policy, 61 Fed. Reg. 16,632 (Apr. 16, 1996).
112. 40 C.F.R. § 164.22(a).
113. *Id.* § 22.24(b) ("Each matter of controversy shall be decided by the Presiding Officer upon a preponderance of the evidence").
114. FIFRA § 16(b), 7 U.S.C. § 136n(b).
115. 40 C.F.R. § 164.80(b).
116. *See* EDF v. EPA, 548 F.2d 998, 1004, 1012–18 (D.C. Cir. 1976), *cert. denied sub nom.* Velsicol Chem. Corp. v. EPA, 431 U.S. 925 (1977); EDF v. EPA, 510 F.2d 1292, 1297, 1302 (D.C. Cir. 1975). *See also* 40 C.F.R. § 164.20(a). *But see* Dir., Office of Workers' Comp., Dep't of Labor v. Greenwich Collieries, 512 U.S. 267 (1994) (finding that the Administrative Procedure Act places the burden of persuasion, not the burden of going forward, on the proponent of an agency rule or order).

FIFRA section 6(b)(2) states that "a person adversely affected by the notice" has a right to request a hearing. Therefore, user groups, as well as registrants, can request a hearing.[117] User groups, however, cannot force a registrant to defend its product in a cancellation proceeding, and cannot disturb a settlement agreement reached between EPA and the registrants.[118] Once a hearing request has been filed, any person may file a request to intervene.[119] Intervenors may not raise matters not pertinent to the issues or matters that unreasonably broaden the issues presented by the hearing requests.[120]

EPA-Initiated Cancellation: Section 6(e)

FIFRA section 6(e) gives EPA the authority to issue a notice of intent to cancel a conditional registration issued under FIFRA section 3(c)(7) at any time it determines that the registrant "has failed to initiate and pursue appropriate action toward fulfilling any condition imposed," or "at the end of the period provided for satisfaction of any condition imposed, . . . [if] that condition has not been met."[121] EPA can provide for the continued sale and use of existing stocks in the cancellation notice.

Unless a person adversely affected by the notice requests a hearing, the cancellation becomes effective at the end of 30 days.[122] The issues at a hearing are limited by statute to (1) whether the registrant has initiated and pursued appropriate action to comply with the conditions of registrations within the time provided; (2) whether the conditions have been satisfied within the time provided; and (3) whether EPA's existing stocks determination is consistent with FIFRA.[123]

Voluntary Cancellation: Section 6(f)

A registrant may, at any time, voluntarily request the cancellation or amendment of its own pesticide registration.[124] A voluntary cancellation applies to all distributor products distributed or sold under the pesticide's registration number, as well as to the registered product. EPA must publish in the *Federal Register* a Notice of Receipt of a Request

---

117. EDF v. Costle, 631 F.2d 922, 937–38 (D.C. Cir. 1980), *cert. denied*, 449 U.S. 1112 (1981).
118. McGill v. EPA, 593 F.2d 631, 636–37 (5th Cir. 1979).
119. 40 C.F.R. § 164.31(a).
120. Id. § 164.31(c).
121. FIFRA § 6(e)(1), 7 U.S.C. § 136d(e)(1).
122. FIFRA § 6(e)(2), 7 U.S.C. § 136d(e)(2).
123. FIFRA § 6(e)(2), 7 U.S.C. § 136d(e)(2).
124. FIFRA § 6(f)(1)(A), 7 U.S.C. § 136d(f)(1)(A).

2.2. Federal Insecticide, Fungicide, and Rodenticide Act (FIFRA) 111

for Voluntary Cancellation and allow a 30-day comment period. For a pesticide registered for a minor agricultural use, EPA must generally allow a 90-day comment period. During this 90-day comment period, registrants can enter into an agreement with users or other interested persons for the transfer of the pesticide registration, and notify EPA of this agreement. An application for the transfer of the registration must be submitted within 30 days of this notification to avoid the cancellation. With regard to pesticides registered for minor agricultural uses, the new registrant assumes the responsibility for any outstanding data requirements pending at the time of the transfer.[125]

Suspension

A suspension is a temporary ban on the sale and distribution of a pesticide during the time required to conduct a cancellation proceeding. EPA can issue a suspension order if it finds that a suspension is necessary to prevent an "imminent hazard" during the time required for the completion of the cancellation proceeding.[126] An "imminent hazard" is a substantial likelihood of serious harm during the duration of a cancellation proceeding. The term is not limited to the concept of a crisis.[127] EPA may issue a suspension notice only if it has issued, or issues at the same time, a notice of intent to cancel.

Following notification of a suspension, affected registrants have five days to request an expedited hearing. Only a registrant can request a suspension hearing. If the registrant does not request a hearing, the suspension takes effect five days after the registrant's receipt of the suspension notice. No court review is permitted if a registrant fails to request an administrative hearing. If an administrative hearing is held, the suspension does not become effective until the issuance of an order favorable to EPA.[128]

A suspension hearing must begin within five days after the hearing request is filed, unless EPA and the registrant agree to extend the time. The sole issue is "whether an imminent hazard exists."[129] Adversely affected nonregistrants can intervene and "file proposed findings and conclusions and briefs."[130] The presiding officer has ten

---

125. FIFRA § 6(f)(3)(D), 7 U.S.C. § 136d(f)(3)(D).
126. FIFRA § 6(c)(1), 7 U.S.C. § 136d(c)(1).
127. See EDF v. EPA, 510 F.2d 1292, 1297 (D.C. Cir. 1975); EDF v. Ruckelshaus, 465 F.2d 528, 540 (D.C. Cir. 1972); Dow Chem. Co. v. Blum, 469 F. Supp. 892, 898–99 (E.D. Mich. 1979).
128. FIFRA § 6(c)(2), 7 U.S.C. § 136d(c)(2).
129. 40 C.F.R. § 164.121(c).
130. Id. § 164.121(e).

days from the conclusion of the testimony to issue recommended findings and conclusions to the Environmental Appeals Board (EAB), who must issue a final order in seven days.[131] Final suspension orders are reviewable in the court of appeals even if cancellation proceedings are still pending.[132] An intervenor is entitled to seek judicial review of a final suspension order.[133]

Emergency Suspension

EPA can issue an emergency suspension if it determines that there is an "emergency" situation sufficiently serious to warrant a suspension prior to a hearing.[134] FIFRA section 25(d) directs EPA to "promptly submit" an emergency suspension action to the SAP "for comment, as to the impact on health and the environment." An emergency suspension order may be issued prior to notification of the registrant, and the order takes effect immediately.[135] Only the registrant can participate in an emergency suspension hearing, with the exception that other adversely affected parties may file briefs in the proceeding.[136]

An emergency suspension order is reviewable in district court to determine if the order is arbitrary, capricious, or an abuse of discretion, or whether it was issued in accordance with legally established procedures.[137] This court action can take place simultaneously with the administrative suspension hearing. Either the registrant or another person with the concurrence of the registrant may seek review of the emergency suspension in district court.

Test Data Development

Authority to Require Testing/Testing Triggers

EPA is authorized under FIFRA section 3(c) to require registrants to develop and submit testing to support and maintain product registrations.

---

131. 40 C.F.R. §§ 164.121(j), 164.122(a).
132. FIFRA § 6(c)(4), 7 U.S.C. § 136d(c)(4); FIFRA § 16(b), 7 U.S.C. § 136n(b).
133. 40 C.F.R. § 164.121(e).
134. FIFRA § 6(c)(3), 7 U.S.C. § 136d(c)(3).
135. FIFRA § 6(c)(3), 7 U.S.C. § 136d(c)(3).
136. FIFRA § 6(c)(3), 7 U.S.C. § 136d(c)(3).
137. FIFRA § 6(c)(4), 7 U.S.C. § 136d(c)(4); Dow Chem. Co. v. Blum, 469 F. Supp. 892, 902–03 (E.D. Mich. 1979).

Chemical Substances/Products That Must Be Tested/Implementing Test Requirements

To facilitate its registration and product labeling decisions, EPA has promulgated elaborate requirements for numerous scientific studies, which it directs pesticide manufacturers to conduct to obtain and maintain registrations for pesticide products.[138] EPA requires a wide range of data in categories that include product chemistry, ecological toxicity, mammalian toxicity, environmental fate, residue chemistry, reentry exposure, and spray drift.[139] Although most data requirements are satisfied by the registrant of a pesticide active ingredient, some data requirements must be satisfied to register each individual end-use pesticide product.[140] EPA also has established test guidelines that contain "standards for conducting acceptable tests, guidance on the evaluation and reporting of data, definition of terms, and suggested study protocols."[141]

Pursuant to EPA regulations, pesticide registration studies must be conducted in compliance with Good Laboratory Practice (GLP) requirements.[142] These regulations are intended to ensure the quality and integrity of all data submitted to EPA to support registration of pesticides under FIFRA and tolerances established under the FFDCA.[143]

Applicants for registrations can satisfy EPA data requirements by, among other means, requesting that EPA waive the submission of otherwise required data.[144] An applicant may claim that a waiver granted previously by EPA applies to a data requirement for its product, which the applicant must document by identifying the prior waiver and explaining the applicability of such waiver to the current request.[145] If an applicant seeks a new waiver of a data requirement, EPA waives data requirements on a case-by-case basis in response to specific written requests of applicants.[146] EPA ensures, however, that sufficient data are available to make the determinations required under applicable statutory standards.[147]

---

138. FIFRA § 3(c)(2)(A), (B), 7 U.S.C. § 136a(c)(2)(A), (B); 40 C.F.R. pt. 158.
139. *See* 40 C.F.R. pt. 158, subpts. D–V.
140. *Id.*
141. 40 C.F.R. § 158.70(c).
142. *Id.* § 160.1(a).
143. *Id.*
144. *Id.* § 152.91.
145. *Id.* § 152.91(a).
146. *Id.* § 158.45.
147. *Id.* § 158.45(a).

FIFRA exempts certain registrants from (1) submitting or citing data pertaining to the active ingredient in the formulated product and (2) offering to pay compensation for such data.[148] The formulator exemption applies only if the applicant purchases a registered pesticide (i.e., active ingredient) from another producer in order to formulate it into the applicant's product. The exemption is limited to uses specifically listed on the MUP label.[149]

To claim the formulator's exemption, an applicant must submit a formulator's exemption statement (Form 8570-27) listing the name and the EPA registration number of ingredients in its pesticide that are eligible for the exemption and must submit a CSF. The formulator's exemption does not apply to data concerning the safety or efficacy of the applicant's end-use product itself, unless the applicant's product is identical to the purchased product. It is difficult, however, for an applicant to show that its product is identical to the purchased product. This can usually be done only if the purchased product lists the names and percentages of all ingredients on its label, or if the applicant's product is merely a repackaging of the purchased product. Unless the applicant's product is identical to the purchased product, the applicant must perform a series of toxicology tests, set forth in EPA's data requirement regulations on the formulated product. EPA approves a registration claiming a formulator's exemption only if the purchased product is a registered MUP whose label does not prohibit its use for making the applicant's end-use product, or a purchased end-use product labeled with each use for which the applicant's product is labeled.

Data Sharing and Compensation

In addition to submitting data to satisfy requirements to obtain or maintain a pesticide registration, FIFRA permits pesticide registrants not relying upon the formulator's exemption to satisfy data requirements by citing, subject to certain limitations, the data submitted by other registrants. There are three main issues regarding data sharing and compensation: (1) exclusive use protection under FIFRA section 3(c)(1)(F)(i) for data generated to support a new pesticide active

---

148. FIFRA § 3(c)(2)(D), 7 U.S.C. § 136a(c)(2)(D); 40 C.F.R. § 152.85.
149. PR Notice 91-8, Revised Policy to Provide Applicants Other Than Basic Manufacturers an Opportunity to Submit Generic Data and Receive Data Compensation for It (Dec. 31, 1991).

2.2. Federal Insecticide, Fungicide, and Rodenticide Act (FIFRA)

ingredient or use;[150] (2) data compensation under FIFRA section 3(c)(1)(F)(iii) for data already submitted to EPA and relied upon by applicants seeking registrations of new or amended pesticide products; and (3) cost sharing and joint data development under FIFRA section 3(c)(2)(B) when EPA seeks additional data from registrants for registered pesticides.

First, FIFRA section 3(c)(1)(F)(i)[151] gives data submitters a ten-year period of exclusive use for data submitted in support of a registration for a new pesticide chemical or new uses of an already registered pesticide. The exclusive use provision applies only to data submitted to support an active ingredient first registered after September 30, 1978. During the ten-year exclusive use period, no other registrant can rely on the exclusive use data without the original registrant's consent. Another company can submit its own data package, however, to obtain a registration. Exclusive use does not apply to "defensive data"[152] (i.e., data that responds to a data call-in (DCI) issued under FIFRA section 3(c)(2)(B) or other data submitted proactively by a registrant to maintain the registration for a pesticide). The FQPA amends FIFRA to extend the time period of exclusive data use by one additional year for "minor uses" that meet certain criteria.

Second, FIFRA section 3(c)(1)(F)(iii)[153] requires subsequent applicants—often referred to as "follow-on" or "me-too" applicants—to pay "compensation" to the data submitter for use of their test data for a period of 15 years from the date the data were originally submitted to EPA. Data compensation applies only to data submitted by an applicant or registrant (1) to support an application for registration, or an EUP, or an amendment adding a new use; (2) to support or maintain in effect an existing registration, or for reregistration; or (3) to support a tolerance or exemption from tolerance under section 409 of the FFDCA. After the expiration of the 15-year compensation period, EPA may consider the data in support of a follow-on application without any need for an offer to pay.

As a condition precedent to EPA's reliance on compensable data in support of the follow-on application, the follow-on applicant must certify in its registration application that all necessary compensation offers

---

150. Prior to the 1990 FIFRA amendments, section 3(c)(1)(F) was designated as section 3(c)(1)(D).
151. FIFRA § 3(c)(1)(F)(i), 7 U.S.C. § 136a(c)(1)(F)(i).
152. FIFRA § 3(c)(1)(F)(i), 7 U.S.C. § 136a(c)(1)(F)(i).
153. FIFRA § 3(c)(1)(F)(iii), 7 U.S.C. § 136a(c)(1)(F)(iii).

have been made (Form 8570-34).[154] There is, however, no requirement for compensation actually to be paid, or for the parties to agree on the proper amount of compensation, for EPA to issue the follow-on registration. If the original and follow-on applicant cannot agree on the amount and terms of compensation within 90 days of the submission of an offer to pay, either party may initiate binding arbitration to resolve the data compensation dispute. Arbitration is conducted by the American Arbitration Association under the FIFRA Arbitration Rules at 29 C.F.R. part 1440. FIFRA does not give any guidance on the appropriate amount of compensation, or the principles that arbitrators should follow for determining compensation. The "public" versions of prior arbitration decisions can be consulted for precedent.

A follow-on registrant who relies on data submitted by other registrants has two choices under EPA's regulations for citing those data. The first option is known as the "cite-all" method.[155] The cite-all method allows (but does not require) a follow-on applicant to rely upon all nonexclusive use data in EPA's files that (1) concern the properties or effects of the applicant's product, or of a product that is identical or substantially similar, or the active ingredient in the applicant's product; and (2) is one of the types of data EPA would currently require if the applicant sought initial registration of a product with composition and uses identical or substantially similar to the applicant's product.[156] To use the cite-all method, the follow-on applicant must certify to EPA that it has (1) notified each affected company on EPA's data submitters list that it intends to apply for registration (notice must include proposed product name and list of product's active ingredients); (2) offered to pay compensation to each data submitter to the extent required by FIFRA section 3(c)(1)(F)(ii); and (3) offered to commence negotiations with each data submitter to determine amount and terms of compensation.[157] The cite-all applicant also must furnish to EPA a "general offer to pay statement," offering and agreeing to pay compensation to any other affected data submitters.[158]

Alternatively, an applicant can use the "selective" method, which requires the applicant to identify and list the data requirements that "would apply to his pesticide," and to demonstrate compliance with

---

154. See 40 C.F.R. §§ 152.80–152.99.
155. 40 C.F.R. § 152.86.
156. Id. § 152.86(d)(2).
157. FIFRA § 3(c)(1)(F), 7 U.S.C. § 136a(c)(1)(F); 40 C.F.R. § 152.86.
158. 40 C.F.R. § 152.86(c).

each data requirement.[159] An applicant can satisfy a data requirement by citing to one or more selected compensable studies or to all the data supporting that requirement. Alternatively, the applicant can demonstrate compliance with a data requirement by submitting his own data, citing to data that are no longer compensable or are in the public literature, or obtaining a waiver from the data requirement. A follow-on applicant can defer its obligation to satisfy a particular data requirement by documenting that there is an unsatisfied data gap that other registrants have not yet been required to fill.[160] An applicant using the selective method must make all required offers to pay for any cited compensable data.[161]

EPA's regulations allow a data owner to petition EPA to deny the application for registration or cancel a registration of a follow-on registrant. Grounds for denying an application canceling a follow-on registration include, but are not limited to (1) failure to offer to pay compensation for a required study; (2) failure to list a data requirement applicable to the follow-on product (selective method); (3) failure to demonstrate compliance with all applicable data requirements (selective method); (4) submission of a study that is not valid or does not satisfy a data requirement; (5) failure to comply with data gap procedures; (6) failure to comply with a data compensation agreement; (7) failure to participate in binding arbitration initiated by a data submitter; or (8) failure to comply with the terms of an arbitration decision.[162]

Lastly, under FIFRA section 3(c)(2)(B), if EPA determines that additional data are required to maintain in effect an existing registration of a pesticide, EPA must notify all existing registrants of the pesticide through a DCI.[163] All registrants of an active ingredient (except those exempted by the formulator exemption) are responsible for

---

159. *Id.* § 152.90.
160. *Id.* § 152.96.
161. *See generally* (1) Pesticide Programs; Pesticide Registration and Classification Procedures; Application Procedures to Ensure Protection of Data Submitters' Rights, 49 Fed. Reg. 30,884 (Aug. 1, 1984) (to be codified at C.F.R. pts. 152 and 162); (2) Certification with Respect to Citation of Data (EPA Form 8570-34); (3) U.S. EPA, General Information on Applying for Registration of Pesticides in the United States (Aug. 1992) (NTIS Document No. PB92-221811); (4) PR Notice 83-4, Interim Procedures for Satisfying Registration Data Requirements under Recent Court Decisions (June 16, 1983).
162. 40 C.F.R. § 152.99(a).
163. FIFRA § 3(c)(2)(B), 7 U.S.C. § 136a(c)(2)(B).

responding to DCIs.[164] FIFRA allows (but does not require) the existing registrants to enter into one of two types of agreements: (1) an agreement to develop data jointly; or (2) an agreement by which the follow-on registrant shares in the cost of the data being developed by the original registrant.[165] An agreement to develop data jointly normally means there is co-ownership of data for all purposes (e.g., joint data development task force agreement).[166] An agreement to share in the cost of developing the data provides the follow-on registrant with the right to "examine and rely" on the data for EPA reregistration, but does not require the follow-on registrant to have equal ownership or "hard copy."[167] If parties agree (explicitly or implicitly) to develop data jointly or to cost share, but cannot further agree upon terms and conditions (e.g., what percentage of costs each should bear), either party may initiate binding arbitration after 60 days.[168] If either party fails to participate in arbitration or comply with the arbitrators' decision, its registration can be suspended by EPA.[169]

REPORTING AND RECORDKEEPING REQUIREMENTS

Reports to Be Submitted to EPA

There are reporting requirements associated with pesticides in addition to the submissions required to register, amend, and maintain product registrations, including both substantive submissions and annual maintenance fees. Importers must file a notice of arrival with the appropriate EPA regional office and receive approval before importing pesticide products, as noted in the "Imports and Exports" section of this chapter. Registrants that obtain information regarding unreasonable adverse effects to human health or the environment of the pesticide must report this information to EPA, as noted in the "Risk Notification" section of this chapter.

---

164. FIFRA § 3(c)(2)(B), 7 U.S.C. § 136a(c)(2)(B).
165. FIFRA § 3(c)(2)(B)(ii), 7 U.S.C. § 136a(c)(2)(B)(ii).
166. FIFRA § 3(c)(2)(B)(ii), 7 U.S.C. § 136a(c)(2)(B)(ii).
167. FIFRA § 3(c)(2)(B)(ii), 7 U.S.C. § 136a(c)(2)(B)(ii).
168. FIFRA § 3(c)(2)(B)(iii), 7 U.S.C. § 136a(c)(2)(B)(iii).
169. FIFRA § 3(c)(2)(B)(iii), 7 U.S.C. § 136a(c)(2)(B)(iii).

In addition, each pesticide "producer operating an establishment" must submit annual pesticide reports.[170] A facility is an establishment, regardless of whether it is owned or operated independently or as part of a larger entity, whether it is located in the United States and producing a pesticide solely for export, or whether it is located in a foreign country and producing a pesticide for import into the United States. Establishments must be registered with EPA by submitting an Application for Registration of Pesticide-Producing and Device-Producing Establishment (EPA Form 3540-8) before pesticides and devices are produced.[171] EPA assigns each registered establishment a number, which must be included on the label of the pesticides and devices produced in that establishment. On or before March 1 of each year, pesticide producers operating establishments must submit pesticide production reports (EPA Form 3540-16) indicating the types and amounts of the following pesticides, active ingredients, or devices: (1) those estimated to be produced at that establishment in the current calendar year; (2) those actually produced at the reporting establishment in the past calendar year; and (3) those sold or distributed in the past calendar year. Foreign-producing establishments are required to report only what they export to the United States.[172] The report must include only those pesticidal products actually produced at the reporting establishment.[173] Reports submitted by foreign-producing establishments must cover only pesticidal products exported to the United States.[174]

The initial report is due no later than 30 days after the first registration of each establishment that a company operates.[175] Thereafter,

---

170. See FIFRA § 7(c)(1), 7 U.S.C. § 136e(c)(1); see also 40 C.F.R. § 167.85(a) ("[e]ach producer operating an establishment" is required to submit a report); id. § 167.3 ("producer" is defined as "any person, as defined by the Act, who produces any pesticide, active ingredient, or device (including packaging, repackaging, labeling and relabeling)").
171. See FIFRA §§ 7, 9, 12(a)(2)(L), 7 U.S.C. §§ 136e, 136g, 136j(a)(2)(L); 40 C.F.R. pt. 167.
172. See FIFRA § 7(c)(1), 7 U.S.C. § 136e(c)(1); 40 C.F.R. § 167.85(b); see also 40 C.F.R. § 167.3 ("amount of pesticidal product" is defined as "quantity, expressed in weight or volume of the product, and is to be reported in pounds for solid or semi-solid pesticides and active ingredients or gallons for liquid pesticides and active ingredients, or number of individual retail units for devices"); id. ("current production [sales or distribution]" is defined as "amount of planned production in the calendar year in which the pesticides report is submitted, including new pesticidal products not previously sold or distributed"); id. ("sold or distributed" is defined as "the aggregate amount of a pesticidal product released for shipment by the establishment in which the pesticidal product was produced").
173. 40 C.F.R. § 167.85(b).
174. Id.
175. See FIFRA § 7(c)(1), 7 U.S.C. § 136e(c)(1); 40 C.F.R. § 167.85(d).

annual reports are due on or before March 1 of each year, even if the establishment has produced no pesticidal product for that reporting year.[176] Should an establishment fail to submit a required report, its establishment registration may be terminated, or it may be subject to civil and/or criminal penalties.[177]

Recordkeeping

All "pesticide producers" are required to maintain certain records.[178] Pursuant to the recordkeeping provisions of the FIFRA regulations, a "producer" is defined as "the person . . . who produces or imports any pesticide or device or active ingredient used in producing a pesticide."[179] These recordkeeping requirements apply to a company as the registrant, even if another company does the actual pesticide production activities, because EPA would consider the other company to be conducting these activities on the registrant's behalf and at the registrant's direction.[180]

A pesticide registrant/producer must keep the following documents for a two-year time period:

- Records showing the product name, EPA registration number, amounts per batch, and batch identification (numbers, letters, etc.) of all pesticide products produced. The batch control number must appear on all production control records.[181]
- Records showing the brand names and quantities of devices produced.[182]

---

176. See 40 C.F.R. § 167.85(d).
177. 40 C.F.R. § 167.20(f).
178. FIFRA § 8(a), 7 U.S.C. § 136f(a); 40 C.F.R. § 169.2.
179. 40 C.F.R. § 169.1(e).
180. FIFRA section 8 requires that all pesticide producers, registrants, and applicants for registration maintain certain records. See FIFRA § 8(a), 7 U.S.C. § 136f(a). Although EPA's regulations discuss only pesticide producers with regard to the recordkeeping requirements, it is commonly understood that the recordkeeping provisions also apply to registrants.
181. 40 C.F.R. § 169.2(a); see also id. § 169.1(b) (defining "batch" as "a quantity of a pesticide product or active ingredient used in producing a pesticide made in one operation or lot or if made in a continuous or semi-continuous process or cycle, the quantity produced during an interval of time to be specified by the producer").
182. 40 C.F.R. § 169.2(b).

2.2. Federal Insecticide, Fungicide, and Rodenticide Act (FIFRA)   121

- Records of its receipt of any product used to produce the registered pesticide product.[183]
- Records showing specified information regarding the shipment of all pesticides and active ingredients used in producing pesticides.[184]
- Copies of domestic advertising, if any, for a RUP.[185]
- For pesticides intended solely for export to any foreign country, copies of various documents.[186]

A pesticide registrant/producer also must keep the following documents:

- Inventory records for the types (e.g., package sizes) and amounts of registered pesticide products produced.[187] The inventory records may be disposed of when a more current inventory record is prepared.[188]
- Copies of guarantees given.[189] These records must be kept for one year after expiration of the guarantee.[190]
- Records regarding the disposal of pesticides or active ingredients and of containers accumulated during production.[191] The records must be retained for 20 years, and may be forwarded to EPA for maintenance after three years.[192]
- Records of tests conducted on human beings whether performed by the registrant/producer or authorized or paid for by the registrant/producer.[193] The records must be retained for

---

183. 40 C.F.R. § 169.2(c). Shipping and receiving documents are deemed satisfactory records. See id.
184. 40 C.F.R. § 169.2(d). "Such records are required regardless of whether any shipment or receipt of shipment is between plants owned or otherwise controlled by the same person." Id. Shipping and receiving documents are deemed satisfactory records. See id.
185. 40 C.F.R. § 169.2(f).
186. Id. § 169.2(h).
187. 40 C.F.R. § 169.2(e).
188. See id.
189. 40 C.F.R. § 169.2(g). Under FIFRA section 12(b), a person who receives a pesticide may escape liability for certain acts unlawful under FIFRA if that person has a guarantee that the pesticide was lawfully registered at the time of sale and complies with FIFRA. Rather, the guarantor would be liable. See FIFRA § 12(b)(1), 7 U.S.C. § 136j(b)(1).
190. 40 C.F.R. § 169.2(g).
191. 40 C.F.R. § 169.2(i).
192. See id.
193. 40 C.F.R. § 169.2(j).

20 years, but may be forwarded to EPA for maintenance after three years.[194]

- Records containing research data relating to registered pesticides, including all test reports submitted to EPA in support of registration or in support of a tolerance petition and all underlying raw data, and interpretations and evaluations thereof, whether the documents are in the possession of the registrant/producer or of an independent testing facility or laboratory.[195]

The records must be kept as long as the registration is valid and the registrant/producer is in business.[196] The requirements discussed above are the minimum requirements. Often registrants keep additional records to ensure a sound defense in the event of any potential alleged violation or potential liability.

Although the list of records that must be kept is expansive, FIFRA explicitly excludes a number of documents from the recordkeeping requirements. Expressly excluded are the following: financial data; sales data, other than shipment data; pricing data; personnel data;[197] and research data, other than data related to registered pesticides or to a pesticide for which an application for registration has been submitted.[198]

A company, along with any entities that distribute, carry, or deal in a company's registered pesticide product, must permit EPA access to the required records.[199] Should a company have an "inability" to produce the required records for inspection, it must provide access to all other records and information regarding the same matters.[200] The definition of "inability" is, however, limited.[201]

---

194. *See id.*
195. 40 C.F.R. § 169.2(k).
196. *See id.*
197. Certain personnel data, such as training logs, may be separately required, pursuant to the GLP standards set forth at 40 C.F.R. part 160.
198. FIFRA § 8(a), 7 U.S.C. § 136f(a).
199. FIFRA § 8(b), 7 U.S.C. § 136f(b); 40 C.F.R. § 169.3(a). The statute and the regulations provide mechanisms for the protection of trade secrets. *See* FIFRA § 10, 7 U.S.C. § 136h; 40 C.F.R. § 169.3(c).
200. FIFRA § 8(b)(2), 7 U.S.C. § 136f(b)(2); 40 C.F.R. § 169.3(d)(1).
201. 40 C.F.R. § 169.1(d) ("The term "inability" means the incapacity of any person to maintain, furnish or permit access to any records under this Act and regulations, where such incapacity arises out of causes beyond the control and without the fault or negligence of such person. Such causes may include, but are not restricted to acts of God or of the public enemy, fires, floods, epidemics, quarantine restrictions, strikes, and unusually severe weather, but in every case, the failure must be beyond the control and without the fault or negligence of said person.").

If there is no such inability, and if a company were to fail to give access to and permit copying of the required records, its failure would "be deemed a refusal to keep records required or a refusal to allow the inspection of any such records or both."[202] Such a refusal could subject a company to civil and/or criminal penalties.[203]

Risk Notification

Under FIFRA section 6(a)(2), if "at any time after the registration of a pesticide the registrant has additional factual information regarding unreasonable adverse effects on the environment of the pesticide, the registrant shall submit such information to the Administrator."[204] "Unreasonable adverse effects on the environment" is defined as "any unreasonable risk to man or the environment, taking into account the economic, social, and environmental costs and benefits of the use of any pesticide" or "a human dietary risk from residues that result from a use of a pesticide in or on any food inconsistent with" certain set standards.[205]

Examples set forth in EPA's regulations of the types of information that must be reported under FIFRA section 6(a)(2) if the criteria for reporting are met and none of the exceptions is satisfied include the following:

- Toxicological studies involving "the toxicity of a pesticide to humans or other non-target domestic organisms" must be reported "if, relative to all previously submitted studies, they show [certain] adverse effect[s]."[206]
- Ecological studies of the toxicity of a pesticide "to terrestrial or aquatic wildlife or plants if, relative to all previously submitted studies, they show an adverse effect" of a specified type.[207]
- Certain information on pesticides in or on food or feed, if the pesticide exceeds established levels.[208]

---

202. 40 C.F.R. § 169.3(d)(2).
203. *See* FIFRA § 14, 7 U.S.C. § 136l.
204. FIFRA § 6(a)(2), 7 U.S.C. § 136d(a)(2).
205. FIFRA § 2(bb), 7 U.S.C. § 136(bb).
206. 40 C.F.R. § 159.165(a). *See also* 62 Fed. Reg. 49,370, 49,390 (Sept. 19, 1997).
207. 40 C.F.R. § 159.165(b).
208. *Id.* § 159.178(a).

- Information showing that a pesticide is present above the "water reference level" in U.S. water, groundwater, or finished drinking water.[209]
- Certain data regarding pesticide metabolites, degradates, contaminants, and impurities.[210]
- Toxic or adverse effect incident reports affecting humans or other nontarget organisms, if specific conditions are met.[211]
- Incidents or studies involving a failure of efficacy for a pesticide that is claimed to work against particular microorganisms that pose a risk to human health.[212]
- Information concerning incidents that involve a pest having developed resistance to any pesticide (both public health and non-public health).[213]
- Other, nonspecified information if the registrant knows, or reasonably should know, that, if correct, EPA might regard the information, alone or in conjunction with other information about the pesticide, as raising concerns about the continued registration of a product or about the appropriate terms and conditions of that registration.[214]

Under EPA regulations, a registrant need not submit certain information where the conditions for exclusion are met, including clearly erroneous information, previously submitted information, published articles or reports containing information otherwise reportable, or information concerning former inerts, contaminants, or impurities.[215] EPA's regulations provide general criteria that must be satisfied for a company to be required to report under section 6(a)(2) (i.e., the registrant must "possess or receive" the information; the information must be relevant to the risks or benefits of a specific pesticide registration; a specific pesticide registration must be at issue; and the registration must be currently or formerly held by the registrant). EPA regulations

---

209. Id. § 159.178(b).
210. Id. § 159.179.
211. Id. § 159.184.
212. Id. § 159.188(a).
213. Id. § 159.188(c).
214. Id. § 159.195(a).
215. Id. § 159.158(b).

and guidance also provides specific criteria to be met for the type of information involved in this case to be reportable.[216]

Depending upon the severity of the incident, different reporting time frames may apply. For example, registrants must report as soon as possible, but in any event, no later than 15 days after learning of a human fatality incident.[217] Other information "must be received by EPA not later than the 30th calendar day after the registrant first possesses or knows of the information."[218] Certain other information may be accumulated for one month and submitted one month thereafter. Finally, information regarding incidents meeting exposure and severity label categories for which no deadline is specified may be accumulated for three months and submitted two months thereafter.[219] For such "five-month" incidents, registrants must provide an aggregated report—a count of incidents and effects for each product— "in lieu of individual reports."[220] According to EPA, this requirement is not discretionary.[221] EPA will reject reports that are not aggregated.[222] A number of industry trade associations have developed an EPA-approved voluntary reporting form for aggregate incident reporting.[223]

IMPORTS AND EXPORTS

Authority to Regulate Imports and Exports

FIFRA section 17(c) provides that the secretary of treasury "shall refuse delivery to the consignee" of any pesticide being imported into the United States "[i]f it appears from the examination of a sample that it is adulterated, or misbranded or otherwise violates the

---

216. 40 C.F.R. pt. 159; 62 Fed. Reg. at 49,370; 63 Fed. Reg. 33,580 (June 19, 1998) (technical corrections); PR Notice 98-3, Guidance on Final FIFRA Section 6(a)(2) Regulations for Pesticide Product Registrants (Apr. 3, 1998); PR Notice 98-4, Additional Guidance on Final FIFRA Section 6(a)(2) Regulations for Pesticide Product Registrants (Aug. 4, 1998).
217. 40 C.F.R. § 159.184(d)(1). See 62 Fed. Reg. at 49,394.
218. 40 C.F.R. § 159.155.
219. 40 C.F.R. § 159.184(d)(3); PR Notice 98-3, at 3 (converting 30-day time limit to one month, 60-day limit to two months, and 90-day limit to three months); 40 C.F.R. §§ 159.155(b), 159.184(d)(2)–(3); PR Notice 98-4, at 7–8.
220. 40 C.F.R. § 159.184(e); see also PR Notice 98-3, at 3.
221. See PR Notice 98-3, at 3 ("The Agency will reject submissions of minor/common incidents if the summaries are not provided.").
222. See id.
223. The forms are also available from five major trade associations and a FIFRA section 6(a)(2) website, http://www.fifra6a2.com.

provisions set forth in [FIFRA], or is otherwise injurious to health or the environment."[224]

FIFRA section 17(a)(1) allows a pesticide, device, or active ingredient used in producing a pesticide to be lawfully exported "when prepared or packed according to the specifications or directions of the foreign purchaser," subject to the producers meeting the labeling requirements of FIFRA sections 2(p), 2(q)(1)(A), (C), (D), (E), (G), and (H), 2(q)(2)(A), (B), (C)(i) and (iii), and (D),[225] the establishment registration requirements in FIFRA section 7, and the books and records provisions of FIFRA section 8.[226]

FIFRA section 17(a)(2) provides that any person exporting any pesticide not registered under FIFRA section 3, or sold under FIFRA section 6(a)(1), shall obtain a "foreign purchaser" acknowledgment statement acknowledging that the purchaser understands that such pesticide is not registered for use in the United States and cannot be sold in the United States under FIFRA.[227]

Export Requirements and Exemptions

EPA's export labeling and procedural requirements are found in 40 C.F.R. part 168, subpart D. Among other requirements, the label warning and caution statements, the ingredient statement, the word "Poison," if required, and the statement of practical treatment in case of poisoning must be in English and in the appropriate foreign languages.[228] The regulations specify that the appropriate foreign language is either the language used to conduct official government business or the predominantly spoken or written language of the country of import.[229] If the country of final destination is known, the label must also be written in the official language of that country or the language that is predominantly spoken in that country.[230]

For exported unregistered pesticides, the foreign purchaser acknowledgment statement requirement can be satisfied annually or

---

224. FIFRA § 17(c), 7 U.S.C. § 136o(c).
225. FIFRA section 2(p) defines the terms "label" and "labeling"; FIFRA sections 2(q)(1)(A), (C), (D), (E), (G), and (H) and 2(q)(2)(A), (B), (C)(i) and (iii), and (D) specify various ways in which a pesticide product can be misbranded.
226. FIFRA § 17(a)(1), 7 U.S.C. § 136o(a)(1).
227. FIFRA § 17(a)(2), 7 U.S.C. § 136o(a)(2).
228. 40 C.F.R. § 168.65(a), (b)(4)(i).
229. Id. § 168.65(b)(4)(ii).
230. Id.

2.2. Federal Insecticide, Fungicide, and Rodenticide Act (FIFRA)

on a per-shipment basis, and such foreign purchaser statements must be filed with EPA according to deadlines specified by regulation.[231] An unregistered pesticide may be transferred within the United States solely for export if the pesticide meets the labeling and packaging requirements of 40 C.F.R. §§ 168.75 and 168.65(b)(1)(iii), and the foreign purchaser has signed an acknowledgment statement stating that the purchaser understands that the pesticide is not registered for use, and cannot be sold, in the United States.[232] One of the labeling requirements is that the unregistered pesticide prominently display the label statement "Not Registered for Use in the United States of America" in English and in the appropriate foreign languages.[233]

Import Requirements and Exemptions

Pursuant to U.S. Customs Service regulations implementing section 17,[234] pesticides or devices may be imported into the United States if two conditions are met:

- The pesticide must be "registered" under FIFRA section 3 (or in the case of devices, the device is compliant with FIFRA section 3); and
- A notice of arrival must be submitted to EPA prior to the shipment's arrival in the United States.[235]

EPA must review the notice of arrival and approve the import before the pesticides may enter the United States.[236] The form must be submitted for approval to the EPA regional office with jurisdiction over the customs port of entry of the shipment. EPA, in reviewing the form, may direct the U.S. Customs Service to release the shipment, refuse delivery of the shipment, or detain the shipment pending EPA's examination to determine if the shipment complies with FIFRA. EPA returns the form to the importer, who must present it to Customs

---

231. Id. § 168.75.
232. See FIFRA § 17(a)(2), 7 U.S.C. § 136o(a)(2); 40 C.F.R. § 168.75(a), (c).
233. 40 C.F.R. § 168.65(b)(1)(iii).
234. 19 C.F.R. §§ 12.110–12.117.
235. 19 C.F.R. §§ 12.111, 12.112. Section 12.111 provides that "[a]ll imported pesticides" must be "registered under the provisions of section 3 of [FIFRA], and under the regulations . . . promulgated thereunder . . . before being permitted entry into the United States." Section 12.112 requires the completion by EPA of a NOA form.
236. See FIFRA § 17(c), 7 U.S.C. § 136o(c); 19 C.F.R. §§ 12.110–12.117.

when the pesticides arrive in the United States. Pesticides detained for examination may be released to the consignee under bond, but may not be used or disposed of until EPA determines if the pesticides comply with FIFRA. Importers of multiuse chemicals imported for nonpesticide uses may request a waiver from the notice of arrival requirements for nonpesticidal imports of those chemicals.

CONFIDENTIAL AND TRADE SECRET INFORMATION

Authority Relating to Confidential Business Information

Data submitted to support an application for a pesticide registration are protected from release until 30 days after a registration is issued.[237] FIFRA section 10 affords certain protections to trade secret and confidential business information (CBI) after the issuance of a registration.

FIFRA section 10(b) protects trade secret or commercial or financial information submitted by an applicant from public release, with a number of exceptions. First, EPA may reveal the formula of a product to another federal agency or at a public hearing or in findings of fact issued by EPA when necessary to carry out the provisions of FIFRA. Second, FIFRA section 10(d)(2) allows EPA to release section 10(b) CBI concerning the production, sale, or inventories of a pesticide in connection with a public proceeding to determine whether the pesticide or any of its ingredients cause unreasonable adverse effects on health or the environment.

FIFRA section 10(d) allows EPA to make health and safety data publicly available, with three exceptions. These exceptions are (1) data concerning pesticide manufacturing or quality control processes; (2) the details of any methods for testing, detecting, or measuring the quantity of any deliberately added inert ingredient in a pesticide product; and (3) the identity or percentage quantity of any deliberately added pesticide inert ingredient.[238] Disclosure of such protected information can only occur if EPA determines that its release is necessary to protect against an unreasonable risk of injury to health or the environment. Before it can release any protected information, EPA must notify the data submitter by certified mail of its intent to release the data, and

---

237. FIFRA § 3(c)(2)(A), 7 U.S.C. § 136a(c)(2)(A); 40 C.F.R. § 152.119.
238. FIFRA § 10(d)(1), 7 U.S.C. § 136h(d)(1).

provide the data submitter with a 30-day period in which to take action in district court to enjoin the release of the information.[239]

FIFRA section 10(g) prohibits EPA from knowingly disclosing information submitted by an applicant or registrant to any employee or agent of a foreign or multinational business without the consent of the registrant or applicant. The statute requires EPA to obtain an affirmation from any person who intends to inspect health and safety data that such person will not deliver the information, or offer it for sale to a foreign or multinational business.[240] In effect, the provisions of section 10(g), prohibiting the release of data to foreign or multinational corporations, often override the section 10(b) disclosure authorization for health and safety data, and protect data from release.[241]

FIFRA Provisions Relating to Confidential Business Information

EPA specifies in its regulations and associated guidance how a submitter should mark and segregate into confidential attachments information eligible for protection as confidential. In responding to Freedom of Information Act (FOIA) requests for information claimed as confidential, EPA follows the procedures in its general FOIA rules.[242] Specifically, EPA notifies the data submitter and gives the data submitter an opportunity to submit comments on, among other things, (1) measures taken by the company to guard against the undesired disclosure of the information to others; (2) the extent to which the information has been disclosed to others; and (3) whether the business claims that disclosure will likely result in substantial harmful effects on the company's competitive position, and if so, what those harmful effects would be, why they should be viewed as substantial, and an explanation of the causal relationship between disclosure and such harmful effects.[243] EPA must give the business at least 15 working days after such notification to submit comments.[244]

---

239. FIFRA § 10(c), 10(d)(3), 7 U.S.C. § 136h(c), 136h(d)(3).
240. FIFRA § 10(g), 7 U.S.C. § 136h(g).
241. EPA can disclose information otherwise protected by FIFRA section 10(g) in connection with a cancellation hearing if such information is relevant to whether the pesticide or its ingredients cause unreasonable adverse effects on health or the environment.
242. *See* 40 C.F.R. §§ 2.100–2.215, 2.307.
243. 40 C.F.R. § 2.204(c).
244. *Id.* § 2.204(e)(2).

If EPA determines that the data are entitled to confidential treatment, it will maintain the information as confidential and notify the data submitter of its determination.[245] If EPA denies the claim of confidentiality, it must notify the data submitter of its denial and state the basis for its determination. A denial of a claim of confidentiality is a final agency action subject to judicial review.[246] The denial notice must state that EPA will make the information public on the 31st day after the data submitter's receipt of the notice, unless the data submitter has notified EPA's legal office within that time that the company has initiated a legal action seeking judicial review of the disclosure determination or an injunction prohibiting disclosure.[247]

Enforcement and Penalties

Unlawful Activities

FIFRA section 12 lists prohibited acts. It is unlawful for any person to sell or distribute (1) an unregistered pesticide or (2) a registered pesticide (a) with claims that differ from those approved by EPA; (b) that differs in composition from the composition approved by EPA; (c) that has not been colored or discolored, if coloration is required; or (4) that is adulterated or misbranded.[248] It is also unlawful to sell or distribute a misbranded pesticide device. A pesticide or device is misbranded if, among other things, its labeling bears any false or misleading statement; does not bear the establishment or product registration number; does not prominently display any required word, statement, or other information; does not contain directions for use that are adequate to protect health and the environment; does not contain a warning or caution statement adequate to protect health and the environment; or does not contain the use classification.[249]

Under FIFRA, a person may not "offer for sale" a pesticide that is unregistered or that makes claims at the time of sale that are substantially different from the claims made for it in its registration application. EPA interprets the term "offer for sale" to extend to advertising,

---

245. See 40 C.F.R. § 2.208.
246. 40 C.F.R. § 2.205(f)(1), (2).
247. Id. § 2.307(e)(3).
248. FIFRA § 12(a)(1), 7 U.S.C. § 136j(a)(1).
249. See FIFRA § 2(q)(1)–(2) for a complete list of the various ways a pesticide product can be misbranded. 7 U.S.C. § 136(q)(1)–(2).

2.2. Federal Insecticide, Fungicide, and Rodenticide Act (FIFRA) 131

defined as promotional material in a medium to which pesticide users or the general public have access, such as television, radio, newspapers, trade journals, industry magazines, and billboards.[250] Thus, advertising claims must be consistent with a product's registration claims. In addition, EPA regulations prohibit, with some exceptions, advertisements that recommend purchasing or using a pesticide for a use authorized by an EUP, an emergency exemption, or a special local needs registration. The regulations also prohibit advertisements that recommend purchasing or using an unregistered pesticide or a registered pesticide for an unregistered use.

A pesticide is "adulterated" if its strength or purity differs from that stated on the label, any substance has been substituted in whole or in part for the pesticide, or any valuable component of the pesticide is not present.[251]

Other prohibited acts for registered pesticides include, among other things, detaching or altering the labeling of a pesticide; refusing to prepare, maintain, or submit required records, or to submit required reports; refusing to allow an EPA inspector to enter or inspect a facility, copy records, or conduct sampling;[252] selling a RUP to a noncertified applicator;[253] using a pesticide in a manner inconsistent with its labeling; using an EUP for a nonauthorized use;[254] violating any "stop sale, use, or removal" order issued under FIFRA section 13;[255] violating any suspension or cancellation order; violating the requirements for filing FIFRA section 7 reports;[256] knowingly submitting a false application for registration, report, or other information, or falsifying any required records; failing to file required reports; improperly adding or removing any substance from a pesticide; conducting human testing of pesticides without informed consent; submitting false test data or information; and violating any regulation issued under FIFRA sections 3(a) or 19 pertaining to the sale or use of an unregistered pesticide or the storage, disposal, transportation, and recall of pesticides.

---

250. See 40 C.F.R. § 168.22. See also FIFRA § 2(p), 7 U.S.C. § 136(p); EPA, The Sale of Pesticides over the Internet, http://www.epa.gov/compliance/assistance/bystatute/fifra/pestecom.html.
251. FIFRA § 2(c), 7 U.S.C. § 136(c).
252. FIFRA § 12(a)(2)(A)–(B), 7 U.S.C. § 136j(a)(2)(A)–(B).
253. FIFRA § 12(a)(2)(F), 7 U.S.C. § 136j(a)(2)(F).
254. FIFRA § 12(a)(2)(H), 7 U.S.C. § 136j(a)(2)(H).
255. FIFRA § 12(a)(2)(I), 7 U.S.C. § 136j(a)(2)(I).
256. FIFRA § 12(a)(2)(N), 7 U.S.C. § 136j(a)(2)(N).

The precursor to an enforcement action is likely to be either an inspection or a subpoena from EPA or a state. FIFRA section 9(a) authorizes EPA or state officers or employees to inspect facilities where pesticides are held for distribution or sale or where canceled or suspended pesticide products are being held.[257]

In the landmark 3M case decided in 1994, the court of appeals held that civil administrative violations under the Toxic Substances Control Act (TSCA) are subject to the general federal five-year statute of limitations in 28 U.S.C. § 2462 because TSCA does not have its own statute of limitations.[258] The court also held that the period of limitations begins to run at the time the violation first occurred, rather than at the time EPA discovered the violation, in the absence of fraudulent concealment.[259] FIFRA also does not contain its own statute of limitations regarding administrative penalty actions for violations under 7 U.S.C. § 136l(a), and the federal five-year statute of limitations has been applied to FIFRA violations as well. Following the 3M decision, EPA has taken the position that the statute of limitations does not apply to "continuing" violations under the theory that a continuing violation restarts the statute of limitations.[260]

Civil Penalties

Penalties for violations of FIFRA section 12 prohibited acts are set forth in FIFRA section 14, as adjusted by 40 C.F.R. part 19 (Adjustment of Civil Monetary Penalties for Inflation) for violations occurring after January 30, 1997.[261] In determining the size of the penalty, EPA must consider the appropriateness of the penalty to the size of the business, the effect on the entity's ability to continue in business, and the gravity of the violation.[262] Each offense committed by a registrant, commercial applicator, wholesaler, dealer, retailer, or other

---

257. *See* EPA, FIFRA INSPECTION MANUAL, http://www.epa.gov/compliance/resources/publications/monitoring/fifra/manuals/fiframanual.pdf.
258. *See* 3M Co. v. Browner, 17 F.3d 1453 (D.C. Cir. 1994).
259. 17 F.3d at 1462.
260. *See* EPA Memorandum, Guidance on the Application of Five-Year Statute of Limitations to Administrative and Judicial Civil Penalty Proceedings, at 2 (July 14, 1994).
261. FIFRA § 14(a)(1), 7 U.S.C. § 136l(a)(1).
262. FIFRA § 14(a)(4), 7 U.S.C. § 136l(a)(4).

distributor is subject to a civil penalty of up to $7,500 using the following matrix:[263]

	SIZE OF BUSINESS		
LEVEL OF VIOLATION	I: Over $10 million	II: $1 million–$10 million	III: Under $1 million
LEVEL 1	$7,500	$7,150	$7,150
LEVEL 2	$7,150	$5,670	$4,250
LEVEL 3	$5,670	$4,250	$2,830
LEVEL 4	$4,250	$2,830	$1,420

Once a base penalty is determined, the amount can be adjusted upward or downward, depending on the actual circumstances of each violation. The factors that can adjust a penalty amount include any economic benefit gained, whether the company self-confessed the violations, and other good-faith efforts by the company.[264] The statute authorizes EPA to issue a notice of warning instead of assessing a penalty when a violation occurred despite the exercise of due care or did not cause significant harm to health or the environment.[265] EPA cannot assess a penalty without giving the person charged an opportunity to request a hearing.[266]

Criminal Penalties

FIFRA section 14(b) allows EPA to bring criminal charges against a registrant, applicant, or producer who knowingly violates any provision of FIFRA.[267] In *United States v. Corbin Farm Service*, the court interpreted the word "knowingly" in FIFRA section 14(b) to require only a general intent to do the acts constituting the violation, not a specific intent to violate the law.[268] The maximum criminal penalty for a registrant, applicant, or producer is $50,000 and/or one

---

263. Office of Compliance Monitoring, Office of Pesticides and Toxic Substances, FIFRA Enforcement Response Policy (Dec. 2009), at 19 [hereinafter FIFRA ERP].
264. FIFRA ERP, at 19–28.
265. FIFRA § 14(a)(4), 7 U.S.C. § 136l(a)(4).
266. FIFRA § 14(a)(3), 7 U.S.C. § 136l(a)(3).
267. FIFRA § 14(b), 7 U.S.C. § 136l(b). *See also* FIFRA INSPECTION MANUAL, *supra* note 257, at ch. 18.
268. United States v. Corbin Farm Svc., 444 F. Supp. 510, 517–20 (E.D. Cal.), *aff'd*, 578 F.2d 259 (9th Cir. 1978); FIFRA ERP, *supra* note 263, at 14.

year in prison.[269] A commercial applicator of a RUP or a distributor who knowingly commits a violation is subject to a maximum criminal penalty of $25,000 and/or one year in prison.[270] The maximum criminal penalty for a private applicator who knowingly violates the statute is $1,000 and/or 30 days in prison.[271] Any person who uses or reveals information about product formulas with the intent to defraud is subject to a maximum criminal penalty of $10,000 and/or three years in prison.[272]

Both the person knowingly committing a violation and the employer of such person can be held criminally liable.[273] Criminal charges are prosecuted by the Department of Justice (DOJ). The court in *United States v. Orkin Exterminating Co.*[274] ruled that the DOJ retains the authority to prosecute FIFRA criminal use violations even though the statute gives primary enforcement authority for use violations to the states.

Specific Enforcement

EPA has the authority under FIFRA section 13(a) to issue "stop sale, use, or removal" orders (SSURO) to prevent the sale, distribution, or use, or require the removal of a pesticide that is in violation of FIFRA.[275] EPA may issue a SSURO if it has reason to believe that the pesticide or device is in violation of FIFRA or has been or is intended to be distributed or sold in violation of FIFRA. EPA may also issue a SSURO if a pesticide's registration has been canceled or suspended. After a SSURO has been issued, any movement of the pesticide is a violation of FIFRA. Removal has been defined broadly in this context. For example, an administrative law judge, in *In re Schattner*, held that shipments to a laboratory for tests regarding the efficacy of the pesticide are prohibited once a SSURO has been issued. EPA often issues SSUROs in conjunction with an administrative penalty action.

---

269. FIFRA § 14(b)(1)(A), 7 U.S.C. § 136l(b)(1)(A).
270. FIFRA § 14(b)(1)(B), 7 U.S.C. § 136l(b)(1)(B).
271. FIFRA § 14(b)(2), 7 U.S.C. § 136l(b)(2).
272. FIFRA § 14(b)(3), 7 U.S.C. § 136l(b)(3).
273. FIFRA § 14(b)(4), 7 U.S.C. § 136l(b)(4).
274. United States v. Orkin Exterminating Co., 688 F. Supp. 223 (W.D. Va. 1988).
275. *See* FIFRA § 13(a), 7 U.S.C. § 136k(a); FIFRA INSPECTION MANUAL, *supra* note 257, at ch. 10.

2.2. Federal Insecticide, Fungicide, and Rodenticide Act (FIFRA)

Stop sale orders can be challenged in federal district court under the general review provisions in FIFRA section 16(a). FIFRA section 13(b) authorizes EPA to bring a seizure action in federal district court against a pesticide that is adulterated or misbranded, unregistered, improperly labeled, not colored or discolored as required, or where the claims made for the pesticide differ from the approved claims, or otherwise where the pesticide causes unreasonable adverse effects on the environment.

KEY BUSINESS ISSUES

Commonly Encountered Issues

A FIFRA registration is a dynamic license. Accordingly, a pesticide registrant must have a sound management-of-change program in place. With few exceptions, any change to the product label or formulation requires prior approval by EPA. Even changes in suppliers or sources of inert ingredients require a submission to EPA. Also, through its periodic review cycles or other initiatives, EPA may require registrants to develop and submit additional data to support their registrations or make label changes. Registrants must *maintain* their registrations or risk noncompliance.

A registrant should coordinate marketing efforts with its regulatory experts. Under FIFRA it is unlawful for a person to distribute or sell any registered pesticide if any claims made for it as part of its distribution or sale substantially differ from any claims approved as part of its registration. Additionally, for documents that could fit the FIFRA definition of labeling, EPA may conclude that the product is misbranded if the labeling contains false or misleading statements. If a company's website is listed on its label, EPA views the website as labeling. Even if the website is not listed on the product label, product claims made on the website must be consistent with label-approved claims. It is easy for an aggressive marketing campaign to overreach and make statements that substantially differ from approved claims. Many registrants include long lists of possible marketing claims on their master labels, from which they can then pick and choose for marketing purposes.

Supplemental distribution is a mechanism frequently used by registrants to leverage more effectively their registrations. Likely because the process for notifying EPA of supplemental distributors requires little effort and the requirements are otherwise self-implementing, registrants often fail to comply with all applicable requirements, especially the requirement that the supplemental distribution product must be produced and packaged in the same registered establishment as the registrant's product. EPA has come to appreciate that this is an area of compliance weakness and has targeted enforcement efforts accordingly. As registrants and supplemental distributors are jointly liable for the compliance status of the supplemental distributor product, contractual language and best management practices should be used to support compliance.

Practical Tips

Practical management of FIFRA compliance requires its own book, but illustrative tips include the following:

Pesticide product registrations are handled by EPA product management teams, who are responsible for all communications regarding a pesticide registration application or an amendment. The product management team is headed by a team leader or product manager (PM) who has a number of assistants for coordinating the review within the Office of Pesticide Programs. All communication regarding a registration or an amendment application should be directed through the assigned product management team leader. It is important to establish good working relationships with the EPA staff members responsible for a company's pesticide products, given the heavy regulation of these products and thus frequent interaction with EPA that product registrants must have.

For registration of a new active ingredient or a product that raises one or more novel issues, registrants should engage EPA prior to data development or submission of an application through a preregistration meeting. All meetings should be scheduled through the appropriate product management team leader unless otherwise directed by EPA. Well-researched questions are more likely to yield helpful responses. EPA should be provided with an agenda and possibly other information, such as product formulation and intended claims, prior to the meeting.

If a new pesticide product contains one or more components with a proprietary formulation, such as a fragrance, the applicant should confirm that EPA has on file the formulation of the substance and that the product is approved for use as an ingredient in a pesticide product. The supplier may provide proprietary information directly to EPA. For further guidance, see inert ingredient guidance documents available on EPA's website.[276]

Registrants must recognize that pesticide products are highly regulated and registrants are responsible for product compliance. To the extent that registrants use toll manufacturers, or have supplemental distributors, the contractual relationships with these parties must address federal and state regulatory compliance.

EPA allows a registrant to distribute or sell a product with labeling that contains any subset of the approved directions for use, provided that in limiting the uses listed on the label, no changes are needed to the precautionary statements, use classification, or product packaging. A registrant thus may obtain an approved master label containing all the known uses for a pesticide that the application can support and then distribute or sell the same registered product into different markets with different "market labels" without separate EPA approval. The registration fee is the same for a pesticide product registration regardless of the number of uses.

As noted in the "Commonly Encountered Issues" section of this chapter, it is unlawful for a person to distribute or sell any registered pesticide if any claims made for it as part of its distribution or sale substantially differ from any claims approved as part of its registration. Additionally, it is unlawful to advertise an unregistered pesticide. Companies must integrate regulatory review and compliance with its product development and product marketing efforts or invite enforcement risk.

Through its registration review program, EPA reviews all registered pesticides every 15 years to confirm that the products continue to meet registration standards. EPA publishes a schedule for beginning reviews in approximately three-year increments. Registrants should monitor EPA's schedule and note when registration review will begin for pertinent active ingredients. EPA increasingly is using the mechanism of focus meetings early in the registration review process to engage with affected registrants and other stakeholders to help ensure that EPA has the best available information for making decisions during the process, which could lead to fewer data requirements and EPA's reliance on avoiding overly conservative assumptions. If a registrant has not received an invitation letter from EPA to a focus

---

276. EPA, Pesticide Inert Ingredients, http://www.epa.gov/opprd001/inerts.

meeting, it may contact the chemical review manager to inquire whether a focus meeting has been scheduled.

Trends

15-Year Registration Review Cycle

Often overlooked, but one of the more significant amendments to FIFRA as part of FQPA, is the requirement that EPA reexamine all pesticide registrations on a 15-year cycle to determine if the pesticide continues to meet the standards for continued use. Because FQPA required a review of all then-existing tolerances as of 1996, and those reviews began to be completed in 1997, the 15-year cycle has begun in earnest.

This review is known as registration review. EPA has now established the schedule and review path for active ingredients subject to the *second* round of FQPA/FIFRA review. The data requirements and regulatory criteria are continually being updated and thus are always, to some degree, subject to change. The 15-year cycle allows EPA to defer or reserve some degree of the review of any specific pesticide until its next round of review (although EPA always has a right and a duty to ensure that any registered product meets FIFRA requirements at all times). For example, as EPA has been pressed via litigation in recent years to comply clearly with the requirements of the Endangered Species Act (ESA), ESA reviews for many pesticides have been incorporated into the registration review schedule.

Digital Records/Electronic Labels

Regardless of policy changes or shifts in program priorities, there is an increasing reliance on digital records, to allow greater ease in both public access to, and registrant submission of, applications and other data. To the extent that digital or electronic submissions make for more accurate transmission and more rapid acceptance of the information, there is broad support by all stakeholders. Further, all stakeholders benefit from the additional ease provided by Internet access to information related to FIFRA registrations (e.g., dockets, study reviews, and progress on PRIA 3 deadlines).

Of more controversy is EPA's initiative promoting electronic labels. Such labels would help to facilitate immediate effect to changes made to any label language, but they raise numerous issues of concern to registrants, and to EPA, that have not been resolved, such as the status of the paper label attached to the actual product in the supply chain, liability issues, and how to ensure that label instructions specific to any particular use situation are current. These issues present difficult challenges that EPA and the registrants need to address.

Putting aside the complex issues associated with electronic labeling, in general, making information supporting pesticide decisions (e.g., decision memos and support documents) digitally available is consistent with and facilitates EPA's goal of greater public input. EPA materials are much more immediately and easily available to the public through the Internet, and the public can also submit comments electronically into the regulatory dockets. These digital trends continue to be enhanced and expanded.

Budget/Resources

Notwithstanding the enactment of PRIA 3 at the end of the 2012 session of Congress, EPA resources, like those of any federal agency, are currently restrained. Although PRIA fees and the other FIFRA maintenance fees provide a supplement to appropriated resources, the appropriations process has become less predictable and more subject to overall partisan wrangling over domestic program spending and priorities. Many operational divisions of the EPA Office of Pesticide Programs currently have significantly reduced resources compared to just a few years ago, and there is little on the horizon indicating that the pesticide program, or EPA, will be likely receiving any significant increases in appropriations. With shrinking resources, PRIA deadlines may begin to be missed more frequently, and, overall, the quality and output of the review staff may suffer. PRIA fees may need to be enhanced in the coming years.

International Activities

EPA's pesticide program continues to be actively engaged in both multilateral approval submissions and international harmonization of data requirements and standards. As global trade in agricultural products continues to grow, multinational approvals become not only more attractive but also more imperative for potential pesticide users to avoid possible trade issues as new products are introduced into the marketplace. Multilateral submissions are attractive to registrants seeking broad introduction of a new product, as well as facilitating trade among the various market partners. EPA has seen an increase in the number of multilateral submissions, and this trend is expected to increase. The goal is to reduce the net time for obtaining approval in those countries, and, as experience is gained, progressing toward harmonization of data requirements and evaluation techniques across governments. It may be a long time before there is absolute harmonization among a large number of countries, but joint submissions are a step toward that goal.

As the effect of the Occupational Safety and Health Administration's adoption of the Globally Harmonized System of Classification and Labelling of Chemicals (GHS) for hazard communication and the potential for conflict with pesticide label requirements remains controversial for registrants, the intermediate problem is for the transition period between full implementation of any international, harmonized approach and compliance with individual country requirements in the meantime. The GHS scheme is expected to eventually facilitate trade among pesticide-producing countries and to improve label communication to all users, but the transition steps are not all in place and some possible dislocations could occur along the way.

"21st-Century Toxicology"/Endocrine Effect Testing
"21st-Century Toxicology" refers to a constellation of efforts to reduce animal testing and needless time and expense while improving the quality of scientific study by supplanting animal testing with more sophisticated science, including in vitro testing and computational science. The pesticide program is currently considered "data rich" because it requires detailed and extensive whole animal (in vivo) data on individual compounds, with costs of these required data estimated to be more than $10 million to $20 million per active ingredient. Applying such requirements to tens of thousands of other chemicals outside the FIFRA world would literally be impossible without the outlay of billions of dollars in regulatory costs. As a result, new techniques designed to meet a number of goals are envisioned as the future of chemical product testing, and some of these may eventually be applied in the FIFRA program. Among the criteria that are being assessed for new chemical testing approaches are tests that can be completed relatively quickly and for low cost; tests that minimize the use of laboratory animals; and tests that can be validated and on which a regulatory determination may be based. These goals are ambitious and are a priority for EPA as a whole.

In the immediate future, pesticides regulated under FIFRA will continue to need the extensive and expensive data sets currently required. As the movement toward new testing and evaluation techniques is developed across EPA, however, one can expect these approaches first to complement and eventually, potentially, to supplement the current FIFRA-imposed requirements.

Similarly, as EPA continues to unfold its endocrine testing program, which was first required under FQPA, those tests, the embedded regulatory concerns, and debates about the meaning of results might first be forged in the pesticide regulatory program. EPA is now moving toward the second tier of required testing, and as it makes progress in the 21st-century toxicology arena, these two program goals may eventually marry into part of the future pesticide testing scheme over the next years.

Resources

Federal Insecticide, Fungicide, and Rodenticide Act (FIFRA), 7 U.S.C. §§ 136–136y.
Federal Food, Drug, and Cosmetic Act (FFDCA), as amended by the Food Quality Protection Act (FQPA), FFDCA §§ 201, 301, 303, 402, 408, 409, 21 U.S.C. §§ 321, 331, 333, 342, 346a, 348; 40 C.F.R. § 152.112(g).
40 C.F.R. parts 152–189.
EPA's Office of Pesticide Programs website is a rich resource of information and tools. The following web links are for some of the most frequently used resources:
    InertFinder Search Engine: http://iaspub.epa.gov/apex/pesticides/f?p=103:1
    Label Review Manual: http://www.epa.gov/oppfead1/labeling/lrm/
    Office of Pesticide Programs main website: http://www.epa.gov/pesticides/
    Pesticide Chemical Search Engine: http://iaspub.epa.gov/apex/pesticides/f?p=chemicalsearch:1
    Pesticide Product Label System: http://iaspub.epa.gov/apex/pesticides/f?p=PPLS:1
    Pesticide Registration Kit: http://www.epa.gov/pesticides/registrationkit/
    Pesticide Registration Manual (Blue Book): http://www.epa.gov/pesticides/bluebook/
    Pesticide Registration (PR) Notices: http://www.epa.gov/PR_Notices/
    Pesticide Registration Review status and documents: http://www.epa.gov/oppsrrd1/registration_review/reg_review_status.htm
    Pesticide Reregistration Status: http://www.epa.gov/pesticides/reregistration/status.htm
    Purdue University's National Pesticide Information Retrieval System (free portion): http://ppis.ceris.purdue.edu/

2.3

State Laws

Carla N. Hutton

EXECUTIVE SUMMARY

Several states have implemented their own chemical regulatory schemes. Perhaps the oldest, and certainly the best known, is California's Safe Drinking Water and Toxic Enforcement Act of 1986 (Proposition 65). Under Proposition 65, the California Office of Environmental Health Hazard Assessment (OEHHA) publishes a list of chemicals known to the state to cause cancer or reproductive toxicity. Businesses must notify Californians about listed chemicals in the products they purchase, in their homes or workplaces, or that are released into the environment. More recently, as part of its Green Chemistry Initiative, California implemented the Safer Consumer Products Regulations (SCPR) effective October 1, 2013. Although Proposition 65 does not restrict or ban the manufacture, processing, or distribution of consumer products containing any listed chemicals, the SCPR requires manufacturers to evaluate the availability of safer alternative ingredients. Depending on the availability of a safer alternative, California could restrict the use of priority products (PP) without the alternative or could even prohibit product sales. California is not the only state to implement its own chemical regulatory scheme, however. In 1989, Massachusetts passed the Toxics Use Reduction Act (TURA). Under TURA, companies that use large quantities of listed chemicals are required to report annually their use of the listed chemicals, pay an annual toxics use fee, and prepare a report examining their use of the listed chemicals and ways to implement toxics use reduction strategies.

2.3.1
CALIFORNIA

2.3.1.1
PROPOSITION 65

PROPOSITION 65 OVERVIEW

California has long been a leader in progressive environmental protection. The Safe Drinking Water and Toxic Enforcement Act of 1986, better known as Proposition 65, revolutionized chemical notification programs and brought "right to know" to a new level. Proposition 65 requires California to publish annually a list of chemicals known to cause cancer or reproductive toxicity.[1] Proposition 65 requires businesses to notify Californians about significant amounts of listed chemicals in the products they purchase, in their homes or workplaces, or that are released into the environment. It also prohibits California businesses from knowingly discharging significant amounts of listed chemicals into sources of drinking water.

---

1. CAL. HEALTH & SAFETY CODE §§ 25249.5–.13.

SCOPE OF PROPOSITION 65

Scope of Substances Subject to Proposition 65

Proposition 65 targets exposure to and discharge of chemicals listed as known to cause cancer or reproductive toxicity, rather than products containing listed chemicals.[2] An individual may be exposed to a listed chemical through water, air, food, consumer products, and any other environmental exposure, as well as through occupational exposures.[3]

Scope of Activities Subject to Proposition 65

Proposition 65 prohibits any person in the course of doing business from knowingly discharging or releasing a listed chemical into water or onto or into the land where such chemical passes or is likely to pass into any source of drinking water.[4] Proposition 65 exempts any discharge or release that takes place within 20 months of the chemical being listed.[5] Proposition 65 also exempts any discharge or release that (1) will not cause any significant amount of the discharged or released chemical to enter any source of drinking water; and (2) is in conformity with all other laws and every applicable regulation, permit, requirement, and order.[6] The regulations clarify that discharge into water or onto or into land also includes a discharge or release to air that is "directly and immediately" deposited into water or onto land.[7] In addition, under the regulations, if the listed chemical is an ingredient or degradation product of a pesticide, and use of the pesticide is otherwise in compliance with the law, then the discharge or release will be presumed not to pass into any source of drinking water.[8]

Under Proposition 65, any person in the course of doing business is prohibited from knowingly and intentionally exposing any individual to a listed chemical without first providing "clear and reasonable warning."[9] The warning need not be provided separately to each potentially exposed individual, but instead may be provided through

---

2. *Id.* §§ 25249.5, 25249.6.
3. CAL. CODE REGS. tit. 27, § 25102(i).
4. CAL. HEALTH & SAFETY CODE § 25249.5.
5. *Id.* § 25249.9(a).
6. *Id.* § 25249.9(b).
7. CAL. CODE REGS. tit. 27, § 25102(f).
8. *Id.* § 25405.
9. CAL. HEALTH & SAFETY CODE § 25249.6.

consumer product labels, notices in mailings to water customers, posted notices, and notices in public news media.[10] The regulations include provisions for consumer product warnings,[11] occupational exposures,[12] and environmental exposures.[13]

Exemptions from the warning requirement are provided for the following:

- An exposure for which federal law governs warning in a manner that preempts state authority;[14]
- An exposure that occurs within 12 months of the listing of the chemical in question;[15]
- An exposure that the person responsible can show poses no significant risk assuming lifetime exposure at the level in question for substances known to cause cancer, and that will have no observable effect assuming exposure at 1,000 times the level in question for substances known to cause reproductive toxicity.[16]

Scope of Persons Subject to Proposition 65

Proposition 65 defines person as "an individual, trust, firm, joint stock company, corporation, company, partnership, limited liability company, and association."[17] Businesses with fewer than ten employees are exempt from the warning requirement and discharge prohibition,[18] as are governmental agencies and public water systems.[19]

---

10. Id. § 25249.11(f).
11. Cal. Code Regs. tit. 27, §§ 25603–25603.3.
12. Id. §§ 25604–25604.2.
13. Id. §§ 25605–25605.2.
14. Cal. Health & Safety Code § 25249.10(a).
15. Id. § 25249.10(b).
16. Id. § 25249.10(c).
17. Id. § 25249.11(a).
18. Id. § 25249.11(b).
19. Id.

Notification/Registration/Approval Requirements and Compliance

Summary of Authority Governing the Requirements for Notification/Registration/Approval

Section 25249.8 requires California to publish annually a list of chemicals known to cause cancer or reproductive toxicity.

Standard of Review/Burden of Proof

Under Proposition 65, a chemical is known to the state to cause cancer or reproductive toxicity if, in the opinion of the state's qualified experts, it has been "clearly shown through scientifically valid testing" to cause cancer or reproductive toxicity, or if a body considered to be authoritative has formally identified it as causing cancer or reproductive toxicity, or if a California or federal agency has formally required it to be labeled or identified as causing cancer or reproductive toxicity.[20]

Listing Procedures

Under the statute, the list includes at a minimum those substances identified by reference in Labor Code section 6382(b)(1) and those substances identified additionally by reference in Labor Code section 6382(d).[21] This is also referred to as the Labor Code mechanism.[22] The statute also provides three other listing mechanisms: listing via the state's qualified experts, the Carcinogen Identification Committee

---

20. *Id.* § 25249.8(b).
21. *Id.* § 25249.8(a).
22. This method of adding chemicals to Proposition 65 has been the subject of recent litigation. Most recently, on October 31, 2012, the Court of Appeals of the State of California, Third Appellate District, issued a decision affirming the decision of the trial court, which found that OEHHA failed to provide sufficient evidence that styrene and vinyl acetate are "known" to cause cancer and thus these substances could not be listed as carcinogens. Styrene Info. and Research Ctr. v. OEHHA, No. C064301 (Cal. Ct. App. 3d Dist. 2012). OEHHA based its decision on a 2002 monograph by the International Agency for Research on Cancer (IARC) that classified styrene and vinyl acetate as Group 2B, "possibly carcinogenic to humans." The Group 2B category applies to "agents for which there is limited evidence of carcinogenicity in humans and less than sufficient evidence of carcinogenicity in experimental animals." The court held that because chemicals may be included in IARC Group 2B based on less than sufficient evidence of carcinogenicity, they may not qualify for Proposition 65 listing on that basis alone.

(CIC) and Developmental and Reproductive Toxicant Identification Committee (DARTIC); listing via the authoritative body mechanism; and listing via the formally required mechanism.[23] If CIC or DARTIC determine that a chemical "has been clearly shown through scientifically valid testing according to generally accepted principles to cause cancer or reproductive toxicity," then the chemical will be added to Proposition 65.[24] If a body considered to be authoritative by CIC or DARTIC has formally identified a chemical as causing cancer or reproductive toxicity, then it may be listed.[25] Authoritative bodies for the identification of chemicals as causing cancer include the International Agency for Research on Cancer (IARC); National Institute for Occupational Safety and Health (NIOSH); National Toxicology Program (NTP); U.S. Environmental Protection Agency (EPA); and U.S. Food and Drug Administration (FDA).[26] Authoritative bodies for the identification of chemicals as causing reproductive toxicity are IARC, solely as to transplacental carcinogenicity; NIOSH; NTP, solely as to final reports of NTP's Center for Evaluation of Risks to Human Reproduction (CERHR); EPA; and FDA.[27] If a California or federal agency formally requires a chemical to be labeled or identified as causing cancer or reproductive toxicity, then it may be listed.[28]

OEHHA Review and Response

Under Proposition 65, listing chemicals as known to cause cancer or reproductive toxicity is not considered to be adopting or amending a regulation within the meaning of the California Administrative Procedure Act.[29] The regulations explicitly provide that OEHHA will publish a notice of intent to list only in the case of adding a chemical to Proposition 65 through the authoritative body mechanism.[30] In that case, OEHHA must publish at least 60 days prior to adding a chemical a notice of intent to list that identifies the authoritative body and chemical.[31] In practice, OEHHA provides at least a 30-day comment

---

23. CAL. HEALTH & SAFETY CODE § 25249.8(b).
24. Id.; CAL. CODE REGS. tit. 27, § 25305(a)(1), (b)(1).
25. CAL. HEALTH & SAFETY CODE § 25249.8(b); CAL. CODE REGS. tit. 27, § 25306(a).
26. CAL. CODE REGS. tit. 27, § 25306(m).
27. Id. § 25306(l).
28. CAL. HEALTH & SAFETY CODE § 25249.8(b); CAL. CODE REGS. tit. 27, § 25902.
29. CAL. HEALTH & SAFETY CODE § 25249.8(e).
30. CAL. CODE REGS. tit. 27, § 25306(i).
31. Id.

period when listing via the state's qualified experts mechanism,[32] formally required mechanism,[33] and Labor Code mechanism.[34]

CONFIDENTIAL AND TRADE SECRET INFORMATION

Because businesses need not provide information to OEHHA, Proposition 65 and its implementing regulations do not generally address confidential business information. Under Proposition 65, a business may request OEHHA make a safe use determination.[35] Under the regulations, the requester may claim that information should not be available for public inspection by specifically identifying the information and the basis for the claim.[36] If OEHHA determines that information claimed confidential must be released to the public under the Public Records Act, it will notify the requester of its determination and provide the requester an opportunity to submit additional justification for the claim or to contest OEHHA's determination.[37]

ENFORCEMENT AND PENALTIES

Unlawful Activities

Under Proposition 65, it is unlawful to contaminate drinking water with a listed chemical.[38] This prohibition includes the discharge or release of a listed chemical both directly into water and onto or into land where such a discharge "passes or probably will pass into any source of drinking water."[39] It is also unlawful to fail to provide the required warning before exposure to a listed chemical.[40]

---

32. OEHHA, Listing via the State's Qualified Experts (SQEs) Mechanism; Health and Safety Code § 25249.8(b) and 27 Cal. Code Regs. § 25305 (June 2008), http://www.oehha.ca.gov/prop65/policy_procedure/pdf_zip/LDfig1.pdf.
33. OEHHA, Listing via the Formally Required Mechanism; 27 Cal. Code Regs. § 25902 (June 2008), http://www.oehha.ca.gov/prop65/policy_procedure/pdf_zip/LDfig3.pdf.
34. OEHHA, Listing Chemicals Identified Via Labor Code § 6382(b)(1) or (d); Health and Safety Code § 25249.8(a) (Apr. 2007), http://www.oehha.ca.gov/prop65/policy_procedure/pdf_zip/LDfig4.pdf.
35. CAL. CODE REGS. tit. 27, § 25204.
36. Id. § 25204(c)(7).
37. Id. § 25204(c)(7)(B).
38. CAL. HEALTH & SAFETY CODE § 25249.5.
39. Id.
40. Id. § 25249.6.

Civil Penalties

Violations of the discharge prohibition or warning requirement provisions are subject to a civil penalty not to exceed $2,500 per day for each violation.[41] In assessing the amount of the penalty, the court may consider factors such as the extent of the violation, its severity, the economic effect of the penalty on the violator, whether the violator took good-faith measures to comply with Proposition 65, the willfulness of the violator's misconduct, and the deterrent effect of the penalty.[42]

Proposition 65 is enforced through civil litigation. Under Proposition 65, the California attorney general, any district attorney, or certain city attorneys may bring suit,[43] as well as private parties acting in the public interest.[44] Private parties must first provide notice of the alleged violation to the attorney general, the appropriate district attorney and city attorney, and the business accused of the violation.[45] A private party may not pursue an enforcement action if one of the governmental officials noted begins an action concerning the alleged violation within 60 days of the notice.[46]

If the private party alleges a failure to provide the required warning before exposure to a listed chemical, then the private party must include a certificate of merit. The certificate must state that the private party or its attorney has consulted with at least one expert "who has reviewed facts, studies, or other data" concerning the alleged exposure and that, based on that information, there is a "reasonable and meritorious case for the private action."[47] The basis for the certificate of merit is not discoverable, except as discussed below.[48]

A business found to be in violation of Proposition 65 is subject to civil penalties of up to $2,500 per day for each violation.[49] In assessing the amount of the civil penalty, the court shall consider the following factors: the nature and extent of the violation; the number of, and severity of, the violations; the economic effect of the penalty on

---

41. Id. § 25249.7(b)(1).
42. Id. § 25249.7(b)(2).
43. Id. § 25249.7(c).
44. Id. § 25249.7(d).
45. Id. § 25249.7(d)(1); CAL. CODE REGS. tit. 27, § 25903.
46. CAL. HEALTH & SAFETY CODE § 25249.7(d)(2).
47. Id. § 25249.7(d)(1).
48. Id. § 25249.7(h)(1).
49. Id. § 25249.7(b)(1).

the violator; whether the violator took good-faith measures to comply with Proposition 65 and the time those measures were taken; the willfulness of the violator's misconduct; and the deterrent effect that the penalty would have on both the violator and the regulated community as a whole.[50]

At the end of an action brought by a private party, if the court determines that there was no actual or threatened exposure to a listed chemical, the court may review the basis for the belief of the person executing the certificate of merit that an exposure to a listed chemical had occurred or was threatened.[51] If the court finds that there was no credible factual basis for the certifier's belief that an exposure to a listed chemical has occurred or was threatened, then the action will be deemed frivolous.[52]

Key Business Issues

Commonly Encountered Issues

While OEHHA adopts safe harbor levels for listed chemicals, it has done so for less than half the listed chemicals.[53] Given California's budgetary constraints, businesses should not rely on OEHHA to adopt a safe harbor level for a particular chemical of concern. Under the regulations, a business may request OEHHA to consider the applicability of Proposition 65 to its business activities and issue a safe use determination.[54] OEHHA will not issue safe use determinations for hypothetical situations or for alternative scenarios in a proposed activity.[55]

In the absence of a safe harbor level or safe use determination, businesses may choose to calculate whether the anticipated level of exposure would pose no significant risk assuming lifetime exposure

---

50. *Id.* § 25249.7(b)(2).
51. *Id.* § 25249.7(h)(2).
52. *Id.*
53. According to OEHHA, as of February 2010 it had established safe harbor levels "for nearly 300 chemicals." OEHHA, Proposition 65 in Plain Language (updated Feb. 2010), at 3, http://oehha.ca.gov/prop65/pdf/P65Plain.pdf. No significant risk levels for carcinogens are listed at Cal. Code Regs. tit. 27, § 25705(b)(1), (c)(2), (d)(3). Specific regulatory levels for chemicals causing reproductive toxicity are listed at Cal. Code Regs. tit. 27, § 25805(b). Exposure at 1,000 times the listed level has observable effect.
54. Cal. Code Regs. tit. 27, § 25204.
55. *Id.* § 25204(b)(5).

at the level in question for substances known to cause cancer,[56] or have no observable effect assuming exposure at 1,000 times the level in question for substances known to cause reproductive toxicity.[57] Should the business be challenged for not providing a warning to a listed chemical, however, the burden of proof is on the business to prove the warning was not required.[58] Under the statute, businesses may choose to determine whether the exposure exceeds the no significant risk level for a carcinogen or 1/1,000 of the no observable effect level for a chemical listed as causing birth defects or reproductive harm.[59] If the exposure does not exceed those levels, no warning is necessary. Faced with the prospect of negative publicity and the expense of defending a suit brought by an overeager bounty hunter, a business may choose to follow the warning provisions, despite an exposure that falls well below the level requiring them.

Practical Tips

Businesses in California need to be familiar with their business operations and the chemicals used and closely monitor OEHHA's rulemaking process to ensure that they can participate fully in any proposed listing of their chemicals. If a business chooses to evaluate the risk posed by exposure to a listed chemical to determine whether a warning is necessary, it should ensure that an experienced toxicologist or similar professional conducts the exposure assessment.

Regarding OEHHA's use of the Labor Code mechanism to list chemicals, OEHHA stated that the court's ruling in *Styrene Information & Research Council v. OEHHA* does not affect its duty to list chemicals identified by IARC as possibly carcinogenic to humans where the determination is based on sufficient evidence of carcinogenicity in either humans or laboratory animals.[60] OEHHA has determined that the language in the court's ruling applies to ten chemicals, including four chemicals listed on Proposition 65 due to an IARC Group 2B designation with less than sufficient animal and human

---

56. Id. § 25721(c).
57. Id. § 25801(a).
58. CAL. HEALTH & SAFETY CODE § 25249.10(c).
59. CAL. CODE REGS. tit. 27, §§ 25721(c), 25801(a).
60. OEHHA, Regarding Certain IARC (International Agency for Research on Cancer) 2B Chemicals (Jan. 4, 2013), http://www.oehha.ca.gov/prop65/CRNR_notices/010413notice.html.

evidence at the time they were listed.[61] OEHHA will review the basis for listing these chemicals.

Another potential issue concerns NTP's CERHR, which is now the Office of Health Assessment and Translation (OHAT). There was controversy when OEHHA designated NTP's CERHR as an authoritative body for the identification of chemicals as causing reproductive toxicity. Should OEHHA propose a listing based on an evaluation prepared by OHAT, it is likely that a similar controversy would recur.

RESOURCES

OEHHA, Proposition 65, http://www.oehha.ca.gov/prop65.html.
OEHHA, Proposition 65: Frequently Asked Questions, http://www.oehha.ca.gov/prop65/background/P65QA.pdf.
OEHHA, Proposition 65 in Plain Language (2010), http://www.oehha.ca.gov/prop65/pdf/P65Plain.pdf.
OEHHA, Fact Sheet on Proposition 65 Safe Use Determination (SUD) Process (2010), http://www.oehha.ca.gov/prop65/CRNR_notices/safe_use/pdf_zip/SUDfacts081809.pdf.

2.3.1.2
SCPR

SCPR Overview

The California Department of Toxic Substances Control (DTSC) promulgated the Safer Consumer Products Regulations (SCPR), which took effect October 1, 2013, as part of its implementation of California's Green Chemistry Initiative. In 2008, two bills intended to promote green chemistry were passed by the legislature and signed into law: A.B. 1879, which provides DTSC greater authority to regulate toxins in consumer products; and S.B. 509, which creates an online Toxics Information Clearinghouse. A.B. 1879 required DTSC

---

61. Id.

by January 1, 2011, to adopt regulations establishing a process to identify and prioritize chemicals in consumer products for consideration as being chemicals of concern (COC).[62] The bill also required DTSC by January 1, 2011, to establish a process by which COCs in products, as well as their alternatives, are evaluated to determine how best to limit exposure.[63]

After A.B. 1879 was signed into law by then Governor Arnold Schwarzenegger (R), DTSC issued several sets of draft regulations, informal draft regulations, proposed regulations, and revised proposed regulations.[64] The final SCPR no longer includes certain controversial lists upon which the COCs, now called candidate chemicals (CC) would be derived, provides more time for responsible entities to respond to DTSC, eliminates inventory recalls, and requirements pertinent to certified assessors and accreditation bodies. The regulatory process remains complicated, however, beginning with the development of the CC list through DTSC's regulatory responses.

Scope of SCPR

Scope of Products Subject to and/or Exempt from SCPR

The SCPR applies to all consumer products that are sold, offered for sale, distributed, supplied, or manufactured in or for use in California.[65] An exemption is provided for products excluded from the definition

---

62. CAL. HEALTH & SAFETY CODE § 25252(a).
63. Id. § 25253.
64. See June 23, 2010, draft regulations, http://www.dtsc.ca.gov/PollutionPrevention/GreenChemistryInitiative/upload/Safer-Product-Alternative-Regulations-6-23-10.pdf; September 17, 2010, proposed regulations, http://www.dtsc.ca.gov/LawsRegsPolicies/upload/SCPA-Regs_APA-format-9-07-10-rev-9-12.pdf; November 16, 2010, revised proposed regulations, http://www.dtsc.ca.gov/LawsRegsPolicies/upload/SCPA_Regs_15Day_Revisions_11162010.pdf; October 31, 2011, informal draft regulations, http://www.dtsc.ca.gov/LawsRegsPolicies/Regs/upload/SCP-Regulations-Informal-Draft-10312011.pdf; May 18, 2012, revised informal draft regulations, http://www.dtsc.ca.gov/LawsRegsPolicies/Regs/upload/SCP-Regulations_REVISED-Informal-Draft-5-18-2012.pdf; July 27, 2012, proposed regulations, http://www.dtsc.ca.gov/LawsRegsPolicies/Regs/upload/SCP-Proposed-Text-underlined-7-23-2012.pdf; January 29, 2013, revised proposed regulations, http://www.dtsc.ca.gov/LawsRegsPolicies/Regs/upload/SCP-Revised-Text.pdf; and April 10, 2013, revised proposed regulations, http://www.dtsc.ca.gov/LawsRegsPolicies/Regs/upload/2-SCP-REVISED-Proposed-Regulations_APA-MARKUP-April-2013.pdf. On August 23, 2013, DTSC released additional revisions to certain proposed sections, http://www.dtsc.ca.gov/LawsRegsPolicies/Regs/upload/SCP-Final-Regs_August-2013-15-Day-Revisions-Text-8-22-13.pdf.
65. CAL. CODE REGS. tit. 22, § 69501(b)(1).

of "consumer product," as specified in Health and Safety Code section 25251.[66] Section 25251 excludes from the definition of consumer products the following: dangerous prescription drugs and devices; dental restorative materials; medical devices; packaging associated with dangerous prescription drugs and devices, dental restorative materials, and medical devices; food; and pesticides.[67] Consumer products that DTSC determines are regulated by one or more federal or California regulatory programs, and/or applicable treaties or international agreements with the force of domestic law, are also exempt.[68]

Scope of Activities Subject to SCPR

Under the SCPR, manufacture means to make or produce.[69] Acts meeting the definition of assemble ("to fit, join, put, or otherwise bring together components to create, repair, refurbish, maintain, or make non-material alterations to a consumer product") are specifically excluded from the definition of manufacture.[70]

SCPR includes the following within its definition of placing into the stream of commerce: to sell, offer for sale, distribute, supply, or manufacture in or for use in California as a finished product or as a component in an assembled product.[71] Selling or offering for sale includes but is not limited to transactions and offers made through sales outlets, catalogs, the Internet, or other electronic means.[72]

Scope of Persons Subject to SCPR

The SCPR defines responsible entities to include manufacturers, importers, assemblers, and retailers.[73] The manufacturer makes a product subject to the SCPR or "controls the manufacturing process for, or specifies the use of chemicals to be included in, the product."[74] An importer imports a product that is subject to the requirements of

---

66. Id. § 69501(b)(2).
67. Cal. Health & Safety Code § 25251.
68. Cal. Code Regs. tit. 22, § 69501(b)(3)(A).
69. Id. § 69501.1(a)(43).
70. Id.
71. Id. § 69501.1(a)(50)(A).
72. Id. § 69501.1(a)(50)(B).
73. Id. § 69501.1(a)(60).
74. Id. § 69501.1(a)(44).

the SCPR.[75] The definition of importer specifically excludes a person who imports a product solely for use in that person's workplace if the product is not sold or distributed to others.[76] The retailer sells or distributes the product to a consumer.[77] While the SCPR states that the "principal duty" for compliance lies with the manufacturer, if the manufacturer fails to comply, then the duty falls on the importer.[78] Under the SCPR, a retailer is required to comply only if the manufacturer and importer fail to comply, and only after DTSC posts this information on its website, on the Failure to Comply List.[79] Retailers who become responsible for complying with the SCPR requirements may choose to opt out by ceasing to order the PP and notifying DTSC.[80] The SCPR allows for a consortium, trade association, public-private partnership, nonprofit organization, or other entity to fulfill the role of the responsible entities.[81]

NOTIFICATION/REGISTRATION/APPROVAL REQUIREMENTS AND COMPLIANCE

Summary of Authority Governing the Requirements for Notification/Registration/Approval

Health and Safety Code section 25252 directs DTSC to adopt regulations by January 1, 2011, establishing a process to identify and prioritize COCs.[82] The final SCPR was effective October 1, 2013.

Listing Procedures

Under the SCPR, within 30 days after the effective date, DTSC must establish an informational list of CCs.[83] For the initial list of CCs, a chemical is designated as a CC if it is identified as exhibiting a hazard

---

75. Id. § 69501.1(a)(39).
76. Id.
77. Id. § 69501.1(a)(61).
78. Id. § 69501.2(a)(1)(A).
79. Id.
80. Id. § 69501.2(b)(2).
81. Id. § 69501.2(a)(2).
82. CAL. HEALTH & SAFETY CODE § 25252.
83. CAL. CODE REGS. tit. 22, § 69502.3(a). DTSC released the list on September 28, 2013. DTSC, Safer Consumer Products (SCP), http://www.dtsc.ca.gov/SCP/index.cfm.

trait and/or an environmental or toxicological endpoint on one or more of 15 different lists already selected by other state, federal, and international agencies and organizations[84] or it is identified by one or more of eight state, federal, or international lists of certain types of chemicals.[85] According to DTSC, the broad list of CCs includes approximately 1,200 chemicals.[86]

Going forward, DTSC will periodically update the CC list "to reflect changes to the underlying lists and sources from which it is drawn."[87] DTSC is also able to add or delete chemicals, first making its proposed revisions available for review and comment.[88] The SCPR also includes a procedure for any person to petition DTSC to evaluate a claim that a chemical should be added or removed from the CC list.[89] In addition, a person may petition DTSC to remove from or add to the CC list "the entirety of an existing chemicals list."

REGULATION OF SUBSTANCES/PRODUCTS

Authority to Regulate Inventory/Listed Substances or Products

Health and Safety Code section 25253 directs DTSC to adopt regulations by January 1, 2011, to establish a process to evaluate COCs and their potential alternatives to best limit exposure or to reduce the level of hazard posed by a COC.[90]

Prioritization and Risk Assessments

Now that DTSC has identified the initial list of CCs, it will develop a list of product-chemical combinations identified as PPs for which alternatives analyses (AA) must be conducted. The SCPR requires that any product-chemical combination identified and listed as a PP must meet both of the following criteria: there must be potential public and/or aquatic, avian, or terrestrial animal or plant organism exposure to the CCs in the product; and there must be the potential

---

84. Id. § 69502.2(a)(1).
85. Id. § 69502.2(a)(2).
86. DTSC, Safer Consumer Products, http://www.dtsc.ca.gov/SCP/index.cfm.
87. CAL. CODE REGS. tit. 22, § 69502.3(a).
88. Id. §§ 69502.2(b), 69502.3(b).
89. Id. § 69504(a).
90. CAL. HEALTH & SAFETY CODE § 25253.

for one or more exposures to contribute to or cause significant or widespread adverse impacts.[91] To develop the list of PPs, DTSC will evaluate the product-chemical combination to determine its associated potential adverse impacts, potential exposures, and potential adverse waste and end-of-life effects.[92] DTSC will also consider the availability of information;[93] and other state, federal, and international regulatory programs under which the product or CC(s) in the product is regulated.[94]

The SCPR limits the initial list of PPs, which must be released no later than 180 days after the effective date of the SCPR, to five PPs.[95] On March 13, 2014, DTSC released the first draft PPs, including the following three draft PPs: (1) children's foam-padded sleeping products containing tris(1,3-dichloro-2-propyl) phosphate (TDCPP); (2) spray polyurethane foam (SPF) systems containing unreacted diisocyanates; and (3) paint and varnish strippers and surface cleaners containing methylene chloride. DTSC held three public workshops in May and June of 2014 to discuss the draft PPs with stakeholders. DTSC expects to initiate rulemaking to codify the initial PP list in regulations in the latter part of 2014, and the process could take up to one year.

DTSC will review and revise the PP list at least once every three years.[96] Under the SCPR, within 60 days after a product-chemical combination is listed as a PP, responsible entities are required to submit to DTSC a PP notification.[97] If applicable, the notification should include an indication that a removal/replacement notification or AA threshold notification is being submitted concurrently with the PP notification, or will be submitted by the due date of the preliminary AA report.[98] Manufacturers or other responsible entities must perform an AA to determine how best to limit exposures to, or the level of adverse public health and environmental impacts posed by, the COC(s) in the product.[99] An AA is not required if the PP is no longer placed into the stream of commerce, or if the PP meets the

---

91. CAL. CODE REGS. tit. 22, § 69503.2(a).
92. Id. § 69503.2(b).
93. Id. § 69503.2(b)(1)(C).
94. Id. § 69503.2(b)(2).
95. Id. § 69503.6(b), (c).
96. Id. § 69503.5(e).
97. Id. § 69503.5(f).
98. Id. § 69503.7(a); see also id. §§ 69505.2(a), 69505.3(a).
99. Id. § 69505.1(b)(1).

2.3. State Laws 159

AA threshold exemption criteria.[100] To be eligible for an AA threshold exemption, the concentration of each COC in the PP must not exceed the practical quantitation limit (PQL) for that chemical or the COC(s) must not exceed the applicable AA threshold(s) specified by DTSC.[101]

Responsible entities that submit a PP notification must then conduct an AA.[102] AAs are conducted in two stages. In the first stage, the responsible entity must (1) identify the product requirements and function(s) of the COC(s); (2) identify alternative(s); (3) identify factors relevant for comparison of alternatives; (4) evaluate and screen alternative replacement chemicals; (5) consider other relevant information and data not specifically identified in the SCPR; and (6) develop a work plan and proposed implementation schedule and a preliminary AA report.[103] Responsible entities may prepare an abridged AA report if they determine, after completing the first five steps of the first stage of the AA, that a functionally acceptable and technically feasible alternative is not available.[104] The preliminary AA report is due no later than 180 days after the date the product is listed on the final PP list, unless DTSC specifies a different date.[105]

In the second stage of the AA, responsible entities must (1) identify factors relevant for comparison of alternatives; (2) compare the PP and alternatives; (3) consider additional information; (4) decide whether to select an alternative or retain the existing PP; and (5) prepare the final AA report.[106] The final AA report is due within 12 months after the date DTSC issues a notice of compliance for the preliminary AA report, unless the responsible entity requests, and DTSC approves an extended due date.[107] A responsible entity may request a one-time extension of up to 90 days to the submission deadline for the AA report, "based on circumstances that could not reasonably be anticipated or controlled by the responsible entity."[108]

---

100. *Id.* § 69505.1(a).
101. *Id.* § 69505.3(a)(4).
102. *Id.* § 69505.1(b)(1).
103. *Id.* § 69505.5.
104. *Id.* § 69505.4(b).
105. *Id.* § 69505.1(b)(2)(A).
106. *Id.* § 69505.6.
107. *Id.* § 69505.1(b)(2)(B).
108. *Id.* § 69505.1(c)(1).

Use Restrictions/Chemical or Product Bans

Once a final AA report is submitted, DTSC will post the final report on its website and accept public comments for at least 45 days.[109] Within 30 days of the close of the comment period, DTSC will notify the submitter of the report of any issues it has determined must be addressed in an AA report addendum. DTSC will include a due date, taking into account the scope and complexity of issues the submitter must address.[110] After DTSC determines that a final AA report and addendum, if any, are compliant (i.e., not deficient), DTSC will issue a notice of proposed determination that one or more regulatory responses are required.[111] DTSC will post the notice and accept comments for at least 45 days, as well as hold a public workshop to provide an opportunity for comment.[112] DTSC will then issue a notice of final determination.[113] If any regulatory responses are required, DTSC will consider the complexity of implementing the regulatory response in assigning a due date for implementation.[114] DTSC may require no regulatory response, or a combination of the following:

- *Product information for consumers:* In certain cases, information must be made available to the consumer prior to product purchase. This information includes the COCs in the product and/or any replacement CCs and known hazards traits and/or environmental or toxicological endpoints for those chemicals; a statement that the product must be disposed of or otherwise managed as a hazardous waste at the end of its useful life, if applicable; any safe handling and storage procedures and/or other information necessary to protect public health or the environment; end-of-life management programs or requirements; and a website address to obtain additional information about the product, the adverse impacts posed by the product, and proper end-of-life disposal or management of the product. The entity must post the information on its website and use at least one of the following means to inform consumers at the point of sale of the information: (1) providing the required

---

109. Id. § 69505.8(a).
110. Id. § 69505.8(b).
111. Id. § 69506.1(c).
112. Id. § 69506.1(d).
113. Id. § 69506.1(e).
114. Id. § 69506.1(g).

information on the product packaging or in accompanying written material that is accessible without breaking the product seal; and/or (2) posting the information in a prominent place at the point of retail display.[115]

- *Use restrictions on chemicals and consumer products:* DTSC may impose restrictions on the use of one or more COCs or replacement CCs in a selected alternative, or COCs in a PP for which an alternative is not selected, or restrictions on the use of the product itself. Use restrictions may include (1) restrictions on the amount or concentration of the COC or replacement CC permitted in a product; (2) restrictions on the settings in which a product may be sold or used; (3) restrictions regarding the form in which a product is sold; (4) restrictions on who may purchase and/or use a product; (5) requirements for training of product purchasers and/or users; and/or (6) any other use restriction that reduces the amount of any COC or replacement CC in the product, or reduces the potential for the product to contribute to or cause an exposure to the COC(s) or replacement CC(s) in the product.[116]
- *Product sales prohibition:* If a responsible entity decides in a final AA report to retain an existing PP or select an alternative that still contains a COC or replacement CC, DTSC may effectively override a responsible entity based on a determination that a safer alternative exists that does not contain a COC and is functionally acceptable, technically feasible, and economically feasible.[117] DTSC also has the ability in certain circumstances to issue a notification of its determination that a product containing a COC or replacement CC may no longer be placed into the stream of commerce in California, notwithstanding that there are no currently identified safer alternatives that are functionally acceptable, technically feasible, and economically feasible, unless the responsible entity demonstrates to DTSC's satisfaction that (1) the overall beneficial public health and/or environmental impacts and/or social utility of the product significantly outweigh the overall adverse impacts of the product; and (2) administrative and/or engineering restrictions on the

---

115. *Id.* § 69506.3.
116. *Id.* § 69506.4.
117. *Id.* § 69506.5(a).

nature and/or use of the product will adequately protect public health and the environment.[118] Under any of those circumstances, the responsible entity will have an opportunity to revise its final AA report to select an alternative that does not contain a COC or replacement CC and thus not be subject to the product sales prohibition.[119]

- *End-of-life management:* A manufacturer must establish and maintain an end-of-life management program if the alternative product (or the PP, if the manufacturer chooses to retain the PP) is required to be managed as a hazardous waste in California at the end of its useful life.[120]
- *Other regulatory responses:* Other responses that DTSC may impose include requiring engineered safety measures or administrative controls to control access to or limit exposure to the COC or replacement CC,[121] and requiring the responsible entity to initiate a research and development project pertinent to the PP.[122]

Test Data Development

Authority to Require Testing/Testing Triggers

Under Health and Safety Code section 25253, the range of responses that DTSC may take following completion of an AA includes "imposing requirements to provide additional information needed to assess a chemical of concern and its potential alternatives."[123] The SCPR defines information to include data, as well as "documentation, records, graphs, reports, or any other depiction of specific pieces of knowledge."[124]

---

118. *Id.* § 69506.5(b).
119. *Id.* § 69506.5(c).
120. *Id.* § 69506.7(a).
121. *Id.* § 69506.6.
122. *Id.* § 69506.8.
123. Cal. Health & Safety Code § 25253(b)(2).
124. Cal. Code Regs. tit. 22, § 69501.1(a)(40).

Chemical Substances/Products That Must Be Tested/Implementing Test Requirements

Responsible entities submitting AA threshold exemption notifications must submit the laboratory analytical testing methodology and quality control and assurance protocols used to detect and measure the COC or replacement concentration in the PP.[125] The submission must identify the testing laboratory.

A responsible entity submitting a COC removal or replacement notification must provide the laboratory analytical testing methodology and quality control and assurance protocols used to confirm the COC(s) and/or replacements in the PP that ensures the COC(s) have been removed or replaced.[126] The submission must identify the testing laboratory.

While the AA process does not require a responsible entity to conduct any testing to fill any data gaps, DTSC may require a responsible entity to provide supplementary information that it "determines is necessary to select and ensure implementation of one or more regulatory responses" under the SCPR.[127] In addition, DTSC may require a responsible entity to provide information to fill one or more of the information gaps identified in the final AA report.[128]

Data Sharing and Compensation

The SCPR does not address data-sharing or compensation issues.

REPORTING AND RECORDKEEPING REQUIREMENTS

Reports to Be Submitted to DTSC

Within 60 days after a product-chemical combination being listed as a PP, responsible entities are required to submit to DTSC a PP notification.[129] If applicable, the notification should include an indication that a removal/replacement notification or AA threshold notification is being submitted concurrently with the PP notification, or will

---

125. Id. § 69505.3(a)(8).
126. Id. § 69505.2(b)(9)(B), (E).
127. Id. § 69506.2(a).
128. Id. § 69506.2(b).
129. Id. § 69503.7(a).

be submitted by the due date of the preliminary AA report.[130] Responsible entities that submit a PP notification must then conduct an AA, which is conducted in two stages. The preliminary AA report is due no later than 180 days after the date the product is listed on the final PP list, unless DTSC specifies a different date.[131] The final AA report is due within 12 months after the date DTSC issues a notice of compliance for the preliminary AA report, unless the responsible entity requests, and DTSC approves, an extended due date.[132]

Recordkeeping

A person subject to a requirement to obtain or prepare information, but who is not required to submit the information to DTSC or has not yet been requested to submit information, must retain the information for a period of three years from the date the person was required to obtain or prepare the information.[133]

Risk Notification

As discussed above under "Use Restrictions/Chemical or Product Bans," in certain cases, information must be made available to the consumer prior to product purchase.[134] This information includes the COCs and/or any replacement CCs in the product; safe handling and storage procedures; end-of-life management programs or requirements; and a website address to obtain additional information about the product, the adverse impacts associated with the product, and proper end-of-life disposal or management of the product.[135] The responsible entity must post the information on the manufacturer's website and the importer's website and use at least one of the following means to inform consumers at the point of sale of the information: (1) providing the required information on the product packaging or in accompanying written material that is accessible without breaking the product seal; and/or (2) posting the information in a prominent place at the point of retail display.[136]

---

130. Id.; see also id. §§ 69505.2(a), 69505.3(a).
131. Id. § 69505.1(b)(2)(A).
132. Id. § 69505.1(b)(2)(B).
133. Id. § 69501.3(e).
134. Id. § 69506.3(b).
135. Id.
136. Id. § 69506.3(c).

Confidential and Trade Secret Information

Authority Relating to Confidential Business Information

Section 25257 of the Health and Safety Code provides for the submission of trade secret information.[137]

SCPR Provisions Relating to Confidential Business Information

With respect to any documents or information submitted to DTSC, a person may assert a claim of trade secret protection.[138] Claims must be substantiated by providing certain information to DTSC and by providing a redacted copy of the documentation being submitted with the trade secret information removed.[139] In general, a submitter may not claim trade secret protection for any hazard trait submission or any chemical identity information associated with a hazard trait submission.[140] The SCPR does allow a claim for trade secret protection to mask temporarily the identification of a chemical that is the subject of a hazard trait submission if the subject of claim is a proposed alternative to a COC in a PP and a patent application is pending for the chemical or its contemplated use in the product.[141] The masking will be authorized only until the information subject to the trade secret claim is made public through any means.

Enforcement and Penalties

The SCPR does not address criminal or civil penalties for failure to meet the notification and other requirements. Instead, the SCPR provides for a Failure to Comply List. If DTSC determines that one or more requirements have not been met for a specific product, it will issue a notice of noncompliance to the manufacturer and importers for the product.[142] If the manufacturer or importer fails to remedy the noncompliance to DTSC's satisfaction, DTSC will post information concerning its determination of noncompliance on its website, on

---

137. Cal. Health & Safety Code § 25257.
138. Cal. Code Regs. tit. 22, § 69509.
139. Id. § 69509(a), (c).
140. Id. § 69509(f).
141. Id. § 69509(g)(1).
142. Id. § 69501.2(c)(1)(A).

the Failure to Comply List.[143] The Failure to Comply List will include a statement placing retailers and, if applicable, assemblers on notice of the failure of the manufacturer and/or importers to comply with the SCPR requirements, and identifying the requirement with which they must comply and the time frame for compliance.[144]

Key Business Issues

Manufacturers, processors, and retailers in California should monitor closely how DTSC implements the SCPR. As required under final SCPR, DTSC has identified less than five PPs in its initial list, which now allows industry stakeholders an opportunity to see how DTSC will implement the regulations, how the AA process will work, and what changes may be necessary. The SCPR allows for a consortium or trade association to fulfill the role of a responsible entity with only a few exceptions, and smaller manufacturers or processors may find that working through a consortium presents less of an economic burden.

Resources

DTSC, Green Chemistry, http://www.dtsc.ca.gov/PollutionPrevention/GreenChemistryInitiative/index.cfm.
DTSC, Safer Consumer Products Regulations, http://www.dtsc.ca.gov/SCPRegulations.cfm.
DTSC, Safer Consumer Products Resources, http://www.dtsc.ca.gov/SCPResources.cfm.
DTSC, Safer Consumer Products main page, http://www.dtsc.ca.gov/SCP/index.cfm.

---

143. Id. § 69501.2(c)(2).
144. Id. § 69501.2(c)(4)(C).

2.3.2
MASSACHUSETTS

TURA Overview

In 1989, then Governor Michael Dukakis (D) signed the Toxics Use Reduction Act (TURA), which was intended to promote the reduction of the use of toxic or hazardous substances within Massachusetts.[145] Under TURA, companies in Massachusetts that use more than a certain amount of listed toxic or hazardous substances must report annually the quantities of listed substances manufactured, processed, or otherwise used in amounts equal to or exceeding the designated thresholds.[146] The first even-numbered year after filing an annual report, companies are required to submit an initial toxics use reduction (TUR) plan, examining how and why they use listed toxic or hazardous substances and evaluating options for reducing their use.[147] The plan must be updated every even calendar year.[148] Companies must also pay an annual toxics use fee.[149] In 2006, then Governor Mitt Romney (R) signed legislation amending TURA to categorize toxic or hazardous substances as high hazard or low hazard, with different reporting thresholds.[150]

Scope of TURA

Scope of Substances/Products Subject to and/or Exempt from TURA

Under TURA and its implementing regulations, reporting and planning requirements apply to chemicals listed under section 313 of the Emergency Planning and Community Right to Know Act (EPCRA) and the Comprehensive Environmental Response, Compensation, and

---

145. Mass. Code Regs. tit. 310, § 50.02.
146. Mass. Gen. Laws ch. 21I, § 10; Mass. Code Regs. tit. 310, § 50.32(1).
147. Mass. Gen. Laws ch. 21I, § 11; Mass. Code Regs. tit. 310, § 50.40.
148. Mass. Gen. Laws ch. 21I, § 11(D); Mass. Code Regs. tit. 310, § 50.41(1).
149. Mass. Gen. Laws ch. 21I, § 19(E); Mass. Code Regs. tit. 310, § 40.03(1).
150. Mass. Gen. Laws ch. 21I, § 9(A)–(B); Mass. Code Regs. tit. 310, § 41.05.

Liability Act (CERCLA), unless specifically excluded.[151] The Administrative Council on Toxics Use Reduction (Council) has the authority to update annually the list of toxic or hazardous substances consistent with changes to EPCRA and CERCLA.[152] In addition, in any one calendar year, the Council may add up to ten substances and delete up to ten substances.[153] The Toxics Use Reduction Institute (TURI) and its Science Advisory Board (SAB) offer recommendations to the Council concerning additions or deletions to the list of toxic or hazardous substances.[154] The regulations include a list of substances that are explicitly excluded from the toxic or hazardous substances list.[155]

The Council may also, in any one calendar year, designate up to ten listed toxic or hazardous substances as higher hazard substances and ten listed substances as lower hazard substances.[156] TURI advises the Council on which toxic or hazardous substances should be designated as higher or lower hazard substances.[157] TURA requires TURI to base its advice on recommendations from the SAB, "taking into consideration the policy implications of such recommendations."[158] When the 2006 TURA amendments took effect, substances identified by EPA as persistent, bioaccumulative, and toxic (PBT) were designated as higher hazard substances.[159] Currently, only a limited number of substances are listed as higher or lower hazard substances.[160]

The threshold amount for toxics users that manufacture or process a toxic or hazardous substance is 25,000 pounds at any one facility, and for higher hazard substances, the threshold is 1,000 pounds.[161] The threshold amount for toxics users that otherwise use a toxic or hazardous substance is 10,000 pounds at any one facility, and 1,000 pounds for higher hazard substances.[162]

---

151. Mass. Gen. Laws ch. 21I, § 9(A)–(B); Mass. Code Regs. tit. 301, § 41.03(1)–(4).
152. Mass. Gen. Laws ch. 21I, § 9(A)–(B); Mass. Code Regs. tit. 301, § 41.04(2).
153. Mass. Gen. Laws ch. 21I, § 9(C); Mass. Code Regs. tir. 301, § 41.04(1).
154. Mass. Gen. Laws ch. 21I, § 9(C); Mass. Code Regs. tit. 301, § 41.04(3).
155. Mass. Code Regs. tit. 301, § 41.03(1)–(4).
156. Mass. Gen. Laws ch. 21I, § 9(D); Mass. Code Regs. tit. 301, § 41.05(1).
157. Mass. Gen. Laws ch. 21I, § 6(J); Mass. Code Regs. tit. 301, § 41.05(2).
158. Mass. Gen. Laws ch. 21I, § 6(J).
159. Id. § 9(D).
160. Mass. Code Regs. tit. 301, §§ 41.06–41.07.
161. Mass. Gen. Laws ch. 21I, § 9A(A); Mass. Code Regs. tit. 310, § 50.10.
162. Mass. Gen. Laws ch. 21I, § 9(B); Mass. Code Regs. tit. 310, § 50.10.

Scope of Activities Subject to TURA

TURA applies to companies that manufacture, process, or otherwise use toxic or hazardous substances. TURA defines manufacture as "produce, prepare, import or compound a toxic or hazardous substance."[163] Manufacture also includes the coincidental production of a toxic or hazardous substance during the manufacture, processing, use, or disposal of another substance or mixture of substances.[164] Under TURA, process is "the preparation of a toxic or hazardous substance, after its manufacture, for distribution in commerce."[165] This definition includes distribution as part of an article containing the toxic or hazardous substance.[166]

The following substances are explicitly excluded from the definition of a toxic or hazardous substance: "(1) present in an article; (2) used as a structural component of a facility; (3) present in a product used for routine janitorial or facility grounds maintenance; (4) present in foods, drugs, cosmetics or other personal items used by employees or other persons at a facility; (5) present in a product used for the purpose of maintaining motor vehicles operated by a facility; (6) present in process water or non-contact cooling water as drawn from the environment or from municipal sources, or present in air used either as compressed air or as part of combustion; (7) present in a pesticide or herbicide when used in agricultural applications; (8) present in crude, lubricating or fuel oils or other petroleum materials being held for direct wholesale or retail sale; or (9) present in crude or fuel oils used in combustion to produce electricity, steam or heat except when production of electricity, steam or heat is the primary business of a facility."[167]

Scope of Persons Subject to TURA

TURA applies to companies that

- Manufacture, process, or otherwise use a listed toxic or hazardous substance at or above any one of the following specific reporting thresholds:

---

163. Mass. Gen. Laws ch. 21I, § 2; Mass. Code Regs. tit. 310, § 50.10.
164. Mass. Gen. Laws ch. 21I, § 2; Mass. Code Regs. tit. 310, § 50.10.
165. Mass. Gen. Laws ch. 21I, § 2; Mass. Code Regs. tit. 310, § 50.10.
166. Mass. Gen. Laws ch. 21I, § 2; Mass. Code Regs. tit. 310, § 50.10.
167. Mass. Gen. Laws ch. 21I, § 2; Mass. Code Regs. tit. 310, § 50.10.

- 25,000 pounds for a toxic or hazardous substance that was manufactured or processed;
- 10,000 pounds for a toxic or hazardous substance that was otherwise used; or
- 1,000 pounds for a higher hazard substance;[168]
• Employ the equivalent of ten or more full-time workers;[169] and
• Fall within at least one of the following Standard Industrial Classification (SIC) codes or equivalent North American Industry Classification System (NAICS) codes:
  - 10–14: Mining
  - 20–39: Manufacturing
  - 40, 44–49: Transportation
  - 50 and 51: Wholesale
  - 72, 73, 75, and 76: Certain Services[170]

Companies that meet the above criteria are large quantity toxics users (LQTUs).[171] TURA exempts from the annual reporting requirements activities in laboratories, to the extent such activities are also exempt from EPCRA section 313 reporting.[172]

Any toxics user that is not a LQTU is a small quantity toxics user (SQTU).[173] SQTUs in user segments designated as priority segments may be required to meet the reporting requirements applicable to LQTUs.[174] Under TURA, within four years from designating a toxic or hazardous substance as a higher hazard substance, the Council may designate user segments that it considers to be priorities for achieving TUR.[175] The designations shall be based on recommendations from the Office of Technical Assistance and Technology (OTA), in consultation with TURI and the Massachusetts Department of Environmental Protection (DEP).[176] User segments would include similar production units in all facilities, regardless of threshold amounts.[177]

---

168. Mass. Gen. Laws ch. 21I, § 9A(A)–(B); Mass. Code Regs. tit. 310, § 50.10.
169. Mass. Gen. Laws ch. 21I, § 10(D)(1)(a); Mass. Code Regs. tit. 310, § 50.31(3).
170. Mass. Gen. Laws ch. 21I, § 10; Mass. Code Regs. tit. 310, §§ 50.10, 50.31.
171. Mass. Gen. Laws ch. 21I, § 2; Mass. Code Regs. tit. 310, § 50.10.
172. Mass. Gen. Laws ch. 21I, § 10(D)(1)(b); Mass. Code Regs. tit. 310, § 50.20(9).
173. Mass. Gen. Laws ch. 21I, § 2; Mass. Code Regs. tit. 310, § 50.10.
174. Mass. Gen. Laws ch. 21I, § 10(I); Mass. Code Regs. tit. 310, § 50.31(4).
175. Mass. Gen. Laws ch. 21I, § 14(A).
176. Id.
177. Id.; Mass. Code Regs. tit. 310, § 50.71(1).

2.3. State Laws 171

The Council may designate up to three priority user segments in any calendar year, and there may be no more than 15 priority user segments at any time.[178] The priority user segment designation will expire after five years, unless it is renewed.[179]

Regulation of Substances/Products

TURA and its implementing regulations do not include a provision to ban the use of a listed toxic or hazardous substance. Toxic or hazardous substances may be designated as higher hazard substances, which are subject to a lower reporting threshold of 1,000 pounds each year at any one facility.[180] In addition, upon recommendation of TURI and SAB, the Council may lower the reporting threshold for a higher hazard substance even further.[181]

Reporting and Recordkeeping Requirements

Reports to Be Submitted to the Massachusetts DEP

Each LQTU is required to submit to the Massachusetts DEP an annual report concerning each toxic or hazardous substance manufactured, processed, or otherwise used in an amount equal to or exceeding the threshold amount.[182] The annual reports should include the information submitted to the U.S. Environmental Protection Agency (EPA), as required under EPCRA section 313,[183] as well as any significant change in toxics use and byproduct generation compared to the previous years, including TUR techniques employed.[184]

LQTUs are required to prepare a TUR plan for each facility for which they filed an annual toxic or hazardous substance report.[185] The TUR plan is due by July 1 of the first subsequent even-numbered year

---

178. Mass. Gen. Laws ch. 21I, § 14(B).
179. Id. § 14(C).
180. Id. § 9A(A)–(B); Mass. Code Regs. tit. 310, § 50.10.
181. Mass. Gen. Laws ch. 21I, § 9A(D); Mass. Code Regs. tit. 310, § 50.10.
182. Mass. Gen. Laws ch. 21I, § 10; Mass. Code Regs. tit. 310, § 50.32(1).
183. Mass. Gen. Laws ch. 21I, § 10(A); Mass. Code Regs. tit. 310, § 50.33(1).
184. Mass. Gen. Laws ch. 21I, § 10(C)(1)(c); Mass. Code Regs. tit. 310, § 50.33(6)(d).
185. Mass. Gen. Laws ch. 21I, § 11(A)(1); Mass. Code Regs. tit. 310, § 50.41(1).

in which an annual report is required.[186] For each production unit in which a toxic or hazardous substance is manufactured, processed, or otherwise used, the TUR plan must include

1. A comprehensive economic and technical evaluation of appropriate TUR technologies for each covered toxic or hazardous substance;
2. An analysis of current and projected toxics use, byproduct generation, and emissions;
3. An evaluation of the types and amounts of listed toxic or hazardous substances used;
4. An identification of each technology, procedure, or training program to be implemented for TUR purposes, its anticipated implementation costs, and the expected savings; and
5. A schedule for implementation of TUR technologies, procedures, and training programs.[187]

Each TUR plan must be certified by a TUR planner as meeting the Massachusetts DEP's criteria for acceptable plans.[188] LQTUs must also evaluate their efforts and update their TUR plans two years after the initial TUR plan, and then according to a schedule specified by the Massachusetts DEP.[189] After a LQTU has completed an initial TUR plan and two updates, it may choose to complete any one of the following in subsequent years:

1. A TUR plan update;
2. An alternative resource conservation plan in accordance with requirements established by the Massachusetts DEP; or
3. Implementation of an environmental management system that meets the requirements for an environmental management system established under TURA and by the Massachusetts DEP.[190]

Although LQTUs must prepare TUR plans and updates, only the summary must be submitted to the Massachusetts DEP.[191] The Mas-

---

186. Mass. Gen. Laws ch. 21I, § 11(A)(1); Mass. Code Regs. tit. 310, § 50.41(1).
187. Mass. Gen. Laws ch. 21I, § 11(A)(3)(a)–(f) [sic]; Mass. Code Regs. tit. 310, §§ 50.42–50.46A.
188. Mass. Gen. Laws ch. 21I, § 11(B); Mass. Code Regs. tit. 310, § 50.42(3).
189. Mass. Gen. Laws ch. 21I, § 11(D); Mass. Code Regs. tit. 310, § 50.41(5).
190. Mass. Gen. Laws ch. 21I, § 11(D); Mass. Code Regs. tit. 310, § 50.41(6).
191. Mass. Gen. Laws ch. 21I, § 11(F); Mass. Code Regs. tit. 310, § 50.47.

sachusetts DEP may require that SQTUs in user segments designated as priority segments also prepare TUR plans.[192]

Recordkeeping

LQTUs are required to maintain at the facility documentation substantiating their annual reports.[193] This includes documentation of the amount of each toxic or hazardous substance used in each production unit, as well as the quantity generated as byproduct by each production unit.[194] Documentation must be kept for at least five years after the date the report was due.[195]

CONFIDENTIAL AND TRADE SECRET INFORMATION

If a toxics user believes that disclosing information in an annual report or TUR plan summary or other document will reveal a trade secret, the user may file a trade secret claim.[196] Under TURA, trade secrets are defined as any "formula, plan, pattern, process, production data, device, information, or compilation of information" used in a toxics user's business that provides the toxics user "an opportunity to obtain an advantage over competitors who do not know or use it."[197] A toxics user may conceal the specific chemical identity of any toxic or hazardous substance by, in the nonconfidential copy where the chemical identity would typically be included, using the generic class or category of the toxic or hazardous substance.[198]

Any Massachusetts resident may petition for the disclosure of information claimed as a trade secret by specifying the information sought to be disclosed.[199] If the Massachusetts DEP commissioner has reason to believe that the information concealed may not be a trade secret, he may initiate action without such a petition.[200] In either case, the Massachusetts DEP commissioner shall review the claimant's

---

192. Mass. Gen. Laws ch. 21I, § 11(G).
193. Id. § 10(C)(2); Mass. Code Regs. tit. 310, § 50.36(1).
194. Mass. Gen. Laws ch. 21I, § 10(C)(2); Mass. Code Regs. tit. 310, § 50.36(1)(b)–(c).
195. Mass. Code Regs. tit. 310, § 50.36(4).
196. Mass. Gen. Laws ch. 21I, § 20(A).
197. Id. § 2; Mass. Code Regs. tit. 310, § 50.10.
198. Mass. Gen. Laws ch. 21I, § 20(A).
199. Id. § 20(E).
200. Id.

explanation and determine whether it presents assertions that, if true, support a finding that the information concealed is a trade secret.[201] If the Massachusetts DEP commissioner determines that the information is not a trade secret, the claimant may request an adjudicatory hearing, and, if the Massachusetts DEP's final decision upon the adjudicatory hearing fails to find that the information is a trade secret, the claimant may seek judicial review.[202]

Enforcement and Penalties

Civil Penalties

The Massachusetts DEP will impose an additional administrative fee of $1,000 for failure to file timely a complete and accurate annual report or to pay the toxics use fee in a timely manner.[203] Late payment fees shall apply if a toxics use report is filed or the toxics use fee is paid more than 30 days late.[204]

Violations of TURA are subject to civil penalties not to exceed $25,000 per day for each violation.[205] Failure to comply with the requirements concerning annual reporting, TUR plans, and trade secret protection could result in a fine of not less than $2,500 and not more than $25,000 for each violation, or by imprisonment for not more than one year, or both.[206]

A trade secret claimant will be subject to a civil penalty not to exceed $25,000 per claim if the Massachusetts DEP commissioner determines that

(1) (i) An explanation submitted by a trade secret claimant presents insufficient assertions to support a finding that the information concealed is a trade secret; or

(ii) After receiving supplemental supporting detailed information that the information concealed is not a trade secret; and

(2) That the trade secret claim is frivolous.[207]

---

201. Id. § 20(F)(1).
202. Id. § 20(F)(4), (6).
203. Id. § 19(F); Mass. Code Regs. tit. 301, § 40.04.
204. Mass. Gen. Laws ch. 21I, § 19(F); Mass. Code Regs. tit. 301, § 40.04.
205. Mass. Gen. Laws ch. 21I, § 21(A).
206. Id. § 21(B).
207. Id. § 21(C).

Specific Enforcement

Under TURA, the Massachusetts DEP may order a toxics user who violates any standard limiting a release of toxic or hazardous substances to the environment to prepare a TUR plan for the production unit at which the violation occurred. The TUR plan would have to demonstrate maximum TUR opportunities available to that user.[208]

At the request of the Massachusetts DEP, the attorney general may bring action for injunctive relief against any person violating a provision of TURA or its implementing regulations.[209] The superior court in equity shall have jurisdiction to enjoin such violation and to grant such further relief as it may deem appropriate.[210]

▬ KEY BUSINESS ISSUES ▬

Commonly Encountered Issues

The annual plans for the most part require information already submitted to EPA under EPCRA section 313. The more difficult aspect of TURA concerns the initial TUR plans and updates. Given that TURA is an established rather than a "new" statute, however, a company should be able to determine whether its submission meets the statutory and regulatory requirements.

Industry may feel, and the results prove, that TURA has been effective in reducing the use of toxic or hazardous substances. Filing the required annual reports and initial TUR plan and updates, in addition to paying the annual toxics use fee, burden companies choosing to operate in Massachusetts when compared to the lack of similar requirements in other states. Other parties may disagree, however, that TURA offers sufficient protection to consumers. In 2011, a bill was introduced in the House that would amend TURA so that, rather than reduce the use of toxics substances, it required the substitution of safer alternatives.[211] The bill failed to pass.

---

208. *Id.* § 16.
209. *Id.* § 22.
210. *Id.*
211. H.B. 1136, S.B. 2079 (Mass. 2011).

Practical Tips

Companies manufacturing, processing, or otherwise using toxic or hazardous substances in Massachusetts should monitor the Massachusetts DEP's website to monitor for any proposed or final additions or deletions to the list of toxic or hazardous substances, as well as changes to the lists of higher and lower hazard substances. To date, the Massachusetts DEP has not designated any user segment as a priority segment, but, should it occur, companies would likely benefit by working with other stakeholders within the user segment to ensure that the Massachusetts DEP obtains the information it seeks. TURI has a robust website updated to provide the most current information.[212] TURI offers workshops and courses for a variety of audiences, including businesses and industry. OTA also offers education and outreach programs for business and industry, in addition to providing services such as compliance assistance and on-site technical support.

Resources

Massachusetts DEP, Toxics Use Reduction Act (TURA), http://www.mass.gov/dep/toxics/toxicsus.htm.

OTA main page, http://www.mass.gov/eea/grants-and-tech-assistance/guidance-technical-assistance/agencies-and-divisions/ota/.

TURI, TURA Overview, http://www.turi.org/About/Toxics_Use_Reduction_Act2/TURA_Overview2.

TURI, http://www.turi.org/.

2.3.3
Trends Overview

In recent years, several states implemented more limited chemical regulatory schemes focused on protecting children's health. In 2008, Washington enacted the Children's Safe Product Act, which required the Washington Department of Ecology to identify by January 1, 2009, high

---

212. *See* Toxics Use Reduction Inst., http://www.turi.org/.

priority chemicals that are of high concern for children.[213] Although the Department missed the January 1, 2009, deadline, it adopted in July 2011 a list of 66 Chemicals of High Concern to Children.[214] Manufacturers must report if their products contain a listed chemical.[215]

In 2008, Maine enacted legislation intended to protect children's health and the environment from toxic chemicals in toys and children's products. Under the amended statute, by January 1, 2010, the Maine DEP, in concurrence with the Maine Department of Health and Human Services and Maine Center for Disease Control and Prevention, was required to publish a list of chemicals of high concern, referred to as "the list of chemicals of concern" after September 1, 2011.[216] In 2011, the Maine DEP published a list of over 1,300 chemicals of concern.[217] From this list, and as required by statute,[218] in July 2012, Maine published a list of 49 "chemicals of high concern."[219] Chemicals of high concern may then be designated as priority chemicals.[220] Maine manufacturers and distributors of children's products containing priority chemicals must provide certain information to Maine DEP.[221]

Minnesota enacted in 2009 the Toxic Free Kids Act, which required the Minnesota Department of Health, after consultation with the Minnesota Pollution Control Agency, to generate a list of chemicals of high concern by July 1, 2010.[222] From the list of chemicals of high concern, the Minnesota Department of Health, in consultation with the Minnesota Pollution Control Agency, may designate priority chemicals.[223] The Minnesota Department of Health acknowledges that, at this time, there are no requirements related to either list. It notes, however, that other states with similar lists, such as Maine and

---

213. WASH. REV. CODE § 70.240.030(1).
214. WASH. ADMIN. CODE § 173-334-130.
215. WASH. REV. CODE § 70.240.040; WASH. ADMIN. CODE § 173-334-080.
216. ME. REV. STAT. ANN. tit. 38, § 1693.
217. Me. Dep't Envtl. Prot., Chemicals of Concern, http://www.maine.gov/dep/safechem/concern/index.html.
218. ME. REV. STAT. ANN. tit. 38, § 1693-A.
219. Me. Dep't Envtl. Prot., Chemicals of High Concern, http://www.maine.gov/dep/safechem/highconcern/.
220. ME. REV. STAT. ANN. tit. 38, § 1694.
221. Id. § 1695.
222. MINN. STAT. § 116.9402. See Minn. Dep't Health, Chemicals of High Concern, http://www.health.state.mn.us/divs/eh/hazardous/topics/toxfreekids/highconcern.html.
223. MINN. STAT. § 116.9403. See Minn. Dep't Health, Priority Chemicals, http://www.health.state.mn.us/divs/eh/hazardous/topics/toxfreekids/priority.html.

Washington, may require manufacturers to report if listed chemicals are within consumer products.[224]

Manufacturers and distributors should follow closely chemical legislation proposed in any state where they have facilities. Given the delay in reforming the federal Toxic Substances Control Act, more states are considering regulating chemicals themselves, resulting in a confusing and potentially conflicting series of local requirements.

---

224. Minn. Dep't Health, Priority Chemicals—Frequently Asked Questions, http://www.health.state.Minn.us/divs/eh/hazardous/topics/toxfreekids/pclist/faq.html.

3

Canada

Lisa R. Burchi

EXECUTIVE SUMMARY

CEPA Overview

The Canadian Environmental Protection Act, 1999 (CEPA 1999) is the legal framework for Canada's regulation of chemical substances, comparable to the U.S. Toxic Substances Control Act (TSCA). CEPA 1999's implementing regulations include those affecting "new" substances (i.e., those not listed on Canada's Domestic Substance List (DSL)) and "existing" substances (i.e., those listed on the DSL). CEPA 1999 New Substance Notification Regulations (NSN Regulations) were established first in 1994 and significantly revised in October 2005.[1] In general, more detailed exposure information is now required for new substances, in recognition of the important role of this factor in determining whether a chemical is "toxic" in Canada. Implementation of CEPA 1999 is carried out primarily through the Minister of the Environment (Environment Canada), which also engages the Minister of Health (Health Canada) to review pertinent parts to human health.[2]

---

1. 139 C. Gaz. pt. II 1864–1928 (Sept. 21, 2005) [hereinafter NSN Regulations].
2. *Id.* at 1869.

Basic Provisions of CEPA

The main components of CEPA 1999 and its implementing regulations include the following:

- *Inventories:* There are two main inventories in Canada:
  - *DSL:* This is the Domestic Substance List, which is the inventory of existing chemical substances that can be commercially manufactured or imported, comparable to the U.S. version of the TSCA Inventory.
  - *NDSL:* The Nondomestic Substance List comprises a list of substances on the TSCA Inventory minus those on the DSL; it represents chemicals that are present in commerce in the United States but not yet in Canada. Registration and data requirements for substances on the NDSL are less onerous than other "new" substances in recognition of the fact that they are already in commerce in the United States. Substances not already listed on the TSCA Inventory can be notified to the NDSL through submission of a notification dossier.
- *Regulation of "new" substances:* A substance that is not on the DSL is not subject to an exemption, and exceeding volume triggers cannot be commercially manufactured or imported until a new substance notification (NSN), which corresponds to a U.S. premanufacture notice (PMN), is submitted and approved following a joint review by Environment Canada and Health Canada. The standard NSN form requires varying degrees of information, including physical-chemical properties, and testing of mammalian and aquatic toxicity, depending on the volume of manufacture of the substance; whether the substance is a "chemical," a polymer, or a biotech organism; and whether the substance is on the NDSL. Multiple reporting schedules are used to determine how much information and test data must be provided, based on factors including the nature of the substance and volume triggers.
- *"Toxic" as the term is used under CEPA:* The term "toxicity" as used in the United States under TSCA is the inherent toxicity of a substance to humans or the environment. This factor must be combined with an exposure term, or the extent to which humans or the environment are exposed to the substance, to define risk. In Canada, the term "toxic" means risk in the sense

that it encompasses both hazard and exposure. Thus, identifying a substance as "toxic" in Canada is very significant.
- *Confidential business information and masked names:* Canada has a mechanism to protect confidentiality in submissions, although such submissions must be substantive upfront. This includes a much stricter generic name requirement for a substance, known as a masked name.
- *Regulation of existing substances:* In addition to the NSN program under CEPA, Environment Canada and Health Canada implemented the Chemical Management Plan (CMP) to assess risks to the environment and human health associated with existing chemicals. Through a categorization process, the full DSL of about 26,000 chemicals was narrowed to about 4,000 chemicals designated for future assessment. Industry was then "challenged" to provide any data that may refute the findings published by Environment Canada and Health Canada. The process bears resemblance to the High Production Volume Chemicals Program in the United States.
- *SNAc:* SNAc stands for significant new activity. A SNAc requires that adequate additional information is provided by the notifier who wishes to manufacture, import, or use the substance for activities indicated in the order. The additional information allows Environment Canada and Health Canada to assess the potential environmental and human health risks associated with the new activities before they are undertaken. Although Canada may not view it as such, comparisons have been made between SNAcs and Significant New Use Rule under TSCA.

SCOPE OF CEPA 1999

Scope of Substances/Products Subject to and/or Exempt from CEPA 1999

CEPA 1999 regulates "substances," defined as

> any distinguishable kind of organic or inorganic matter, whether animate or inanimate, and includes
>
> > (a) any matter that is capable of being dispersed in the environment or of being transformed in the environment into matter that is capable of

being so dispersed or that is capable of causing such transformations in the environment,

(b) any element or free radical,

(c) any combination of elements of a particular molecular identity that occurs in nature or as a result of a chemical reaction, and

(d) complex combinations of different molecules that originate in nature or are the result of chemical reactions but that could not practicably be formed by simply combining individual constituents.[3]

The following categories are considered "substances" and are subject to CEPA 1999 and its regulations but are not subject to provisions regarding new substances:[4]

- *Mixtures:* The definition of substance includes "any mixture that is a combination of substances and does not itself produce a substance that is different from the substances that were combined."[5]
- *Manufactured item:* The definition of substance also includes what is defined as a "manufactured item" which is similar but not identical to the definition of "article" under TSCA. Specifically, a substance, but for the new substances provisions, includes "any manufactured item that is formed into a specific physical shape or design during manufacture and has, for its final use, a function or functions dependent in whole or in part on its shape or design."[6]
- *Wastes:* The definition of substance can also include, but for the new substances provisions, "any animate matter that is, or any complex mixtures of different molecules that are, contained in effluents, emissions or wastes that result from any work, undertaking or activity."[7]

---

3. CEPA 1999 § 3(1).
4. The sections under which mixtures, manufactured items, and animate matter are not considered substances are 66, 80 to 89, and 104 to 115.
5. CEPA 1999 § 3(1). *See* ENV'T CANADA & HEALTH CANADA, GUIDELINES FOR THE NOTIFICATION AND TESTING OF NEW SUBSTANCES: CHEMICAL AND POLYMERS § 3.2.1 (Role of the DSL) (2005), http://publications.gc.ca/collections/Collection/En84-25-2005E.pdf [hereinafter NSN GUIDELINES].
6. CEPA 1999 § 3(1). *See* NSN GUIDELINES, *supra* note 5, § 3.2.2.
7. CEPA 1999 § 3(1). *See* NSN GUIDELINES, *supra* note 5, § 3.2.3.

Other substances that are not subject to NSN Regulations include the following:

- *Substances subject to other acts:* NSN Regulations do not apply to a substance that is manufactured or imported for a use that is regulated under any other act or regulations listed in Schedule 2 to the Act.[8]
- *In transit:* NSN Regulations do not apply to a substance that is loaded on a carrier outside Canada and moved through Canada to a location outside Canada, whether or not there is a change of carrier during transit.[9] If a substance is brought into Canada and stored for subsequent distribution, the substance is subject to the NSN Regulations.[10]
- *Polymers subject to the 2 percent rule:* A polymer that is listed on the DSL and is modified only by adding reactions, none of which constitute more than 2 percent by weight of the polymer, does not require the specific substance name to be changed and thus is not subject to the NSN Regulations.[11]
- *Proteins subject to the 2 percent rule:* A protein that is manufactured by modifying a protein that is listed on the DSL may not be subject to NSN Regulations provided certain criteria are met.[12]
- *Transient reaction intermediates:* Transient reaction intermediates that are not isolated and are not likely to be released into the environment can be exempt from NSN Regulations.[13]
- *Impurities:* Impurities, contaminants, and partially unreacted materials, the formation of which is related to the preparation of a substance, can be exempt from NSN Regulations.[14]
- *Incidental reaction products:* Substances produced when a substance undergoes a chemical reaction that is incidental to the use to which the substance is put or that results from storage or from environmental factors.[15]

---

8. CEPA 1999 § 81(6)(a); NSN Regulations, *supra* note 1, § 3(1). *See also* NSN GUIDELINES, *supra* note 5, § 3.3.1.
9. NSN Regulations, *supra* note 1, § 3(2).
10. NSN GUIDELINES, *supra* note 5, § 3.2.4.
11. *Id.* § 3.2.5.
12. *Id.* § 3.2.6.
13. CEPA 1999 § 81(6)(b). *See also* NSN GUIDELINES, *supra* note 5, § 3.3.2.
14. CEPA 1999 § 81(6)(c). *See also* NSN GUIDELINES, *supra* note 5, § 3.3.3.
15. CEPA 1999 § 81(6)(d). *See also* NSN GUIDELINES, *supra* note 5, § 3.3.4.

- *Low volume:* A substance that is manufactured, used, or imported in a quantity that does not exceed the maximum quantity prescribed as exempt from this section.[16]
- *Substances occurring in nature:* Substances not subject to the NSN Regulations include those that are naturally occurring and must be unprocessed; processed only by manual, gravitational, or mechanical means, by dissolution in water, by flotation, or by heating solely to remove water; or extracted from air by any means.[17]

Nanotechnology is not defined or mentioned in CEPA 1999. In 2007, Environment Canada and Health Canada issued a Proposed Regulatory Framework for Nanomaterials under the Canadian Environmental Protection Act, 1999.[18] Environment Canada and Health Canada state: "Although there is no internationally recognized definition of this type of substance, nanomaterials can be described generally as substances having one or more dimensions in a nanoscale range, typically between 1–100 nanometers."[19] Environment Canada and Health Canada also state the following regarding the Canadian government's policy for when a nanomaterial will be considered "new" or "existing" for purposes of DSL listing:

What nanomaterials are subject to the Regulations?

Nanomaterials which are manufactured in or imported into Canada that are not listed on the DSL are considered new. The nanoscale form of a substance on the DSL is considered a "new" substance if it has unique structures or molecular arrangements. New nanomaterials are subject to notification under the Regulations. For example, the nanomaterial fullerene (CAS No. 99685-96-8) is not listed on the DSL and is considered a "new" substance under the Regulations.

What nanomaterials are not subject to the Regulations?

Substances listed on the DSL whose nanoscale forms do not have unique structures or molecular arrangements are considered existing. Existing

---

16. CEPA 1999 § 81(6)(e); NSN Regulations, *supra* note 1, § 4. *See also* NSN GUIDELINES, *supra* note 5, § 3.3.5.
17. NSN GUIDELINES, *supra* note 5, § 3.3.6.
18. Env't Canada & Health Canada, Proposed Regulatory Framework for Nanomaterials under the Canadian Environmental Protection Act, 1999 (Sept. 10, 2007), https://www.ec.gc.ca/subsnouvelles-newsubs/default.asp?lang=En&n=FD117B60-1.
19. *Id.* at app. A (New Substances Program Advisory Note 2007-06).

nanomaterials are not subject to the Regulations and do not require notification. For example, titanium dioxide (CAS No. 13463-67-7) is listed on the DSL and since its nanoscale form does not have unique structures or molecular arrangements, it is not subject to the Regulations.

In addition, incidentally produced or naturally occurring nanomaterials are not subject to notification.[20]

There are three "special category substances" under CEPA 1999. A research and development (R&D) substance is a special category substance that includes both chemicals and polymers. Under CEPA 1999, "research & development" includes both scientific and technological R&D. Specifically, R&D is defined as

> a substance that is undergoing systematic investigation or research, by means of experimentation or analysis other than test marketing, whose primary objective is any of the following:
>
> (a) to create or improve a product or process;
>
> (b) to determine the technical viability or performance characteristics of a product or process; or
>
> (c) to evaluate the substance prior to its commercialization, by pilot plant trials, production trials, including scale-up, or customer plant trials, so that technical specifications can be modified in response to the performance requirements of potential customers.[21]

Another special category substance is a site-limited intermediate substance, defined as

> a substance that is consumed in a chemical reaction used for the manufacture of another substance and that is
>
> (a) manufactured and consumed at the site of manufacture;
>
> (b) manufactured at one site and transported to a second site where it is consumed; or
>
> (c) imported and transported directly to the site where it is consumed.[22]

To be classified as a site-limited intermediate, the substance must "at all times during its existence (manufacture, importation, storage,

---

20. Id.
21. NSN Regulations, *supra* note 1, § 1(1).
22. Id. See id. for definitions of "consumed" and "contained."

transport, handling, use and disposal) be contained . . . to prevent any significant environmental release."[23]

The last special category is for contained export-only substances, which are limited to new substances manufactured in or imported into Canada that are destined solely for foreign markets and that are contained.[24]

For substances that are subject to NSN Regulations, the substances are divided into three categories: chemicals, polymers, and inanimate and animate biotechnology products.

Scope of Activities Subject to CEPA

Environment Canada has responsibility for CEPA 1999 and works in equal partnership with Health Canada in areas where CEPA 1999 relates to human health. To meet CEPA 1999 responsibilities, Health Canada may gather information and conduct investigations and evaluations, including human health screening assessments, for the purpose of assessing whether a substance is entering or may enter the environment under conditions that constitute or may constitute a danger in Canada to human life or health.[25]

CEPA 1999 does not place any burdens of compliance or necessarily define each of the entities that may be affected by CEPA 1999. Most activities subject to CEPA 1999 apply to the manufacture and import of substances, although other activities can apply to processing, distribution, use, and disposal.

Environment Canada and Health Canada will publish in the *Canada Gazette*, and in any other manner that the ministers consider appropriate, a notice requiring any person described in the notice to provide the ministers with any information that may be in the possession of that person or to which the person may reasonably be expected to have access, including information regarding the following:

 (a) substances on the Priority Substances List;

 (b) substances that have not been determined to be toxic under Part 5 because of the current extent of the environment's exposure to

---

23. NSN GUIDELINES, *supra* note 5, § 3.5.2.
24. *Id.* § 3.5.3.
25. Health Canada, Environmental and Workplace Health, Screening Health Assessment of Existing Substances, http://www.hc-sc.gc.ca/ewh-semt/contaminants/existsub/screen-eval-prealable/index-eng.php.

them, but whose presence in the environment must be monitored if the Minister considers that to be appropriate;

(c) substances, including nutrients, that can be released into water or are present in products like water conditioners and cleaning products;

(d) substances released, or disposed of, at or into the sea;

(e) substances that are toxic under section 64 or that may become toxic;

(f) substances that may cause or contribute to international or interprovincial pollution of fresh water, salt water or the atmosphere;

(g) substances or fuels that may contribute significantly to air pollution;

(h) substances that, if released into Canadian waters, cause or may cause damage to fish or to their habitat;

(i) substances that, if released into areas of Canada where there are migratory birds, endangered species or other wildlife regulated under any other Act of Parliament, are harmful or capable of causing harm to those birds, species or wildlife;

(j) substances that are on the list established under regulations made under subsection 200(1);

(k) the release of substances into the environment at any stage of their life-cycle;

(l) pollution prevention; and

(m) use of federal land and of aboriginal land.[26]

Scope of Persons Subject to CEPA

CEPA 1999 section 81 prohibits any "person" from manufacturing or importing a substance not listed on the DSL (and not subject to any exemption) until a notification for a new substance has been submitted and approved.[27] A "person" is not defined within CEPA 1999, but it is common in the individual programs/notices (e.g., CEPA part 4)[28] for "persons" to mean manufacturers and importers and to include

---

26. CEPA 1999 § 46(1).
27. Id. § 81.
28. Id. § 70 ("Where a person 1. (a) imports, manufactures, transports, processes or distributes a substance for commercial purposes, or 2. (b) uses a substance in a commercial manufacturing or processing activity. . . .").

the manufacturers, processors, and in some cases users of chemical substances in Canada. If the person is not a Canadian resident (or Canadian entity), then there are NSN Regulation requirements to identify a Canadian resident as the agent who is authorized to act on the person's behalf.[29]

With regard to toll manufacturing, which occurs when a company contracts a manufacturer to process its raw materials and create a new substance, the NSN Guidelines provide that "ownership of the raw materials and resulting substance remains with the contracting company throughout the activity." The contracting company, and not the toll manufacturer, must be designated as the notifier, must be responsible for complying with the necessary regulations, and must submit the appropriate NSN package.[30]

NOTIFICATION/REGISTRATION/APPROVAL REQUIREMENTS AND COMPLIANCE

Summary of Authority Governing the Requirements for Notification/Registration/Approval

CEPA 1999, section 80 to 89, and its implementing regulations address substances and activities new to Canada.

Standard of Review/Burden of Proof

Once a company submits a NSN of an intent to manufacture or import a new substance, Environment Canada and Health Canada will conduct a joint assessment to determine whether the substance has the potential to cause adverse effects on the environment and human health. Specifically, an important concept in Environment Canada and Health Canada's review is a determination of whether a substance is "toxic" or "capable of becoming toxic."[31] Under CEPA 1999 section 64, a substance is "toxic" if it is entering or may enter the environment in a quantity or concentration or under conditions that

---

29. NSN Regulations, *supra* note 1, § 14(3).
30. NSN GUIDELINES, *supra* note 5, § 1.4.3.
31. NSN Regulations, *supra* note 1, § 2(1).

(a) have or may have an immediate or long-term harmful effect on the environment or its biologic diversity;

(b) constitute or may constitute a danger to the environment on which life depends; or

(c) constitute or may constitute a danger in Canada to human life or health.

A substance may be capable of becoming toxic if either the adverse effects of a substance or potential exposure to a substance is of concern. NSN Guidelines state: "For example, substances with considerable potential for exposure because of continuous release of high quantities or persistence in the environment may be suspected of being toxic, although there may be uncertainty regarding any biological or environmental hazard from the information available for the initial assessment."[32]

Environment Canada/Health Canada's actions to protect the environment and health are guided by the precautionary principle, which provides that "where there are threats of serious or irreversible damage, lack of full scientific certainty shall not be used as a reason for postponing cost-effective measures to prevent environmental degradation."[33] CEPA 1999 section 76.1 specifically directs Environment Canada and Health Canada to apply a weight-of-evidence approach and the precautionary principle when conducting and interpreting the results of assessments of existing substances under section 74, 75(3), or an assessment whether a substance specified on the Priority Substance List is toxic or capable of becoming toxic.[34]

CEPA 1999's preamble also "recognizes the integral role of science," adopts the "polluter pays" principle, and commits to ensuring that the government of Canada's operations are carried out "consistent with the principles of pollution prevention." Environment Canada states the following regarding CEPA 1999's guiding principles:

- *Sustainable development:* The government of Canada's environmental protection strategies are driven by a vision of environmentally sustainable economic development. This vision depends on a clean, healthy environment and a strong, healthy economy that meets the needs of the present

---

32. NSN Guidelines, *supra* note 5, § 9.4.2.
33. CEPA 1999 pmbl.
34. CEPA 1999 § 76.1.

generation without compromising the ability of future generations to meet their own needs.
- *Pollution prevention:* CEPA 1999 shifts the focus away from managing pollution after it has been created to preventing pollution. Pollution prevention is "the use of processes, practices, materials, products, substances or energy that avoid or minimize the creation of pollutants and waste and reduce the overall risk to the environment or human health."
- *Virtual elimination:* CEPA 1999 requires the virtual elimination of releases of substances that are persistent (take a long time to break down in the environment), bioaccumulative (collect in living organisms and end up in the food chain), toxic (according to CEPA 1999 section 64), and primarily the result of human activities. Virtual elimination is the reduction of releases to the environment of a substance to a level below which its release cannot be accurately measured.
- *Ecosystem approach:* Based on natural geographic units rather than political boundaries, the ecosystem approach recognizes the interrelationships among land, air, water, wildlife, and human activities. It also considers environmental, social, and economic elements that affect the environment as a whole.
- *Precautionary principle:* The government's actions to protect the environment and health are guided by the precautionary principle, which states that "where there are threats of serious or irreversible damage, lack of full scientific certainty shall not be used as a reason for postponing cost-effective measures to prevent environmental degradation."
- *Intergovernmental cooperation:* CEPA 1999 reflects that all governments have the authority to protect the environment and directs the federal government to endeavor to act in cooperation with governments in Canada to ensure that federal actions are complementary to and avoid duplication with other governments.
- *National standards:* CEPA 1999 reinforces the role of national leadership to achieve ecosystem health and sustainable development by providing for the creation of science-based, national environmental standards.
- *"Polluter pays" principle:* CEPA 1999 embodies the principle that users and producers of pollutants and wastes should bear the responsibility for their actions. Companies or people that pollute should pay the costs they impose on society.

- *Science-based decision making:* CEPA 1999 emphasizes the integral role of science and traditional aboriginal knowledge (where available) in decision making and that social, economic, and technical issues are to be considered in the risk management process.[35]

Inventories or Other Listing Procedures

The DSL includes approximately 24,000 "existing" substances manufactured, imported, or used in Canada for commercial purposes between January 1, 1984, and December 31, 1986 (before CEPA came into existence), at a quantity of greater than 100 kilograms (kg) per year.[36] Substances can be added to the list after satisfying the requirements of the NSN Regulations. There is a confidential portion of the DSL for those substances whose identities are masked under the Masked Name Regulations (see "Provisions Relating to Confidential Business Information").[37] A substance is "new" if it is not listed on the DSL.

The NDSL includes substances that are not present on the DSL but are listed on the TSCA Inventory.[38] As new substances are added to the TSCA Inventory through the PMN process, they become eligible to be added to the NDSL, with a lag time of about one year.[39] This is a substantial improvement over the earlier process prior to 2005, which could take as long as five years.

Registration requirements for new substances under the NSN Regulations are categorized by several factors, including chemical type (chemicals, polymers, and biologically based materials), annual production volumes, use, and whether the substance is present on the NDSL.[40] Environment Canada's NSN Guidelines provide detailed information and flowcharts to assist companies in determining what information is required.[41] As an example, the testing

---

35. *See also* Env't Canada, A Guide to Understanding the Canadian Environmental Protection Act, 1999, ch. 3 (CEPA 1999 Guiding Principles), http://www.ec.gc.ca/lcpe-cepa/default.asp?lang=En&n=E00B5BD8-1&offset=3&toc=show.
36. *See* CEPA 1999 §§ 66, 87.
37. NSN GUIDELINES, *supra* note 5, § 2.1 (Role of the Domestic Substance List).
38. CEPA 1999 § 66(2).
39. ENV'T CANADA & HEALTH CANADA, NEW SUBSTANCES PROGRAM OPERATIONAL POLICIES MANUAL 13 (Apr. 2004) (Operational Policy for Administering the DSL and NDSL), http://www.ec.gc.ca/subsnouvelles-newsubs/66C60DFB-BD97-4F1F-B32B-545CD8A87773/ops_pol_e.pdf. *See also* NSN GUIDELINES, *supra* note 5, § 2.2 (Role of the NDSL).
40. NSN Regulations, *supra* note 1, §§ 5–12, scheds. 1–12.
41. NSN GUIDELINES, *supra* note 5, § 4.

and administrative requirements for chemicals (i.e., substances that are not polymers or biologically based materials) are described in NSN Regulations Schedules 4, 5, and 6. Schedule 4 contains chiefly administrative information, Schedule 5 includes most of the testing requirements, and Schedule 6 requires additional tests. For chemicals not present on the NDSL, a Schedule 4 must be submitted prior to exceeding 100 kg/yr, Schedule 5 before exceeding 1,000 kg/yr, and Schedule 6 before exceeding 10,000 kg/yr.[42]

The listing of a substance on the NDSL does not allow a company to manufacture or import the substance into Canada for commercial purposes. Substances on the NDSL are subject to notification under the NSN Regulations, but authorities can allow for diminished information requirements in the NSN packages in recognition of the fact that the substance is not "new" in world commerce.[43] The benefit of a substance's listing on the NDSL is that a Schedule 4 or 5 may be filed without the need to file a Schedule 6. This means that fewer data are required, but a notification must nonetheless be made. By filing a Schedule 5 that is subsequently approved by Environment Canada and Health Canada, the substance(s) will be placed on the DSL provided that it is already on the NDSL. If it is not, a Schedule 6 is required to achieve DSL status.

The information required is different for biochemicals and polymers.[44] With regard to polymers, certain polymers were designated as "low concern" in the original regulations; qualifying criteria were similar, but not identical, to those for the polymer exemption under TSCA. In the new scheme under the NSN Regulations, the term "low concern" has been replaced by "reduced regulatory requirements," with essentially the same qualifying criteria.[45]

An R&D provision allows greater quantities to be manufactured or imported into Canada prior to submission of an NSN. As an example, Schedule 1 information is required for any R&D chemical at least 30 days prior to exceeding 1,000 kg/yr.[46] At least 30 days prior to an R&D chemical exceeding 10,000 kg/yr, the person must update and

---

42. NSN Regulations, *supra* note 1, § 8.
43. CEPA 1999 § 66.
44. NSN Regulations, *supra* note 1, §§ 9–12; NSN GUIDELINES, *supra* note 5, §§ 4.7, 4.8.
45. NSN Regulations, *supra* note 1, § 9; NSN GUIDELINES, *supra* note 5, § 4.9.
46. NSN Regulations, *supra* note 1, § 5(1). *See also* NSN GUIDELINES, *supra* note 5, § 4.2.1.

submit the section 1 information or indicate that there have been no changes to the information previously submitted.[47]

Environment Canada has developed a NSN reporting form to aid persons in providing the necessary information.[48] There are four main sections of the NSN reporting form: Part A, Administrative and Substance Identity Information; Part B, Technical Information; Part C, Biochemical or Biopolymer Information Requirements; and Part D, Additional Information Requirements. There also is Appendix I to provide information on "Manufacture, Import, Use, Exposure and Release Information (Known and Anticipated)" and Appendix II for information on the fee payment.

The fees required to be provided with a NSN range from $50 to $3,500, depending on several factors, including annual sales, the specific schedule of information being submitted, if other services are being requested (e.g., masked name application), and if the notifier qualifies for fee reductions (i.e., annual sales in Canada are less than $40 million).[49] There are currently no fees for R&D substances, organisms regulated under the NSN Regulations for Organisms, and substances that are manufactured or imported for a use that is regulated under any other Act of Parliament.[50]

Environment Canada/Health Canada Review and Response

When Environment Canada receives a NSN notification, it will conduct an initial assessment to confirm compliance with the NSN Regulations, including that the necessary information required by the NSN Regulations is included, fees have been paid, masked names are acceptable (if applicable), and claims for confidential business information have been substantiated.[51]

The assessment periods of NSN notifications range from five to 75 calendar days, depending on the type and amount of substance

---

47. NSN Regulations, *supra* note 1, § 5(5).
48. NSN GUIDELINES, *supra* note 5, § 6 & app. 2.
49. New Substances Fees Regulations (SOR/2002-374), *available at* http://laws-lois.justice.gc.ca/PDF/SOR-2002-374.pdf. *See also* NSN GUIDELINES, *supra* note 5, § 1.5.2 & app. 3; NEW SUBSTANCES PROGRAM OPERATIONAL POLICIES MANUAL, *supra* note 39, at 21–25 (Operational Policy for Processing Fees and Refunds).
50. New Substances Fees Regulations § 2.
51. NEW SUBSTANCES PROGRAM OPERATIONAL POLICIES MANUAL, *supra* note 39, at 19 (Operational Policy for Processing New Substances Notifications).

being manufactured or imported.[52] As noted, the crux of Environment Canada and Health Canada's review is to determine whether a substance is "toxic" or "capable of becoming toxic."[53]

There are three potential outcomes following the joint review and assessment of a new substance:

- A determination that the substance is not suspected to meet or be capable of meeting the criteria of "toxic" as defined under CEPA section 64;
- A suspicion that the new substance meets or is capable of meeting the "toxicity" criteria, or
- A suspicion that current activities associated with the substance are not suspected to meet or be capable of meeting the "toxicity" criteria, but a significant new activity in relation to the substance may result in the new substance meeting the "toxicity" criteria.[54]

If there is no suspicion that the substance is "toxic" or capable of becoming "toxic," the notifier may proceed with import or manufacture after the prescribed assessment period has expired, although CEPA 1999 does provide some circumstances when an assessment can be terminated such that a substance can be manufactured or imported earlier.[55]

The last task to be completed to place a new substance on the DSL is for the notifier to submit a notice of manufacture or import (NOMI) or notice of excess quantity (NOEQ). A NOEQ must be submitted within 30 days of exceeding the manufacture or import trigger quantities.[56] Alternatively, a notifier can submit a NOMI for chemicals and polymers. This notice, comparable to the notice of

---

52. NSN Regulations, *supra* note 1, § 16; NSN GUIDELINES, *supra* note 5, § 1.5.1.
53. NSN Regulations, *supra* note 1, § 2(1). For a discussion of Health Canada's "toxic" review, see Health Canada, Determination of "Toxic" for the Purposes of the New Substances Provisions (Chemicals and Polymers) under the Canadian Environmental Protection Act: Human Health Considerations, http://www.hc-sc.gc.ca/ewh-semt/pubs/contaminants/toxic-toxique/index-eng.php.
54. NSN GUIDELINES, *supra* note 5, § 9.5; NEW SUBSTANCES PROGRAM OPERATIONAL POLICIES MANUAL, *supra* note 39, at 37–38 (Operational Policy for Taking Action after a Risk Assessment).
55. CEPA 1999 §§ 83(6), 108(6). See NEW SUBSTANCES PROGRAM OPERATIONAL POLICIES MANUAL, *supra* note 39, at 27 (Operational Policy for Early Termination of Assessment Periods).
56. CEPA 1999 § 81(14); NSN GUIDELINES, *supra* note 5, § 10.1.3.

commencement (NOC) under TSCA and added as part of 2005 NSN Regulations, is an alternative to the submission of a NOEQ for those who want to list a substance on the DSL without tracking of manufacture or import quantities.[57] Once the NOMI or NOEQ is submitted to Environment Canada, the substance will be added to the DSL, but only when it is listed in the *Canada Gazette*.

If there is a suspicion that the substance is "toxic" or capable of becoming "toxic," the notifier will be informed prior to the end of the assessment period of the results of the risk assessment and the risk management measures under consideration. A new substance suspected of being toxic can be controlled by several measures, ranging from the imposition of ministerial conditions on the notifier regarding the import, manufacture, use, or disposal of the substance to the prohibition of such import or manufacture (discussed in "Use Restrictions").[58] In these cases, the assessment period is generally extended to provide time to prepare and approve the risk management measures.[59] Substances subject to ministerial conditions are not added to the DSL, so any subsequent manufacturers or importers would have to comply with the NSN process even though the substance has already been reviewed.

A SNAc also could be issued for a new substance if there is no suspicion of toxicity but changes to the exposure profile raise a suspicion that a "significant new activity in relation to the substance may result in the substance becoming toxic."[60] New substances for which SNAc notices are issued must be published in the *Canada Gazette* within 90 days after the end of the assessment period of the information.[61] For substances not yet on the DSL, the SNAc notice applies to the notifier and users of the substance. Once the substance has been added to the DSL with a flag indicating the substance is subject to a SNAc notice, that flag definition and related restriction(s) applies to

---

57. NSN Regulations, *supra* note 1, §§ 17(2)(a), 18(2)(a); NSN Guidelines, *supra* note 5, § 10.1.4.
58. A compilation of all ministerial conditions imposed on chemicals and polymers since 1996 is available at Env't Canada, Ministerial Conditions, http://www.ec.gc.ca/subsnouvelles-newsubs/default.asp?lang=En&n=9EE2EE18-1. A compilation of all ministerial conditions imposed on living organisms since 1996 is available at http://www.ec.gc.ca/subsnouvelles-newsubs/default.asp?lang=En&n=7DCC2769-1.
59. NSN Guidelines, *supra* note 5, § 9.5.3.
60. CEPA 1999 §§ 85(1), 110(1).
61. *Id. See* New Substances Program Operational Policies Manual, *supra* note 39, at 39–41 (Operational Policy for Issuing Significant New Activity Notices).

all manufacturers, importers, and users of that substance.[62] A person who intends to use, manufacture, or import any of these substances for a new activity must provide information prior to initiating the new activity, comparable to a significant new use notice under TSCA.

Regulation of Substances/Products

Authority to Regulate Inventory/Listed Substances or Products

An "existing" substance is one that is listed on the DSL. Substances listed on the DSL are subject to scrutiny under CEPA 1999 as existing substances. In the first instance, under CEPA 1999 section 73, Environment Canada and Health Canada must categorize all substances on the DSL that have not been subject to new chemical assessments. Environment Canada and Health Canada must then under CEPA 1999 section 74 conduct screening assessments for certain substances to determine whether they are toxic or capable of becoming toxic. There are several lists that a substance can be placed on that will subject it to further review and/or risk management measures: (1) the Priority Substance List; (2) Schedule 1 List of Toxic Substances; (3) Virtual Elimination List; and (4) the Export Control List.[63]

Prioritization and Risk Assessments

Categorization for Screening Assessment

The categorization of DSL substances required under CEPA 1999 section 73 was completed in September 2006 (i.e., with seven years of CEPA's Royal Assent). Approximately 4,000 of Canada's 23,000 existing substances were identified during categorization as needing screening assessments.[64] These substances include those that (1) are "inherently toxic," and display either of the characteristics of persistence (take a long time to break down in the environment) or bioaccumulation (collect in living organisms and end up in the food chain), or (2) may present to individuals in Canada the greatest potential for exposure.[65]

---

62. NSN Guidelines, *supra* note 5, § 9.5.2.
63. *See* CEPA 1999 §§ 65(2), 76, 77(3), 100.
64. A summary of the categorization is available at Env't Canada, What Are the Overall Results of DSL Categorization?", http://www.ec.gc.ca/lcpe-cepa/default.asp?lang=En&n=5F213FA8-1&wsdoc=76D45C61-40EC-2CC6-6FE9-AD1576E210C0.
65. CEPA 1999 § 73.

The term "inherently toxic" is distinguished from the word "toxic" as defined under CEPA 1999 section 64. "Inherently toxic" is described in Environment Canada guidance as inherently toxic to the environment ("chemical substances that are known or suspected, through laboratory and other studies, to have a harmful effect on wildlife and the natural environment on which it depends") and inherently toxic to humans ("these are chemical substances that are known or suspected of having harmful effects on humans. Substances were examined for a number of human health effects, including cancer, birth defects and damage to genetic material").[66]

The approximately 4,000 substances identified for screening-level risk assessment have been, or will be, involved under the Chemical Management Plan in a more in-depth analysis of a substance to determine whether the substance is "toxic" or capable of becoming "toxic" as defined in CEPA 1999. This determination of toxicity consists of integrating the assessment of known or potential exposure of a substance with known or potential adverse effects on the environment. All assessments are published. Under CEPA 1999 section 77(2), the screening-level risk assessment will result in one of the following outcomes:

- Take no further action based on a determination that the screening assessment indicates that (1) the substance is not "toxic," or (2) the substance is "toxic" but risk preventive or control activities for the substance are already being adequately addressed in a program outside of CEPA that are sufficient to manage the risks in a timely manner;
- Add the substance to the Priority Substance List based on a decision to conduct an assessment to assess more comprehensively the risks associated with the release of and exposure to the substance; or
- Recommend that the substance be added to the List of Toxic Substances, Schedule 1 of CEPA 1999, and, where applicable, the implementation of virtual elimination under subsection 65, based on a determination that the substance meets the criteria for "toxic" and that regulatory or pollution prevention or environmental emergency planning risk management measures should be taken.[67]

---

66. Env't Canada, How Were Substances on the DSL Categorized?, http://www.ec.gc.ca/lcpe-cepa/default.asp?lang=En&n=5F213FA8-1&wsdoc=6FCF94B3-CD63-CE3A-4A08-7764E4B847C6.
67. CEPA 1999 § 77(2).

Priority Substances List (PSL)

Environment Canada and Health Canada have shared responsibility to assess substances that pose the greatest risk to human health and the environment. In that regard, Environment Canada and Health Canada are required under CEPA 1999 section 76 to establish and maintain a Priority Substances List (PSL) that identifies substances to be assessed on a priority basis to determine whether they are "toxic" as defined under section 64 or capable of becoming toxic. A priority substance may be a chemical, a group or class of chemicals, effluents, or wastes.

There are several ways that a substance can be added to the PSL.[68] A substance can be added to the PSL following an initial screening assessment conducted under CEPA 1999 section 74 where it is determined that a more comprehensive risk assessment is needed to determine if the substance is "toxic." Environment Canada and Health Canada also are required to review decisions of other Organization for Economic Co-operation and Development (OECD) countries and Canadian provinces and territories that ban or substantially restrict a substance and can add that substance to the PSL if the substance is of concern but there is insufficient evidence to determine that it is "toxic."[69] Any person may also file a written request to add a substance to the PSL.[70]

For a substance on the PSL, Environment Canada must conduct an assessment and determine whether no further action is needed, or whether the substance is "toxic" and should be recommended for inclusion in Schedule 1. CEPA 1999 requires that the substance be assessed within five years from the date the substance is added to the list. Regarding these risk assessments, Environment Canada states:

> Risk assessments done under CEPA 1999 consider impacts on human and non-human organisms and the physical environment. These assessments consider not only the hazard posed by a substance, but the exposure or likelihood that a person, organism or the environment will come in contact with that substance. The exposure or potential for exposure of a substance depends on the amount of substance released into the environment

---

68. *Id.* § 76(5).
69. *Id.* § 75(3).
70. *Id.* § 76(3).

and its fate. The conclusion of the assessment is based on the application of the precautionary principle and a weight of evidence approach.[71]

Under CEPA before the 1999 amendments, there were two PSLs developed. The first PSL (PSL1) was published in 1989 and included 44 substances or groups of substances.[72] All the assessments for these substances were completed by February 1994, with 25 substances identified as toxic. The second PSL (PSL2) was published in 1995 and included 25 substances, including single chemicals as well as mixtures and effluents.[73] Eighteen of the PSL2 substances were identified as toxic (with two assessments suspended pending the collection of new or additional information).

Use Restrictions

Significant New Activities

SNAc for existing substances are considered by many to be the Canadian counterpart to Significant New Use Rule under TSCA. CEPA 1999 section 80 defines "significant new activity" to include, in respect of a substance, any activity that results in or may result in the following:

> (a) the entry or release of the substance into the environment in a quantity or concentration that, in the Ministers' opinion, is significantly greater than the quantity or concentration of the substance that previously entered or was released into the environment; or
>
> (b) the entry or release of the substance into the environment or the exposure or potential exposure of the environment to the substance in a manner and circumstances that, in the Ministers' opinion, are significantly different from the manner and circumstances in which the substance previously entered or was released into the environment or of any previous exposure or potential exposure of the environment to the substance.[74]

---

71. Env't Canada, A Guide to Understanding the Canadian Environmental Protection Act, 1999, § 5.2.1, http://www.ec.gc.ca/lcpe-cepa/default.asp?lang=En&n=E00B5BD8-1&offset=5&toc=show. *See also* CEPA 1999 § 76.1 (When the ministers "are conducting and interpreting the results of . . . an assessment whether a substance specified on the Priority Substances List is toxic or capable of becoming toxic, the Ministers shall apply a weight of evidence approach and the precautionary principle").
72. Env't Canada, First Priority Substances List (PSL1), http://www.ec.gc.ca/ese-ees/default.asp?lang=En&n=95D719C5-1.
73. Env't Canada, Second Priority Substances List (PSL2), http://www.ec.gc.ca/ese-ees/default.asp?lang=En&n=C04CA116-1.
74. CEPA 1999 § 80.

A SNAc is required to be published in the *Canada Gazette* within 90 days after the expiration of the review period for assessing the NSN.[75] To assist companies, Environment Canada has provided online a comprehensive listing of all substances subject to a SNAc.[76]

Schedule 1

Under CEPA 1999 section 77(3), Environment Canada and Health Canada can list under Schedule 1 when, based on a screening assessment conducted under section 74, the substance is determined to be toxic or capable of becoming toxic, and Environment Canada and Health Canada are satisfied that

(a) the substance may have a long-term harmful effect on the environment and is

(i) persistent and bioaccumulative in accordance with the regulations, and

(ii) inherently toxic to human beings or non-human organisms, as determined by laboratory or other studies, and

(b) the presence of the substance in the environment results primarily from human activity.

Substances also may be added to the Schedule 1 List of Toxic Substances pursuant to CEPA 1999 section 90(1)—without having gone through a Priority Substances List assessment, a screening assessment, or the review of another jurisdiction's decision—if, on the recommendation of Environment Canada and Health Canada, the Governor in Council is satisfied that a substance is toxic. A substance is "CEPA-toxic equivalent" if it satisfies the definition of "CEPA-toxic" as a result of a systematic, risk-based assessment. Such assessments can include determinations made under other federal statutes or can incorporate appropriate elements of assessments done by or for provinces or territories, international organizations, or other appropriate scientific authorities.[77]

---

75. *Id.* § 85(1).
76. Env't Canada, Comprehensive Listing of Substances That Are Subject to a Significant New Activity Notices, http://www.ec.gc.ca/subsnouvelles-newsubs/default.asp?lang=En&n=0F76206A-1.
77. A list of CEPA-equivalent acts for the purposes of risk assessments of new substances is available at Env't Canada, Factsheet: Regulatory Roadmap for New Substances in Canada, http://www.ec.gc.ca/subsnouvelles-newsubs/25166DE1-D27B-45AA-8B88-C2B8DF97B1C9/Factsheet%20-%20EN.pdf.

Once a substance is determined to be "toxic" and added to the List of Toxic Substances in Schedule 1 of the Act, Environment Canada and Health Canada will propose risk management measures. These instruments and tools may be used to control any aspect of the substance's life cycle, from the design and development stage to its manufacture, use, storage, transport, and ultimate disposal. The proposed risk management regulation or instrument must be developed within two years of a determination that the substance is toxic and published in the *Canada Gazette* for public comment.

The list of substances included in Schedule 1 and the risk management tools developed for each listed substance are available online.[78]

Chemical or Product Bans

Virtual Elimination

CEPA 1999 section 77(4) provides that Environment Canada and Health Canada can propose virtual elimination of a substance based on a determination that

> (a) the substance is persistent and bioaccumulative in accordance with the regulations,
>
> (b) the presence of the substance in the environment results primarily from human activity, and
>
> (c) the substance is not a naturally occurring radionuclide or a naturally occurring inorganic substance.[79]

Under CEPA 1999 section 65, virtual elimination means "in respect of a toxic substance released into the environment as a result of human activity, the ultimate reduction of the quantity or concentration of the substance in the release below the level of quantification specified by the Ministers in the [Virtual Elimination] List." The level of quantification is "the lowest concentration [of a toxic substance] that can be accurately measured using sensitive but routine sampling and analytical methods."[80]

---

78. *See* Env't Canada, Toxic Substances List http://www.ec.gc.ca/lcpe-cepa/default.asp?lang=En&n=0DA2924D-1.
79. CEPA 1999 § 77(4). *See also* Env't Canada, The Canadian Environmental Protection Act, 1999 and Virtual Elimination, http://www.ec.gc.ca/lcpe-cepa/default.asp?lang=En&n=BB1FDE0A-1.
80. CEPA 1999 § 65.1.

If virtual elimination is to be implemented, CEPA 1999 section 65(3) states: "When the level of quantification for a substance has been specified on the [Virtual Elimination] List . . . , the Ministers shall prescribe the quantity or concentration of the substance that may be released into the environment either alone or in combination with any other substance from any source or type of source, and, in doing so, shall take into account any factor or information provided for in section 91, including, but not limited to, environmental or health risks and any other relevant social, economic or technical matters."[81]

As of 2013, there are only two substances on the Virtual Elimination List: (1) hexachlorobutadiene, with the molecular formula C_4Cl_6 (with a level of quantification of 0.06 ng/mL, in a chlorinated solvent; and (2) perfluorooctane sulfonate and its salts (with no identified level of quantification).[82]

Prohibition

When a new substance is reviewed and is suspected to be toxic or capable of becoming toxic, Environment Canada can prohibit manufacture or import of the substance in any quantities to mitigate any risk to human health or the environment.[83] Prohibitions are published in the *Canada Gazette*, part I, after they have been issued to the notifier.[84] The notifier may submit additional information and request a reevaluation. Environment Canada's new substance program reviews and considers this additional information and may amend or rescind the prohibition or take alternative risk management measures. The prohibition stands unless a notice is published in the *Canada Gazette* to amend or rescind the prohibition based on the additional information.

---

81. *Id.* § 65(3).
82. Env't Canada, Virtual Elimination List for Hexachlorobutadiene (Dec. 13, 2006), http://www.ec.gc.ca/lcpe-cepa/eng/Regulations/DetailReg.cfm?intReg=82; Env't Canada, Regulations Adding Perfluorooctane Sulfonate and Its Salts to the Virtual Elimination List (Feb. 4, 2009), http://www.ec.gc.ca/lcpe-cepa/eng/regulations/detailReg.cfm?intReg=164.
83. CEPA 1999 § 84. *See* NSN GUIDELINES, *supra* note 5, §§ 9.5.3.2, 9.5.3.3.
84. A list of chemicals and polymers for which ministerial prohibitions have been issued since 1996 is available at Env't Canada, Ministerial Prohibitions, http://www.ec.gc.ca/subsnouvelles-newsubs/default.asp?lang=en&n=E7C8ED07-1. To date, there are no prohibitions published that pertain to living organisms. *See* Env't Canada, Prohibitions, http://www.ec.gc.ca/subsnouvelles-newsubs/default.asp?lang=En&n=EF56AAE3-1.

Test Data Development

Authority to Require Testing/Testing Triggers

New Substance Notifications

There are data requirements that must be provided, or for which waivers must be sought, when submitting a NSN. Specific data requirements address physical-chemical properties, ecotoxicity, health toxicity, and genotoxicity. Test data that must comply with the practices set out in the OECD's Good Laboratory Practice (GLP) principles have been specifically identified in the regulations.

Existing Substances

Under CEPA 1999 section 68, Environment Canada and Health Canada are provided authority to collect or generate data and conduct investigations for the purpose of assessing whether a substance is toxic or is capable of becoming toxic, or for the purpose of assessing whether to control, or the manner in which to control, a substance, including a substance specified on the List of Toxic Substances in Schedule 1.[85]

CEPA 1999 article 3 sets forth the authority to gather information for environmental monitoring, research, state of the environment reporting, and creating inventories and for the development of objectives, guidelines, codes of practice, and regulations. Importantly, the authority to collect information is limited to information "that may be in the possession of that person or to which the person may reasonably be expected to have access."[86]

Chemical Substances/Products That Must Be Tested/Implementing Test Requirements

Environment Canada may request industrial information under CEPA 1999 section 71 by publishing a notice in the *Canada Gazette* inviting all parties that import, manufacture, sell, or use a substance to provide (i.e., in the person's possession or to which the person may reasonably be expected to have access) or generate any information they have for the purpose of assessing whether the substance is "toxic" or capable of becoming "toxic" or whether to control or the

---

85. CEPA 1999 § 68.
86. *Id.* § 46(1).

manner in which to control the substance.[87] A notice seeking information can ask for "any information and samples, including (a) in respect of a substance, available toxicological information, available monitoring information, samples of the substance and information on the quantities, composition, uses and distribution of the substance and products containing the substance; and (b) in respect of a work, undertaking or activity, plans, specifications, studies and information on procedures."[88] The right to seek information under section 71 only extends if the ministers have reason to suspect that the substance is toxic or capable of becoming toxic or it has been determined under this Act that the substance is toxic or capable of becoming toxic.[89]

Data Sharing and Compensation

Typically, when test data are generated on a substance, the substance owner seeks to maintain full rights to the substance and any data on it. In most circumstances, such testing is sponsored and thus paid for by the substance manufacturer.

REPORTING AND RECORDKEEPING REQUIREMENTS

Reports to Be Submitted to Environment Canada/Health Canada

CEPA 1999 section 71 authorizes Environment Canada to seek the submission of information for the purpose of assessing whether the substance is "toxic" or capable of becoming "toxic" or whether to control or the manner in which to control the substance.[90] Under section 71 authority, on December 1, 2012, Environment Canada published a notice in the *Canada Gazette* issuing a mandatory survey, somewhat similar to EPA's Chemical Data Reporting rule, to be completed and submitted for approximately 2,700 substances in Phase 2 of the DSL Inventory Update.[91] The information sought includes information on production/import volumes and uses. The information will be used to

---

87. Id. § 71.
88. Id. § 71(2).
89. Id. § 72.
90. Id. § 71.
91. *See* Canada Gazette, Notice with Respect to Certain Substances on the Domestic Substances List, http://www.gazette.gc.ca/rp-pr/p1/2012/2012-12-01/html/sup-eng.html.

support subsequent risk assessment and risk management activities. The information is due by September 4, 2013.

Recordkeeping

CEPA 1999 authorizes the establishment of regulations respecting the maintenance of books and records for the administration of certain regulations made under CEPA 1999.[92] CEPA's regulations thus include several provisions requiring companies to maintain records. Section 13 of the NSN Regulations, for example, requires that a notifier who provides information to the minister or the new substance program under the NSN Regulations must keep a copy of that information and any supporting data at the notifier's principal place of business in Canada or at the principal place of business in Canada of a representative of that notifier. The information and the supporting data must be kept for a period of five years after the year in which the information is provided.

Risk Notification

CEPA 1999 section 70 states:

> Where a person
>
> (a) imports, manufactures, transports, processes or distributes a substance for commercial purposes, or
>
> (b) uses a substance in a commercial manufacturing or processing activity, and obtains information that reasonably supports the conclusion that the substance is toxic or is capable of becoming toxic, the person shall without delay provide the information to the Minister unless the person has actual knowledge that either Minister already has the information.[93]

Thus, under section 70, industry is responsible for providing Environment Canada with any new information that may support the conclusion that a substance imported, manufactured, sold, or used by that industry is "toxic" or capable of becoming "toxic" as defined under CEPA 1999. This information must be provided unless the notifier has actual knowledge that Environment Canada already has the information.[94]

In addition, under CEPA 1999 section 95, any person who "owns or has the charge, management or control of a substance immediately

---

92. *See, e.g.*, CEPA 1999 §§ 89(e), 93(t), 114(c).
93. CEPA 1999 § 70.
94. NSN GUIDELINES, *supra* note 5, § 10.1.2.

before its release or its likely release into the environment" or "causes or contributes to the release or increases the likelihood of the release" is obliged to report to an enforcement officer any releases of Schedule 1 listed toxic substances that contravene a regulation or order.[95] Such person also must take all reasonable measures to prevent or remedy release and make "a reasonable effort to notify any member of the public who may be adversely affected by the release or likely release." Whistleblower protection (i.e., no personal civil or criminal) liability is available to any person who makes a voluntary report of a release.

IMPORTS AND EXPORTS

Authority to Regulate Imports and Exports

CEPA 1999 sections 100 to 103 set forth the requirements for the export of substances. In particular, Environment Canada and Health Canada may create an Export Control List and add to that list any substance the use of which is prohibited in Canada, any substance that is subject to an international agreement that requires notification or requires the consent of the country of destination before the substance is exported from Canada, and any substance the use of which is restricted in Canada.[96] The Export Control List is divided into three parts to distinguish between those substances prohibited from use in Canada (part 1), those substances subject to notification or consent pursuant to an international agreement (e.g., the Prior Information Consent procedure of the Rotterdam and Stockholm Conventions) (part 2), and those substances whose use is restricted in Canada (part 3).[97] Details concerning these exports are made public through the CEPA Environmental Registry website.

Regulations can be made for all three categories of substances, addressing:

- Prohibitions on export;
- Conditions under which an export may be made;
- The type of information to be provided to the minister with respect to the export; and

---

95. Id. § 95.
96. Id. § 100.
97. See Env't Canada, Export Control List, http://www.ec.gc.ca/lcpe-cepa/default.asp?lang=En&n=06942923-1.

- The type of information to accompany an export and to be kept by the exporter.[98]

Export Requirements and Exemptions

On May 2, 2013, Environment Canada merged what were previously two separate regulations affecting Canadian exports of substance: (1) the Export Control List Notification Regulations (ECLN Regulations), describing the manner by which the minister must be notified for exports of substances listed in the Export Control List; and (2) the Export of Substances under the Rotterdam Convention Regulations (ESURC Regulations) that apply to exports of substances on the Export Control List that are destined to another party to the Rotterdam Convention.[99] The Export of Substances on the Export Control List Regulations (ESECL Regulations) are intended to merge and streamline (i.e., eliminate duplications) the ECLN Regulations and ESURC Regulations and also include new provisions to ensure that Canada complies with the export obligations under the Stockholm Convention.

The Export Control List is divided into three parts:

- Substances listed in part 1 of the Export Control List are those whose uses are prohibited in Canada and can be exported only in limited circumstances (i.e., they are to be destroyed or to comply with ministerial direction under CEPA section 99(b)(iii)).[100]
- Substances listed in part 2 of the Export Control List are those subject to an international agreement requiring the consent and/or notification of the importing country (i.e., the Rotterdam Convention).[101]

---

98. CEPA 1999 § 102.
99. See Canada Gazette, Export of Substances on the Export Control List Regulations SOR/2013-88, http://www.gazette.gc.ca/rp-pr/p2/2013/2013-05-22/html/sor-dors88-eng.html [hereinafter ESECL Regulations]. See also Canada Gazette, Regulatory Impact Analysis Statement, http://www.gazette.gc.ca/rp-pr/p2/2013/2013-05-22/html/sor-dors88-eng.html; Env't Canada, Guidance Document for Export of Substances on the Export Control List Regulations, http://www.ec.gc.ca/lcpe-cepa/default.asp?lang=En&n=15A D7A69-1 [hereinafter Export Guidance].
100. CEPA 1999 §§ 100(a), 101(2); ESECL Regulations, *supra* note 99, § 9.
101. CEPA 1999 §§ 100(b), 101(3); ESECL Regulations, *supra* note 99, § 3(2).

- Substances listed in part 3 of the Export Control List are those whose use is restricted in Canada and also require notification and/or a permit granted prior to export.[102]

Exporters of a substance on the Export Control List must provide a notice of proposed export to the environment minister in accordance with ESECL Regulations.[103] Specifically, a notice, and certification by a duly authorized representative, must be provided at least 30 days before the export and must contain, for each export, the information specified in Schedule 1 of the ESECL Regulations:

- Name of the substance on the Export Control List;
- Country of destination;
- Expected date(s) of export;
- Estimated quantity of substance to be exported;
- Indication of whether the substance will be destroyed or intended for another use;
- For substances listed in part 2 or 3 of the Export Control List and also on Annex A or B of the Stockholm Convention, further information is required to demonstrate that the export meets at least one condition required under the Stockholm Convention (e.g., the export is for laboratory use, the substance is present incidentally in trace amounts).[104]

The notice and certification may be submitted either in writing or an electronic format compatible with one used by the minister.[105] The exporter must "notify the Minister in writing of any change to the information provided in a notice of proposed export within 30 days after learning of it."[106]

Exporters also are required to submit an export permit. Specifically, with limited exceptions, permit requirements apply to substances on the Export Control List part 1, part 2 (also listed on Annex III of the Rotterdam Convention), and part 3.[107] The permit must be

---

102. CEPA 1999 §§ 100(c), 101(3); ESECL Regulations, *supra* note 99, § 3(2).
103. ESECL Regulations, *supra* note 99, §§ 3–7.
104. Id. §§ 5(3), 6(2), sched. 1. *See also* Export Guidance, *supra* note 99, §§ 4 and 5 for provisions applicable to Stockholm Convention and Rotterdam Convention.
105. ESECL Regulations, *supra* note 99, § 5(4).
106. Id. § 5(5).
107. Exemptions from notice requirements are set forth at *id.* §§ 7(2), 8, and 9.

provided "before the export takes place" and include the information specified in Schedule 2 of the ESECL Regulations:

- Information about the exporter and authorized representative, if applicable (e.g., name, address, telephone number, fax number, e-mail address);
- Name of the substance on the Export Control List, common name and trade name of the substance, if applicable, and Chemical Abstracts Service (CAS) registry number, if applicable;
- Country of destination;
- Expected date(s) of export;
- Estimated quantity of substance to be exported;
- Indication of whether the substance will be destroyed or intended for another use;
- Commodity code for the substance set out in the Harmonized Commodity Description and Coding System prepared by the World Customs Organization;
- If the substance is contained in a product, the name of the product and the concentration of the substance in the product;
- Information, if known or applicable, regarding the customs office through which it is expected to be exported, the countries through which the substance will transit, and the proposed number of exports for the calendar year; and
- The safety data sheet (for the substance or the product containing the substance).[108]

As with notices, a certification must accompany the permit and the permit may be submitted either in writing or an electronic format compatible with one used by the minister.[109] In cases where an exporter would be submitting a notice of proposed export and export permit application at the same time, the ESECL Regulations now allow for exporters to submit a combined application.[110]

For certain exports, there are requirements for exporters to maintain certain liability insurance coverage, include certain documents

---

108. *Id.* § 11; sched. 2.
109. *Id.* § 11(3)–(4).
110. *Id.* § 11(6); Regulatory Impact Analysis Statement § 3. *See also* Env't Canada, Combined Notice of Export and Export Permit Application template, http://www.ec.gc.ca/lcpe-cepa/default.asp?lang=En&n=15AD7A69-1&offset=15&toc=hide#Combined_Notice_Permit.

with each shipment (e.g., export permit, safety data sheet), and meet labeling requirements.[111] Exporters also must keep, at the exporter's principal place of business in Canada, certain documents (e.g., the export permit, proof of liability insurance, labels, shipping documents) for a period of five years.[112]

Import Requirements and Exemptions

The import of "new" chemical substances is regulated as discussed earlier in this chapter. Regarding existing substances, Canada has a range of goods over which it imposes import controls. These goods are listed in the Import Control List (ICL) of the Export and Import Permits Act and include certain toxic substance listed in Schedule 1.

CONFIDENTIAL AND TRADE SECRET INFORMATION

Authority Relating to Confidential Business Information

The authority related to confidential business information (CBI) is found in CEPA 1999 sections 88, 113, and 313–321. Of particular relevance, section 88 relates to confidentiality of chemical identity, by providing:

> Where the publication under this Part of the explicit chemical or biological name of a substance would result in the release of confidential business information in contravention of section 314, the substance shall be identified by a name determined in the prescribed manner.[113]

In addition, CEPA 1999 section 113 extends that confidential protection to biological substances:

> Where the publication under this Part of the explicit biological name of a living organism would result in the release of confidential business information in contravention of section 314, the living organism shall be identified by a name determined in the prescribed manner.[114]

---

111. ESECL Regulations, *supra* note 99, §§ 10(2)(c), 20–22.
112. *Id.* § 19.
113. CEPA 1999 § 88.
114. *Id.* § 113.

CEPA 1999 section 313 sets forth the requirements applicable to CBI claims. It provides that "a person who provides information to the Minister under this Act, or to a board of review in respect of a notice of objection filed under this Act, may submit with the information a request that it be treated as confidential."[115] Generally, Environment Canada shall not disclose any information for which a CBI claim has been made under CEPA 1999 section 313, except in certain circumstances specified in CEPA 1999 sections 315 (disclosure in the public interest), 316 (disclosure for specified purposes), and 317 (when disclosure would not be prohibited under the Access to Information Act, section 20).

Provisions Relating to Confidential Business Information

Canada has a mechanism to protect confidentiality of a substance's identity on the DSL. This mechanism includes generic naming requirements for a substance, known as a masked name, that is much stricter than that provided by TSCA. The Masked Name Regulations provide that the explicit chemical or biological name of a substance shall be masked by:

- Replacing a single distinctive element of the name with a nondescriptive term;
- Replacing each of two or more distinctive elements of the name with a nondescriptive term;
- Removing the locants; or
- Removing the locants and replacing each of one or more other distinctive elements with a nondescriptive term.[116]

There are also requirements for where all or part of the explicit chemical name of a substance can be described with a structural diagram and where all or any part of the name of a substance is an explicit biological name, any of the following is a single distinctive element of that explicit biological name.[117] This approach requires that a CAS name be used, and then only a single element of it is genericized (such as alkyl for butyl). It is possible to mask more than one element

---

115. *Id.* § 313.
116. Env't Canada, Masked Name Regulations (SOR/94-261) § 3(1), http://www.ec.gc.ca/lcpe-cepa/eng/regulations/detailReg.cfm?intReg=12.
117. *Id.* §§ 4–8.

if it can be justified; for this reason, development of a name that protects the technology can become an art. It is important to note that "generic" chemical names accepted in the United States under TSCA may not be acceptable by Environment Canada under CEPA because of the CEPA requirement that allows companies to mask only a single element in the fully specific chemical name. The Guidelines for the Notification and Testing of New Substances: Chemicals and Polymers set forth the specific information and justifications to be provided when a substance's identity is claimed confidential.[118]

Companies making submissions under CEPA can also make CBI claims provided there is upfront documentation and substantiation of those CBI claims. In particular, the confidentiality privilege is satisfied by (1) identifying which particular information is confidential by entering the letter "C" on the NSN form in the spaces provided; (2) providing the substantiation information prescribed in the Guidelines for the Notification and Testing of New Substances for each item claimed CBI; and (3) signing the general CBI Certification Statement on the NSN form.

ENFORCEMENT AND PENALTIES

Unlawful Activities

As an initial matter, it is interesting to note the nonlitigious nature of Canada with respect to CEPA and other environmental laws. Negotiations are preferred to more contentious processes on technical and scientific matters. CEPA 1999 part 10 sets forth the authority to carry out investigations and inspections to ensure compliance with CEPA and its implementing regulations. CEPA 1999 addresses the principles of enforcement tools and penalties, to be carried out by enforcement officers and analysts.

CEPA 1999 provides the authority for enforcement officers to inspect premises, generally with consent or after obtaining a warrant, when, for example, there are reasonable grounds to believe that the place has a substance (or a product containing such a substance) to

---

118. NSN GUIDELINES, *supra* note 5, § 7 & app. 7. *See also* NEW SUBSTANCES PROGRAM OPERATIONAL POLICIES MANUAL, *supra* note 39, at 29–30 (Operational Policy for Maintaining Confidentiality of Substance Identities).

which CEPA 1999 applies, or where any books, records, electronic data, or other documents relevant to the administration of CEPA can be found.[119] When conducting an inspection, enforcement officers have broad authority to examine any substance, examine any books, records, or electronic data, take samples, seize evidence, conduct tests or take measurements, or stop and detain conveyances such as a vehicle, ship, or aircraft for the purpose of conducting an inspection.[120] CEPA 1999 provides enforcement officers with several enforcement tools, ranging from warnings when there is minimal or no threat to the environment or human life or health to environmental protection compliance orders (EPCO) and injunctions when further actions are needed to prevent or stop a violation.

Environment Canada states the following regarding the selection of enforcement tools:

> While each fact situation will be different in relation to alleged violations of CEPA, 1999, probably the most important factor in determining an enforcement response is the effectiveness of the response in securing compliance as quickly as possible with no recurrence of violation. Therefore, except in circumstances where prosecution will always be pursued as described later in this chapter, the enforcement officer will give first consideration to an enforcement response among warnings, directions, Ministerial orders, detention orders for ships and environmental protection compliance orders, as these do not require a court proceeding, and compliance may be restored in a shorter time frame tha[n] would be possible through a court prosecution.[121]

CEPA 1999 also allows for the designation of individuals as CEPA analysts to support enforcement initiatives. CEPA analysts can be chemists, biologists, engineers, forensic accountants, or laboratory personnel. Although CEPA analysts do not have the power to use enforcement tools that enforcement officers do, analysts are entitled to accompany enforcement officers on inspections and in those circumstances have the power to enter premises, open receptacles, take samples, conduct tests and measurements, and require that documents and data be provided to them.

---

119. CEPA 1999 § 218(1).
120. *Id.* § 218(7)–(10).
121. Env't Canada, Compliance and Enforcement Policy for CEPA (1999) (Mar. 2001), http://www.ec.gc.ca/lcpe-cepa/default.asp?lang=En&n=5082BFBE-1.

Civil Penalties

The Environmental Enforcement Act (EEA) is an omnibus legislation passed in June 2009 that amends the fines, sentencing provisions, and enforcement tools of six acts administered by Environment Canada, including CEPA 1999 and three acts administered by Parks Canada. Based on regulations issued on July 4, 2012, EEA increases maximum fine amounts and establishes minimum fines for certain offenses with the following new fine scheme:[122]

Offender	Type of Offense	Summary		Indictment	
		Minimum	Maximum	Minimum	Maximum
Individuals	Most serious offenses	$5,000	$300,000	$15,000	$1 million
	Other offenses	N/A	$25,000	N/A	$100,000
Small Corporations and Ships under 7500 tons	Most serious offenses	$25,000	$2 million	$75,000	$4 million
	Other offenses	N/A	$50,000	N/A	$250,000
Corporations and Ships over 7500 tons	Most serious offenses	$100,000	$4 million	$500,000	$6 million
	Other offenses	N/A	$250,000	N/A	$500,000

All fines are doubled for second and subsequent offenses. All fines go to the Environmental Damages Fund so that they may be used to repair the harm done.

EEA also creates the Environmental Violations Administrative Monetary Penalties Act (EVAMPA), which came into force on December 10, 2010. EVAMPA sets forth an administrative monetary penalty (AMP) system applicable to less serious environmental violations. AMPs are civil or administrative in nature, rather than penal or criminal. They are designed to provide an efficient response to violations where there is no need for denunciation and punishment.

---

122. Env't Canada, New Fine Scheme under the Environmental Enforcement Act, http://www.ec.gc.ca/alef-ewe/default.asp?lang=En&n=7CB7E78A-1.

Criminal Penalties

CEPA section 222.1 provides enforcement officers with the authority to arrest without warrant "a person or ship that the enforcement officer believes, on reasonable grounds, has committed an offence against this Act or the regulations, or that the enforcement officer finds committing or about to commit an offence against this Act or the regulations." Enforcement officers also are authorized to prosecute under the authority of a Crown prosecutor.

Upon conviction of an offender for a CEPA 1999 violation, enforcement officials will provide recommendations for penalties that are proportionate to the nature and gravity of the offense. Potential penalties include fines or imprisonment or both, court orders to accompany a fine or imprisonment, and court orders governing conditional discharge of the offender. When imposing penalties or court orders, courts look to the criteria set forth in CEPA 1999 for guidance. In addition to the sentencing criteria set forth in CEPA 1999, enforcement officers will recommend a penalty and/or court order that will, in their view, be effective in deterring others from committing the same or other violations of the Act.[123] EEA now provides additional sentencing guidance for courts to consider when determining a sentence by establishing principles and purposes of sentencing and by identifying relevant factors, such as prior convictions, the nature of the damage, whether the violation was committed intentionally or recklessly, and whether the perpetrator stood to benefit financially. The courts also would be able to issue orders, upon conviction of an offender, to cancel or suspend licenses and/or permits and to require corporations to inform shareholders about offenses.

Environmental protection alternative measures (EPAMs) are defined as "measures, other than judicial proceedings, that are used to deal with a person who is alleged to have committed an offence under this Act."[124] EPAMs thus are an alternative to court prosecution for a violation of CEPA 1999. To be eligible for an EPAM, the attorney general of Canada must consider the factors set forth at CEPA 1999 section 296 (e.g., consider the protection of the environment and human health, the person's history of compliance, and whether

---

123. Env't Canada, Compliance and Enforcement Policy for CEPA (1999)–March 2001 (Penalties and Court Orders Upon Conviction), http://www.ec.gc.ca/lcpe-cepa/default asp?lang=En&n=5082BFBE-1&offset=7&toc=show#s12.
124. CEPA 1999 § 295.

the offense is a repeat occurrence) and determine whether an EPAM should be offered. If an EPAM is negotiated, it will contain measures to restore compliance (e.g., installation of better pollution control technology or monitoring systems, or cleanup of environmental damage). EPAM agreements are registered in court as public documents, and CEPA 1999 section 301 requires their publication in the CEPA Environmental Registry.

Key Business Issues

Commonly Encountered Issues

CEPA 1999 and its implementing regulations are considered by many to be a relatively straightforward system. Potentially common problems in other regulatory schemes can be avoided in Canada by understanding the laws and regulations and working with Environment Canada and/or Health Canada.

If a substance is "new" to Canada, and is expected to be manufactured or imported in the United States, one issue is whether to add the substance to the DSL or whether to first add the substance to the TSCA Inventory, which would then allow the substance to be added to the NDSL. If the substance needs to be added to both the Canadian and U.S. inventories, there may be some benefit to adding the substance to the TSCA Inventory first and then taking advantage of less onerous registration and testing requirements for NDSL substances.

The last task to be completed to place a new substance on the DSL is the submission of the NOMI or NOEQ. This last requirement suggests interesting options in terms of strategy. If a company wishes to get the substance on the DSL as soon as possible (which is often the case), for example, because of customer demand, then the NOMI approach can be used, just as the NOC is used under TSCA. If, however, the submitter does not care if the substance is listed on the DSL and wishes to exclude other possible manufacturers, the company can wait until the NOEQ annual volume trigger is reached, which may take years, if ever. The 2005 regulations were revised to remove the second volume trigger, which occurred when a cumulative volume was reached. An interesting nuance to the above process is that it does not have to occur in the sequence indicated. For example, it is possible to submit a NOMI at the same time as the notification schedule that represents the highest schedule required. The caveat here is that the substance must have actually been manufactured or imported into Canada, for example, as a R&D substance for which no notification was required.

Practical Tips

Canada has made an effort to make information available to the public by creating the Environmental Registry. Environment Canada calls the Environmental Registry "a comprehensive source of information on a variety of CEPA 1999-related tools, including proposed and existing policies, guidelines, codes of practice, government notices and orders, agreements, permits, and regulations."[125] The resources in the Environmental Registry can be easily consulted to assist in understanding any applicable regulatory requirements.

It can be a complicated process to determine the testing requirements and information required to be submitted with a new substance notification since there are several factors that must be analyzed, including but not limited to the type of substance at issue and the volumes intended to be manufactured in or imported into Canada. It is important for companies considering manufacture or importation of a new chemical substance to have a clear understanding not only of production volumes but also the timing for when certain reporting thresholds may be reached so that any data that need to be developed and other notification information can be timely submitted.

During a new substance review, if Environment Canada believes that a substance may be "toxic" (impose a risk), it may restrict the submitter's manufacture, processing, and use of the substance through a Ministerial Condition. In this case, the substance is not added to the DSL, and other manufacturers or importers have to do their own notification. This is different than a SNAc, since in those cases the substance subject to a SNAc is listed on the DSL but with restrictions. One trap for other manufacturers is to identify the substance as being present on the DSL but not to realize that it is subject to a SNAc.

---

125. Env't Canada, CEPA Environmental Registry, http://www.ec.gc.ca/lcpe-cepa/default.asp?lang=En&n=D44ED61E-1.

Trends

In October 2011, Environment Canada renewed its commitment to pursue its Chemical Management Plan (CMP) over the next five years. The next phase is intended to build on successes and lessons learned from the first phase of the CMP. According to the most recent annual report, the key elements of the next phase of the CMP include the following:

- Completing assessments of approximately 500 substances across nine categories; these categories represent substances that have been grouped based on shared similar characteristics (such as structural or functional similarities);
- Investing in additional research for substances like BPA (bisphenol A), flame retardants, and substances that affect hormone function;
- Addressing approximately 1,000 additional substances in the next five years through other initiatives, including rapid screening of substances;
- Updating the commercial use information of substances through mandatory reporting to inform risk assessment and risk management activities;
- Continuing the assessment and management (where required) of priority substances identified through the first phase of the CMP (e.g., petroleum sector stream substances and the Challenge).[126]

As the United States looks to modernize and reform TSCA, some look to the Canadian system as a model for such reform.[127]

---

126. Env't Canada, Canadian Environmental Protection Act, 1999 Annual Report for April 2011 to March 2012 (§ 5, Controlling Toxic Substances), http://www.ec.gc.ca/lcpe-cepa/default.asp?lang=En&n=0EB06C79-1.
127. See, e.g., Am. Chem. Council, TSCA Modernization, http://www.americanchemistry.com/Policy/Chemical-Safety/TSCA ("Canada's approach to prioritization and review may provide an effective model for some changes to TSCA"); SOCMA, Myth versus Fact about Chemicals in Commerce), http://www.socma.com/GovernmentRelations/index.cfm?subSec=26&articleID=3259 ("If EPA really wants to address this fundamental criticism of its current TSCA program, it then needs to *prioritize all existing chemicals in the same step*, through a comprehensive screening program of the sort that Canada has successfully implemented.").

Resources

Canadian Environmental Protection Act, 1999 (S.C. 1999, c. 33), http://laws-lois.justice.gc.ca/eng/acts/C-15.31/.

New Substances Notification Regulations (Chemicals and Polymers) (SOR/2005-247), *available at* http://laws-lois.justice.gc.ca/eng/regulations/SOR-2005-247/index.html.

GUIDELINES FOR THE NOTIFICATION AND TESTING OF NEW SUBSTANCES: CHEMICALS AND POLYMERS, VERSION 2005, http://publications.gc.ca/collections/Collection/En84-25-2005E.pdf.

Environment Canada, Guidance Document for Export of Substances on the Export Control List Regulations, http://www.ec.gc.ca/lcpe-cepa/default.asp?lang=En&n=15AD7A69-1.

Environment Canada, Compliance and Enforcement Policy for CEPA (1999)—Mar. 2001, http://www.ec.gc.ca/lcpe-cepa/default.asp?lang=En&n=5082BFBE-1.

Environment Canada, Masked Name Regulations (SOR/94-261), http://www.ec.gc.ca/lcpe-cepa/eng/regulations/detailReg.cfm?intReg=12.

ENVIRONMENT CANADA & HEALTH CANADA, NEW SUBSTANCES PROGRAM OPERATIONAL POLICIES MANUAL (Apr. 2004), http://www.ec.gc.ca/subsnouvelles-newsubs/66C60DFB-BD97-4F1F-B32B-545CD8A87773/ops_pol_e.pdf.

Environment Canada, CEPA Environmental Registry, http://www.ec.gc.ca/lcpe-cepa/default.asp?lang=En&n=D44ED61E-1.

4

Central America and Mexico

Michael S. Wenk

EXECUTIVE SUMMARY

Each chemical regulatory framework within the respective countries in Central America functions independently, without reference to those of its neighbors and few, if any, harmonized aspects. Prospective manufacturers or importers must take the initiative to familiarize themselves on a country-specific basis, as needed, with regulatory requirements addressing hazardous substances, notification, labeling and hazard communication, and any other pertinent mandates or programs. Overlapping requirements, such as those found in Mexico, can be difficult to sort out. One basic area in which Central American countries tend to lag is hazard communication. Requirements for warning labels and providing information about chemical properties may not be as comprehensive as in various other jurisdictions. As of this writing, the Globally Harmonized System of Classification and Labelling of Chemicals (GHS) awaits adoption in Central America; Mexico has done so, albeit with compliance at least initially voluntary. Significant practical challenges encountered in some countries, such as Honduras, are that not all relevant legislation is available in electronic form and that amendments are not always promptly consolidated with preexisting versions of the laws that they affect. We summarize below key elements of these regulatory structures in Costa Rica, Panama, Honduras, and Mexico.

4.1
Costa Rica

In Costa Rica, the core document for chemical control regulations is the General Health Law, Law 5395 (Ley General de Salud), administered by the Ministry of Health (Ministério de Salud). The Ministério "determines which substances are 'hazardous,' addresses their registration and permitting, and has broad authority to ban and/or restrict specific substances."[1] Section IV addresses the duties and restrictions on hazardous substances, and article 239 prohibits any chemically related activity without meeting specific requirements.[2]

Costa Rica has implemented the Chemical Information Exchange Network (CIEN), "a network of people involved in the management of chemicals" with support from the United Nations Environment Programme (UNEP).[3] There has been relatively little action, however, in the organization since early 2009. The goals of the CIEN project are to eliminate barriers to information exchange; facilitate access to technical information on chemicals on the Internet; create synergies between national agencies involved in chemical management; and strengthen national capacity for the environmentally sound management of chemicals and participation in international activities and agreements.[4]

---

1. IHS EIATRACK, Restricted Substances Overview in Costa Rica, http://www.eiatrack.org/s/204
2. Ministerio de Salud, Ley General de Salud, http://costarica.eregulations.org/media/L-5395.pdf
3. U.N. Env't Programme, Chem. Info. Exch. Network, http://jp1.estis.net/communities/cien/.
4. Id.

REGULATION OF SUBSTANCES/PRODUCTS

Existing and new chemicals are subject to the registration and permitting requirements of Decree No. 28113-S of September 1999, the Regulation for Registration of Dangerous Products. This version repeals Executive Decree Nos. 24867 of January 31, 1996, and 26805-S of March 11, 1998. The regulation is administered by the Ministry of Health, Department of Toxic Substances and Industrial Medicine. The regulation includes a list of product types that are exempted from regulation, such as human and animal drugs, pesticides, chemicals used as food additives, and laboratory reagents, among other related exemptions.[5] The regulation also addresses the classification and labeling of certain substance types.

The regulation addresses the registration of dangerous products according to criteria specified in Annex I (Guide for the Classification of Dangerous Products), or other goods declared as such (dangerous) by the Minister.[6] Costa Rica has created the Office of Technical Coordination for the Management of Chemical Substances (OTCMCS) via Decree No. 33104 (2006), composed of representatives of the Ministério de Salud, the Ministério de Ambiente, Energía y Telecomunicaciones (MINAET), and others.[7] Although OTCMCS does not have the regulatory authority to impose bans on given substances, it functions as a "regulatory clearing house." For example, the Ministério de Salud administers aspects of the Basel Convention (e.g., hazardous waste), whereas MINAET administers the Stockholm Convention (e.g., persistent organic pollutants).[8]

TRENDS

Costa Rica has not yet adopted the GHS.[9] Costa Rica does have specific requirements for Safety Data Sheets (SDS), mirroring those of the 16-section ISO format, as set forth in ISO 11014-1. If a Costa

---

5. Regulation for Registration of Dangerous Products, Issue No. 194, Alcance No. 74, LA GACETA, DIARIO OFICIAL (Oct. 6, 1999), http://www.glin.gov/download.action?fulltextId=20476&documentId=92061&glinID=92061.
6. Id.
7. Restricted Substances Overview in Costa Rica, *supra* note 1.
8. Id.
9. U.N. Econ. Comm'n for Europe, GHS Implementation, http://www.unece.org/trans/danger/publi/ghs/implementation_e.html.

Rican SDS is produced in a language other than Spanish, a technical data sheet must be completed and submitted along with the alternate SDS language.

4.2
PANAMA

The governing document for chemical regulation in Panama is the General Environmental Law (Ley General del Ambiente) No. 41 (July 1, 1998).[10] As with Costa Rica and other Central American countries, the Ministério de Salud has responsibility for administering the law. Article 60 of chapter II—Hazardous Waste and Hazardous Substances (Desechos Peligrosos y Sustancias Potencialmente Peligrosas)—addresses the registration of hazardous substances before their introduction into commerce in the country. Registration responsibility is delegated to the National Environmental Authority.[11] Registration of a chemical substance may be denied if the substance is banned or otherwise restricted in the importer's home country.[12] Article 2 of Executive Decree No. 305 (2002) expressly forbids the use of substances listed in Annex I of the decree if the entity has not obtained a "special license," whose requirements are set forth in article 7.[13] Panama has implemented a CIEN. Finally, all documents must be in Spanish.[14]

REGULATION OF SUBSTANCES/PRODUCTS

Within Panama, the following categories of chemicals are considered to be hazardous materials, substances, and wastes that must comply with the regulations established in articles 70 to 92 of Executive Decree No. 640:

---

10. Ley General del Ambiente, *available at* http://www.lawyers-abogados.net/es/recursos/Ley-41-1998-Ley-General-Ambiente.htm.
11. Id.
12. Id.
13. Id.
14. Letter from Marco Antonio Ponce, Junior Assoc., Medina, Rosenthal & Fernandez, to author (Oct. 7, 2011) (on file with author).

- explosives
- gases
- inflammable liquids
- inflammable solids
- combustive substances and organic peroxides
- toxic and infectious substances
- radioactive material
- corrosive substances

Decree Law No. 2 (2008), Regulating Safety, Health and Hygiene in the Construction Industry, promulgated by the Ministry of Work and Labor Development, was designed primarily for the construction industry. Its provisions, however, have generally been accepted in other areas. Under article 21, Risk Classification, reference is made to the World Health Organization's (WHO) classification of workplace chemicals.[15] Panama charges industry with making health-based risk assessments for such chemicals and with updating the assessments when conditions change. The risk assessments must be completed on specific forms that the competent authority provides. Articles 373 and 374 of the same law mandate that it is the employer's responsibility to ensure that containers of hazardous materials are labeled with critical information, such as health effects, physical, chemical, and related factors.

TRENDS

With respect to SDS in Panama, the enabling legislation is Decree Law No. 2. Resolution 124, however, speaks to the specific requirements of the SDS. Though Decree Law No. 2 is specific in nature to the construction industry, SDSs and their attendant requirements have generally been incorporated into other aspects of chemical management in the country. Article 368 of Decree Law No. 2 mandates that an SDS must be provided to locations where hazardous substances are delivered, when they are used, or for when they are disposed. All workers who come into contact with the substance for which the SDS has been generated must receive appropriate training

---

15. Decree Law No. 2 (Feb. 22, 2008).

on the information contained therein. The SDS is composed of 12 sections and must be presented in Spanish.

At present, Panama has not adopted GHS.[16]

4.3
HONDURAS

There are two key challenges with accessing Honduran national laws. All legislation and executive decrees are published in the official government daily, *La Gaceta: Diario Oficial de la República de Honduras*, but it is only inconsistently available electronically. Second, many of the principal laws have been amended significantly over a number of years but have not been officially consolidated. This situation makes it challenging to determine the current status of many provisions.[17]

REGULATION OF SUBSTANCES/PRODUCTS

Honduras has regulations currently in place for the control of chemical substances. The country's legislation provides for the control of all chemical substances listed in international conventions (such as the Stockholm Convention, the Montreal Protocol, and the Kyoto Protocol). The country does not have a formal mechanism to monitor and prevent the diversion of controlled chemical substances or an automated information management system for the same.[18] The General Directorate for Health Regulation, through the Department for Health Control of Goods and Services, is the entity responsible for carrying out activities related to the national registry of licensees, control of licenses for manufacture and distribution, inspections, control of distribution and final sale, preexport notifications, and the imposition of sanctions. The Executive Directorate for Revenue

---

16. U.N. Econ. Comm'n for Europe, GHS Implementation, http://www.unece.org/trans/danger/publi/ghs/implementation_e.html.
17. Rule of Law in Armed Conflicts Project, Access to Honduran Nat'l Law, http://www.adh-geneva.ch/RULAC/national_legislation.php?id_state=85.
18. Letter from Marco Antonio Ponce, *supra* note 14.

also interacts with the General Directorate to control the import and export of chemical substances.[19]

Use Restrictions

In Honduras, there is no format for the presentation of information on the chemical product label, as long as the hazard pictograms, the word "WARNING," and hazard statements appear together. The manufacturer or importer may also provide supplementary information for carcinogens, reproductive toxins, or systemically toxic chemicals.[20] The General Rules of Preventive Measures of Occupational Accidents and Diseases, in force since 2004, is a complementary legal instrument to GHS, which also regulates labeling regarding chemicals.[21]

The chemical product label must contain, at a minimum, the following: words of warning; indication of danger; precautionary statements and precautionary pictograms (pictures and colors); product identification (name of the substance or preparation); and supplier identification (name, address, and telephone number of manufacturer and/or provider).[22]

4.4
Mexico

Chemical regulations in Mexico are particularly challenging to deconstruct because there are multiple overlapping laws and regulations. Recognizing this, in 1987 the Comisión Intersecretarial para el Control del Proceso y Uso de Plaguicidas, Fertilizantes y Sustancias Tóxicas (CICOPLAFEST) was created to control the importation of pesticides, fertilizers, and toxic substances.

On January 31, 2013, the Mexican Instituto Nacional de Ecología y Cambio Climático released the National Inventory of Chemical

---

19. Id.
20. Id.
21. Id.
22. Id.

Substances (Inventario Nacional de Sustancias Químicas). Although details are still emerging, the Instituto has commented that the Inventory was "integrated from second hand information sources" as part of its development and that "the inventory is not a compulsory regulation."[23] Thus, at least according to the initial position of the Instituto, registration of chemical products/substances does not appear to be mandatory or part of the national regulation.

Regulation of Substances/Products

There are six agencies that maintain and enforce laws relating to chemicals in Mexico:

- Secretariat of Health (Secretaría de Salud)
- Secretariat of Agriculture, Livestock, Rural Development, Fisheries and Food (Secretaría de Agricultura, Ganadería, Desarrollo Rural, Pesca y Alimentación, or SAGARPA)
- Secretariat of Environment and Natural Resources (Secretaría del Medio Ambiente y Recursos Naturales, or SEMARNAT)
- Secretariat of the Economy (Secretaría de Economía)
- Secretariat of Communications and Transport (Secretaría de Comunicaciones y Transportes)
- Secretariat of Labor and Social Welfare (Secretaría de Trabajo y Previsión Social)

A new CICOPLAFEST agreement was published on December 17, 2001, and became effective on January 17, 2002. In addition, the Federal Commission for the Protection against Sanitary Risk (Comisión Federal para la Protección contra Riesgos Sanitarios, or COFEPRIS), is a decentralized group under the Secretaría de Salud, with responsibility arising out of the General Health Law, tasked with protecting the Mexican population from sanitary, or health, issues. One of the stated objectives of COFEPRIS is the "sanitary control of products, services and their import and export, and establishments dedicated to processing the

---

23. Personal correspondence from Miguel Angel Martínez Cordero, Mexican Nat'l Inst. of Ecology, to author (Feb. 18, 2013) (on file with author).

products," which provides the necessary authority for chemical regulation in instances where it occurs.[24]

Whereas multiple agencies have at least partial responsibility for chemical regulation in Mexico, there are two key items that address the issue of chemical management: the General Health Law (Ley General de Salud) and its attendant regulation, the Regulation under the General Health Law Regarding the Sanitary Control of Activities, Facilities, Products, and Services (Reglamento de la Ley General de Salud en Materia de Control Sanitario de Actividades, Establecimientos, Productos y Servicios). Taken together, these items define "toxic substances," establish a system by which such substances must be licensed and permitted, and set specific labeling standards for chemical substances. Under the authority of the Ley General de Salud, the secretary promulgates the Official Mexican Standards (Normas Oficial Mexicana, or NOM), which "apply to the use, application, development, labeling, storage, packaging, commercialization and distribution of all chemical substances."[25]

Under the Chemical Substances Instruction (Instrucción de Substancias Químicas) published in the *Diario Oficial de la Federación* on December 12, 1988, data on toxic substances, similar in concept to that required for the registration of new pesticides, fertilizers, or vegetable nutrients, must be submitted to CICOPLAFEST. There are currently no premanufacture review requirements for the use, manufacture, or distribution of new toxic substances per se, at least in the vein of other globally recognized schemes, such as the U.S. Toxic Substances Control Act (TSCA) or the Canadian Domestic/Non-Domestic Substances List (DSL/NDSL).[26] In June 2012, Mexico announced its intent to implement a national chemical inventory, although the components and mechanism of such an inventory are currently unclear.

With respect to workplace management of chemicals, the Federal Labor Law and Work Safety Regulation (Ley Federal del Trabajo y Seguridad en el Trabajo) sets out the protection requirements for individuals working in facilities where chemical substances believed to be present as an integral part of the business operations could pose

---

24. Secretaría de Salud, What Is the COFEPRIS?, http://www.cofepris.gob.mx/Paginas/Idiomas/Ingles.aspx.
25. Summary of Environmental Law in Mexico, ch. 11, *available at* http://www.cec.org/law database/MXII.cfm?varlan=english..
26. *Id.*

occupational health risks.²⁷ The Health Secretariat, in conjunction with the Secretariat of Labor and Social Welfare, has oversight of this area, while the Regulation for the Land Transport of Hazardous Materials and Wastes (Reglamento para el Transporte Terrestre de Materiales y Residuos Peligrosos), under the oversight of the Secretariat of Communications and Transport, manages the transport of hazardous materials, wastes, and chemical substances.

Trends

The NOM associated with writing and distributing SDSs in Mexico is NOM-018-STPS-2000, the System for the Identification and Communication of Hazards and Risks for Hazardous Chemical Substances in the Workplace (Sistema para la Identificación y Comunicación de Peligros y Riesgos por Sustancias Químicas en los Centros de Trabajo). Published on March 10, 2000, NOM-018 was initially a draft amendment to the Mexican Official Standard NOM-114-STPS-1994 of the same name. Within NOM-018, appendix C covers the SDSs themselves, and appendix D discusses the specific requirements for completing an SDS.

The stated objective of NOM-018 was "[t]o establish the minimum requirements of a system for the identification and communication of hazards and risks by hazardous chemical substances, that in accordance to their physical, chemical characteristics, toxicity, concentration and time of exposure, can affect the health of workers or damage the workplace."²⁸

Section C.1.1 of the NOM requires: "All work places must have SDS's of each of the dangerous chemical substances handled in it and must be permanently available for all workers involved in their use."²⁹ The SDS must be in Spanish and must be updated as new information is obtained or otherwise known. A compliant SDS must contain, at a minimum, the following information, with no blank information spaces, within the document, in the proscribed 12-section format.³⁰

---

27. Id.
28. Secretaría del Trabajo y Previsíon Social, Sistema para la Identificación y Comunicación de Riesgos por Sustancias Químicas en los Centros de Trabajo (Mar. 10, 2000), http://www.stps.gob.mx/bp/secciones/dgsst/publicaciones/guias/Guia_018.pdf.
29. Id.
30. Id.

On June 3, 2011, Mexico published its GHS standard, NMX-R-019-SCFI-2011, which entered into force on June 4, 2011. The Mexican standard is composed of two parts: the Hazard Communication program and the GHS implementation itself. Although compliance with the NMX is still voluntary at present, the "standard establishes criteria for Mexicans to classify chemicals according to their physical hazards, health and the environment. It also establishes the elements of hazard communication standard chemicals and the requirements for labeling and data sheets for their safety."[31]

---

31. *Mexico Is First in the Region to Put GHS into Practice*, Nexreg, July 14, 2011, http://www.nexreg.com/news/jul-14-eu-mexico-first-region-put-ghs-practice.

5

South America

Michael S. Wenk

Executive Summary

The chemical regulatory frameworks within the countries of South America are often perceived, and correctly so, as largely disjointed with few harmonized elements, even among even neighboring countries or within the same regional trading blocs.

The South American region of the globe consists of two major trading blocs, Mercosur (Mercado Común del Sur), which was formed on March 26, 1991, composed of Argentina, Brazil, Paraguay, and Uruguay; and the Andean Community of Nations (Comunidad Andina), whose roots go back to the Cartagena Treaty of 1969 and whose current members are Bolivia, Colombia, Ecuador, and Peru. These two blocs impose overarching legislative and regulatory constraints on the individual member countries that often do not exist locally, sometimes creating a regulatory disconnect between local (national) law and regional regulations. As an example, the Andean Community has a regional community provision to regulate chemical substances used in the manufacture of illegal narcotic drugs and psychotropic substances; the individual member countries that make up the community generally do not have such comprehensive chemical control regulations.[1] This dichotomy presents an ongoing challenge for regulating entities. We summarize below key elements of these regulatory structures in Argentina, Brazil, Chile, Colombia, Paraguay, Peru, Uruguay, and Venezuela.

5.1
ARGENTINA

NOTIFICATION/REGISTRATION/APPROVAL REQUIREMENTS AND COMPLIANCE

Argentina has no explicit requirements for the notification of new chemical substances. There are well-established laws in force, however, that regulate chemicals used as precursors for the manufacture of illegal drugs, such as potassium permanganate and ephedrine, among others.[2] There are pre-notification requirements for dangerous substances and preparations, as set out in Law No. 26.045 (July 6, 2005)—Registro Nacional de Precursores Químicos (National Registry of Chemical Precursors), as well as requirements for packaging and labeling.[3]

Resolution No. 709/98 of the Ministry of Health and Social Action Overview

The National Administration for Drugs, Food, and Medical Technology (Administración Nacional de Medicamentos, Alimentos y Technologia Médica, or ANMAT) issued Resolution No. 709/98. The resolution entered into force on July 9, 1998; it addresses the registration of domestic-use chemical products and thereby creates the National Register for Domestic Use Products.

---

1. Comunidad Andina, About Us, http://www.comunidadandina.org/ingles/quienes/results.htm.
2. U.S. DEP'T OF STATE, 2012 INTERNATIONAL NARCOTICS CONTROL STRATEGY REPORT (INCSR) (Mar. 7, 2012), http://www.state.gov/j/inl/rls/nrcrpt/2012/vol1/184096.htm.
3. InfoLeg, Registro Nacional de Precursores Químicos, http://www.infoleg.gov.ar/infoleg Internet/anexos/105000-109999/107623/norma.htm.

Scope of Resolution No. 709/98

ANMAT has been delegated responsibility by the Ministry of Health to regulate chemical products and to publish guidelines for them. Such products can be registered with ANMAT for up to five years. Key within this regulation, though, is the provision that failing to reregister a product results in product cancellation without prior notice.[4] Article 9 of the resolution prohibits the use of Group I International Agency for Research on Cancer (IARC)—World Health Organization (WHO) carcinogens in domestic-use products. Risk Category II within the resolution requires specific labeling and a prospectus for attendant products if free samples are to be distributed.

REGULATION OF SUBSTANCES/PRODUCTS

Argentina parallels the European Union's Dangerous Substances Directive (67/548 EEC) with respect to the classification, packaging, and labeling of dangerous substances. In addition, it relies upon the Instituto Argentino de Normalización y Certificación (IRAM) Standard No. 3797, Rotulado de Sustancias Químicas (May 12, 1986) to address the labeling of chemicals for transport.[5] Decree No. 351/79 (Feb. 5, 1979), Higiene y Seguridad en el Trabajo, addresses workplace safety and hygiene requirements, specifically that workers must be educated about safety measures to be employed when working with hazardous chemicals.[6] The Argentine Ministry of the Economy requires product labels for both domestic and imported products to include the following information: the product name, the country of origin, the degree of purity or quality, and the net weight.[7] Article 4 of the Rotulado specifically directs that primary labeling must be in Spanish but that the label may contain other languages as well.[8]

---

4. Resolution No. 709/98 of the Ministry of Health and Social Action, at 33–34.
5. Estrucplan, Rotulado de Sustancias Químicas—Norma IRAM 3797 (Jan. 1, 2000), http://www.estrucplan.com.ar/producciones/entrega.asp?identrega=1742.
6. InfoLeg, Higiene y Seguridad en el Trabajo, Información Legislativa, http://infoleg.mecon.gov.ar/infolegInternet/anexos/30000-34999/32030/texact.htm.
7. Higiene y Seguridad en el Trabajo, Law No. 22802/83, ch. I, § 1 (Apr. 21, 1983), *available at* http://200.69.252.41/hypersoft/Normativa/NormaServlet?id=1506.
8. Argentina Law No. 22802/83 art. 4, *available at* http://200.69.252.41/hypersoft/Normativa/NormaServlet?id=1506.

5.2
CHILE

NOTIFICATION/REGISTRATION/APPROVAL REQUIREMENTS AND COMPLIANCE

Manufacturers and importers of chemicals or chemical substances in Chile have no formal registration requirements. There is no national inventory of existing chemical substances in Chile and "no requirement exists to notify the government about new substances."[9] Importers and exporters of ozone-depleting substances, however, must register under the appropriately named Register of Importers and Exporters of Ozone-Depleting Substances, under the auspices of the National Customs Service (Servicio Nacional de Aduanas or SNA).[10] The purpose of the register is to ensure compliance with the international commitments arising from the Montreal Protocol.[11]

Decree No. 78, issued in September 2010, regulates the storage and labeling of dangerous chemicals, with certain exceptions, including pesticides. With respect to such products, on September 26, 2012, the Chilean Department of Environmental Health began the process of expanding its online database of registered sanitary and domestic disinfectants to include pesticide product registration. Individuals and businesses use an online system (www.ispch.cl) for registration of such products.[12] Under Title XIII of the decree, containers of such chemicals are to be labeled legibly in Spanish, via black lettering presented horizontally on a white background. Labels should include at least information correlating with the Safety Data Sheet (SDS) data, including the chemical name and U.N. number(s), as well as the name, address, and telephone number of the manufacturer/importer.[13]

---

9. *In Recent Years, Chile Has Implemented Wide Reaching Chemical Regulations*, CHEMS. CONTROL WATCH, Dec. 2010.
10. Chile Exempt Resolution No. 5630 of (October 17, 2007).
11. Article 84 of the Chile Customs Ordinance (DFL No. 30).
12. Instituto de Salud Pública, Unidad de Plaguicidas y Desinfectantes Inició el Registro Electrónico de Productos Plaguicidas de Uso Sanitario y Doméstico a Través de Servicios en Línea, http://www.ispch.cl/noticia/16983.
13. Decree No. 78, tit. XIII (Sept. 11, 2010).

Scope of 1968 Chilean Sanitary Code (Decree No. 725)

Within Chile, the Ministerio de Salud y Servicios Sanitarios (Ministry of Health and Health Services) is the agency with authority for managing chemicals and waste. This ministry is administered via articles 90 through 93 of the 1968 Chilean Sanitary Code (Decree No. 725), specifically through regulation on the manufacture, import, distribution, transport, sale, and disposal of such items.[14]

Regulation of Substances/Products

Although Chile does not maintain a labeling and classification regime similar to the European model, Chilean Decree No. 594/2003, Regulation on Basic Sanitary and Environmental Conditions in Workplaces (Reglamento sobre Condiciones Sanitarias y Ambientales Básicas en los Lugares de Trabajo), establishes occupational exposure limits (OEL) for certain chemicals in the workplace that can cause narcotic effects. The decree also regulates SDSs. Two groups share responsibilities for health and safety in the Chilean workplace: the Ministry of Labor and Social Security and the Ministry of Health.[15]

Chilean OEL determinations do not parallel those used by the U.S. Occupational Safety and Health Administration (OSHA). Chilean OELs are based on an eight-hour workday, but with a total of 48—not 40—working hours per week.[16] Chile also uses a unique standard, the temporary permissible limit (TPL), which is the maximum exposure value allowed during a time-weighted average (TWA) period of 14 continuous minutes. This limit may not be exceeded at any time during that 14-minute time frame. The TPL differs from the allowable absolute limit (AAL), the maximum value for airborne concentrations of these chemicals measured at any point during working hours. Exposure limits for the workplace are provided in the Reglamento.[17]

---

14. *In Recent Years, Chile Has Implemented Wide Reaching Chemical Regulations*, supra note 9.
15. Am. Soc'y Safety Eng'rs, Country Overview (Chile), at 17 (Winter 2006), http://www.asse.org/practicespecialties/international/docs/Chile.pdf.
16. Chile Decree 594/2003, *available at* http://www.inh.cl/norma_iso/CALIDAD/DOCUMENTOS%20EXTERNOS/7%20SHYSMAT/8%20DEC.%20SUP.%20N594.pdf.
17. Biblioteca del Congreso Nacional de Chile, Ley Chile, Reglamento sobre Condiciones Sanitarias y Ambientales Básicas en los Lugares de Trabajo, http://www.leychile.cl/Navegar?idNorma=167766.

Several other decrees supplementally regulate the production, use, sale, and distribution of certain organic mixtures known to have specific occupational health effects. Decree No. 144 of July 26, 1985, Organic Solvents Harmful to Health (Los Solvents Orgánicos Nocivos para la Salud), requires that such organic solvents must print the phrase "The prolonged inhalation of this product produces irreparable brain damage" on the product label.[18]

Reporting and Recordkeeping Requirements

Chilean SDSs are governed by Decree No. 594/2003 and the Regulation of Health Conditions and Basic Environmental Workplaces (Regulación de las Condiciones de Salud y Lugares de Trabajo Básico del Medio Ambiente). The *regulación* requires SDSs, in Spanish, to be available placed in all areas in which hazardous substances are stored. A secondary decree, Norma Chilena (NCh) No. 2245 (2003), issued by the Chilean National Standards Institute, established the requirements for the content and the order of sections in both SDSs and labels.[19] These decree requirements parallel the 1994 ISO standard 11014-1194 (Safety Data Sheets for Chemical Products) and the EU's Directive 2001/58/EC.

Chile has expressed its intent to adopt the Globally Harmonized System of Classification and Labelling of Chemicals (GHS) of U.S. OSHA. In 2008, Chile published its National Security Policy for Chemicals, which, among other aspects, sets out goals to comply with GHS.[20] Also in 2008, the Draft Regulation for the Storage of Hazardous Chemicals (Anteproyecto de Reglamento de Almacenamiento de Sustancias Químicas Peligrosas) was also released for public consultation. Finally, during the meeting of the Committee of Experts on the Globally Harmonized System of Classification and Labelling

---

18. Chile Decree 144, Organic Solvents Harmful to Health (May 10, 1985).
19. Chile Decree 594/2003, *available at* http://www.inh.cl/norma_iso/CALIDAD/DOCUMENTOS%20EXTERNOS/7%20SHYSMAT/8%20DEC.%20SUP.%20N594.pdf.
20. *In Recent Years, Chile Has Implemented Wide Reaching Chemical Regulations*, *supra* note 9.

of Chemicals in Geneva July 4–6, 2012, a project to strengthen the ability to implement GHS in the country was approved.[21]

IMPORTS AND EXPORTS

Companies seeking to import or manufacture substances defined as "hazardous substances" under part II, section b of Chilean Circular No. 2/C 152 (1982) must first obtain authorization from the Ministerio de Salud y Servicios Sanitarios prior to the item(s) clearing Chilean customs. "Hazardous substances" are defined as "those substances having the characteristic of corrositivity, irritant, flammability or combustibility, explosivity as well as radioactive substances which represent a risk to the health, safety or well-being of humans or animals."[22] In addition, Chile's Law 18.164 (1982) prohibits toxic or dangerous substances, as well as substances used in food and cosmetics, from clearing Chilean customs unless accompanied by a certificate from the health authority.[23]

5.3
COLOMBIA

Colombia is a member of the Andean Community (Comunidad Andina, or CAN) (Bolivia, Colombia, Ecuador, and Peru). This group is a self-described "Community of four countries that decided voluntarily to join together for the purpose of achieving more rapid, better balanced and more autonomous development through Andean, South American and Latin American integration."[24] Therefore, a member nation subscribes to the directives of the community, in addition to

---

21. U.N. ECON. COMM'N FOR EUROPE, REPORT OF THE SUB-COMMITTEE OF EXPERTS ON THE GLOBALLY HARMONIZED SYSTEM OF CLASSIFICATION AND LABELLING OF CHEMICALS ON ITS TWENTY-THIRD SESSION (July 4–6, 2012), *available at* http://www.unece.org/trans/main/dgdb/dgsubc4/c4rep.html [hereinafter GHS REPORT].
22. Chilean Circular No. 2/C 152 (1982), pt. 2, § b.
23. *In Recent Years, Chile Has Implemented Wide Reaching Chemical Regulations*, *supra* note 9.
24. Comunidad Andina, About Us, http://www.comunidadandina.org/ingles/who.htm.

having its own national legislation. Colombia does not have a formalized national inventory for the management of chemicals.

Regulation of Substances/Products

Authority to Regulate Inventory/Listed Substances or Products

Although Colombia does not have a formalized national inventory for the management of chemicals, the national Health Code (Title III of Law 9 of January 24, 1979) allows the Colombian Ministry of Health (Ministerio de Salud) to prohibit or restrict the import, manufacture, transport, storage, sale, or use of a substance or product if it is believed to pose a public health danger, which is generally seen to encompass chemical substances.[25] Law 9 is administered by the National Institute for the Surveillance of Food and Medicines (Instituto Nacional de Vigilancia de Medicamentos y Alimentos, or INVIMA).

Prioritization and Risk Assessments

With regard to SDSs, Colombia observes Article 8 of Law No. 55/1993, which outlines the requirements for firms handling substances determined to be dangerous, as well as Norma Técnica Colombiana (NTC) 4435, which mirrors the ANSI Z400.1 guidance.[26] Companies subject to the law or to NTC 4435 must develop and provide an SDS containing specific information on the identity, manufacturer, classification, risk, safety precautions, and emergency procedures for the substance(s).[27] In addition, the chemical or common name used to identify the substance must be the same as the one that appears on the label. Spanish is the preferred language, but English is also permitted.[28]

---

25. Instituto Nacional de Vigilancia de Medicamentos y Alimentos, https://www.invima.gov.co/.
26. F. Zanatta, GHS Definitions and Concepts, presentation at the Responsible Care Latin America Symposium (June 17–18, 2010), http://canais.abiquim.org.br/reachpdfs/Reach_7_Fernando_Zanatta.pdf.
27. Chemicals Recommendation, No. 177 of the International Work Conference on Safety in the Use of Chemicals at Work (June 25, 1990). As nationalized in Law 55/1993, *available at* http://www.ilo.org/ilolex/cgi-lex/convde.pl?R177.
28. Id.

Neither Colombia nor any other Andean Community member has adopted GHS.[29] Each country recognizes the importance of the program and is considering its adoption. In concert, the U.N. Committee of Experts on the GHS approved in 2012 a project to strengthen the capacity for implementing GHS in Colombia.[30]

Furthermore, the Ministerio de Salud, acting under the authority of Article 133 of Law No. 9, may regulate the classification of dangerous substances; set requirements concerning information provision, packaging, transport, and labeling; and establish other regulations to prevent harm. Generally, however, the country adheres to the U.N. guidelines for the classification, packaging, and labeling of dangerous goods and applies these requirements to any substance defined in the national legislation as "dangerous" under Law No. 55/1993.[31] According to the law, such labels must be easily understood by those individuals handling the dangerous substances and must provide information regarding classification, hazard, and all necessary safety precautionary measures.[32] Furthermore, toxic products must display the skull and crossbones, the word "poison" in Spanish, and information regarding usage and antidotes.[33] Where no antidotes exist, such products cannot be licensed.

Imports and Exports

Because the production and export of illegal narcotics is of substantial concern in Colombia, the country adheres to Community Decision 602. This decision is intended to "enhance control and surveillance of import, export, transport and any other type of transactions at the Andean level and from other countries, of the chemical substances included in the Basic Harmonized List for Community Use, identified in Annex I."[34]

---

29. U.N. Econ. Comm'n for Europe, GHS Implementation, http://www.unece.org/?id=25735.
30. GHS Report, *supra* note 21.
31. Chemicals Recommendation, No. 177 of the International Work Conference on Safety in the Use of Chemicals at Work.
32. Id.
33. Id.
34. Decision 602, Andean Regulation for the Control of Chemical Substances Used in the Illegal Manufacture of Narcotic Drugs and Psychotropic Substances, http://www.comunidadandina.org/ingles/normativa/D602e.htm.

5.4

VENEZUELA

Venezuela was previously a member of the Andean Community (Comunidad Andina, or CAN) but resigned its membership on June 17, 2006, effective at the close of 2011.[35] The country had planned to exit CAN in favor of Mercosur (Mercado Común del Sur, the "Common Market of the South"), composed of Argentina, Brazil, Paraguay, and Uruguay, but a date for formal entry was delayed as the country waited for the Paraguayan Senate to approve its accession as a full member.[36] This approval occurred on July 31, 2012. Although the country does have key regulations in place regarding pesticides and SDSs, generally the country's chemical regulatory enforcement is famously "fluid."

Venezuela's chemical regulations and policy derive from a combination of global, regional, and domestic regulations. Globally, Venezuela is party to three multilateral environmental agreements (MEA): the Rotterdam Convention (1998, which focused on Prior Informed Consent, or PIC), the Basel Convention (1989, which addressed hazardous waste and e-waste exports), and the Stockholm Convention (2001, which emphasized persistent organic pollutants, or POPs).[37] Specifically, the country is bound to Annex III of the Rotterdam Convention, which requires exporting parties/countries to inform importing parties/countries of potential human or environmental hazards of 40 chemicals: 25 pesticides, four severely hazardous pesticide formulations, and 11 industrial chemicals.[38] Venezuela's Ministry of the Environment has declared that it will enact its own version of the EU's Waste Electrical and Electronic Equipment (WEE) regulation in the near future, but it has yet to put forth any enabling legislation or other proposals.[39]

---

35. *Venezuela: A Weak Regulatory Environment Is Accompanied by a Poor Economic & Political Setting*, CHEMS. CONTROL WATCH, Feb. 2011.
36. *Venezuela's Entry into Mercosur Must Wait*, EL UNIVERSAL (CARACAS), (Dec/21.2011), http://www.eluniversal.com/economia/111221/venezuelas-entry-into-mercosur-must-wait.
37. L. Jinhui, Z. Nana & C. Ying, *Perspective on Synergy of Chemicals and Hazardous Waste Management*, PROC. INT'L CONF. SOLID WASTE TECH. MGMT. 1221–28 (2008), EBSCO Online Database Environment Complete.
38. Rotterdam Convention, Annex III Chemicals, http://www.pic.int/Default.aspx?tabid=1132.
39. K. Ripley, *Perspectives on Electronic Waste in Latin America and the Caribbean* (2009), available at http://www.slideshare.net/ioman01/perspectives-on-electronic-waste-in-latin-america-and-the-caribbean.

REGULATION OF SUBSTANCES/PRODUCTS

Venezuela's Law on Dangerous Substances, Materials, and Wastes (Ley sobre Sustancias, Materiales y Desechos Peligrosos), published on November 13, 2001, is the country's first regulation for the regulation of hazardous substances.[40] Specific authority for handling of substances, hazardous materials, and waste is granted in articles 65 and 66.[41] Although the law is fairly broad in its scope, the key required elements that deal directly with chemical substance management are as follows:

1. Proof of registration (given by the Directorate General of Environmental Quality);
2. A descriptive report of the activity to be developed;
3. A list of substances, materials or hazardous waste, attaching:
    a. For hazardous substances: Material Safety according to 'Venezuelan COVENIN [Comisión Venezolana de Normas Industriales] Standard 3059: Hazardous Materials: Data Sheet Material Safety (HSDM)' [sic];
    b. For materials and hazardous waste: physical and chemical characterization of recyclable materials and hazardous wastes, indicating classes and risk levels performed by a laboratory approved by the MARN (Environment and Natural Resources Ministry);
4. Plan of periodic staff training, in accordance with criteria Norma COVENIN 3061-Hazardous Materials;
5. Provide training to those who handle, store and/or transport hazardous materials;
6. Plan for maintenance of vehicles, materials and equipment to provide the service;
7. Risk analysis of the activities involved in managing substances, materials and hazardous wastes and a specification of the levels of the same;

---

40. EXTRAORDINARY OFFICIAL GAZETTE (VENEZUELA) no. 5.554 (Nov. 13, 2001).
41. Id.

8. Emergency and Contingency Plans (COVENIN Rules 2226 and 2670), should include a system or contract for service and care 24 hours a day for unforeseen emergencies.[42]

Other regulations address the management of chemical substances. Among these regulations are Decree No. 2635, Norms on the Control of Recovery of Hazardous Materials and Handling of Hazardous Waste (Normas sobre el Control de Recuperación de Materiales Peligrosos y Manejo de Residuos Peligrosos, 1998);[43] Resolution No. 40, Requirements for Registration and Authorization of Operators of Dangerous Substances, Materials and Waste (Requisitos para el Registro y Autorización de Operadores de Sustancias Peligrosas, Materiales y Residuos, 2003),[44] which mandates registration of activities that are likely to degrade the environment; and the Organic Law against Illicit Trafficking and Consumption of Narcotic and Psychotropic Substances (La Ley Orgánica contra el Tráfico Illicito y el Consumo de Sustancias Estupefacientes y Psicotrópicas, 2005), whose articles 125 and 131–137 require registration of any "person or entity involved in the production, manufacture, preparation, use, processing, storage, marketing, brokerage, export, import, transport, waste and any other transaction where controlled chemicals are involved."[45] Finally, Venezuela manages the classification and labeling of chemicals via COVENIN 3060 (2002), which incorporates the nine-group U.N. classification system.[46]

KEY BUSINESS ISSUES

Commonly Encountered Issues

Venezuela has a standard for SDSs, initially set forth in COVENIN Standard 3059, Hazardous Materials: Material Safety Data Sheets [sic] (Materiales Peligrosos: Hoja de Datos de Seguridad de Los Materiales

---

42. Id.
43. OFFICIAL GAZETTE (VENEZUELA) no. 5245 (Aug. 3, 1998).
44. OFFICIAL GAZETTE (VENEZUELA) no. 37.700 (May 23, 2003).
45. OFFICIAL GAZETTE (VENEZUELA) no. 38.337 (Dec. 16, 2005).
46. Norma Venezolana, COVENIN 3060:2002, Materiales Peligrosos. Clasificación, Símbolos y Dimensiones de Señales de Identificación (1ra Revisión), *available at* http://www.sencamer.gob.ve/sencamer/normas/3060-02.pdf.

(HDSM) (2002, 2nd rev. 2006).[47] This standard generally follows ISO 11014-1: 1994, and mandates a 16-section format. Appendices A and B are particularly useful to view the formatting that is required for a compliant SDS.[48] COVENIN has been replaced, however, by the Fondo para la Normalización y Certificación de la Calidad (FONDONORMA). FONDONORMA is a nonprofit foundation composed of trade and industry associations, large private companies, and individuals. Nonetheless, Venezuelan standards developed by FONDONORMA still are called "COVENIN norms."[49] This transition has been the source of confusion.

Practical Tips

The general political climate and governmental structure in the country present unique challenges for companies attempting to operate in Venezuela. The government has taken a much more nationalistic role in recent years, requiring companies entering the Venezuelan market to enter into 50/50 joint ventures with the government, and/or completely nationalizing companies such as oil, gas, and electricity. As such, the regulatory environment in general is decidedly challenging to navigate.

Trends

At present, Venezuela has not expressed any intention to implement GHS. The country, however, has asked for GHS capacity building assistance from UNITAR-ILO.[50]

---

47. Norma Venezolana, COVENIN 3059:2002, Materiales Peligrosos Hoja de Datos de Seguridad de Los Materiales (HDSM) (1fra Revisión), *available at* http://www.sencamer.gob.ve/sencamer/normas/3059-02.pdf.
48. *Id.*
49. Jerry Miller, Bureau Veritas Consumer Prod. Servs., Latin American Laws and Regulations (2007), http://www.icphso.org/oldfiles/2007pdf/presentations/LatinAmericaLawsRegs.pdf.
50. UNITED NATIONS INST. FOR TRAINING & RESEARCH, CHEMS. & WASTE MGMT. PROGRAMME. REPORT ON THE PREPARATION FOR GHS IMPLEMENTATION IN NON-OECD COUNTRIES (Nov. 2007), http://www2.unitar.org/cwm/publications/cw/ghs/UNITAR_ILO_OECD_Questionnaire_report_final.pdf.

5.5
URUGUAY

The base law for chemicals management in the República Oriental del Uruguay (Uruguay), a Mercosur member, is the General Law of the Environment. In addition, article 47 of the Uruguayan Constitution states that protection of the environment is in the general interest. Thus, "the protection of the environment is declared to be in the general interest; as are: the production, import, export, transport, packaging, labelling, storage, distribution, commercialization, use and disposal of all chemical substances to ensure adequate levels of environmental protection against the adverse effects derived from normal use, accident or waste."[51] Specific laws also exist for the management of pesticides (Regulatory Dispositions on the Labeling of Phytosanitary Products), explosives, medicines, fuels, ozone-depleting substances, lead batteries, and others.[52] Initially, Decree No. 406/988 (June 3, 1988) set out the general requirements for managing chemical risks, but the Uruguayan authorities subsequently determined that the increase in such risks required more specific and technical regulation.[53] Of particular note, article 10 mandates an information and training system for the workers, which must contain (in part) "clear and simple language" about the risks from the hazardous chemical agents, prevention and protection measures to be taken, the results of any risk assessment(s), and access to any SDSs.

Uruguay is party to a host of international conventions, such as the Montreal Protocol, the Kyoto Protocol, the Stockholm Convention, and the United Nations Framework Convention on Climate Change, among others.

Electronic databases in Uruguay are not well developed. The Parlamento website contains all laws enacted since 1935, legislative

---

51. P. Sandonato de Leon, *A Guide to Uruguay's Legal System and Research*, GLOBALEX (Jan./Feb. 2010), http://www.nyulawglobal.org/globalex/uruguay.htm.
52. National Workshop on the Development of Legal and Institutional Infrastructures and Sustainable Financing Options for the Sound Management of Chemicals in Uruguay (Dec. 15–17, 2010), http://www.unep.org/chemicalsandwaste/Portals/9/Mainstreaming/LIRA-Country%20Workshop/Uruguay_Proceedings.pdf.
53. Uruguay Decree No. 406/988 (June 3, 1988), updating the regulatory provisions on occupational safety, hygiene, and health to bring them into line with the new conditions of the world of work.

procedures, and related information, and most codes can be found (in Spanish) on the website of the Uruguayan Parliament.[54]

Reporting and Recordkeeping Requirements

Article 1 of Decree No. 307/009 "establishes the minimal mandatory provisions in the field of health protection and safety for the workers, against risks related to chemical agents during the workday." Under article 4, it is the responsibility of the employer to identify "hazardous chemical agents" (defined in article 2.5) in the workplace, and if they exist, to conduct a risk assessment and subsequently to develop a prevention plan.[55] The risk assessment must include aspects such as the dangerous properties of the chemical agents; the quantities used and stored; the type, level, and duration of the exposure; and the biological and environmental limit values.[56] The risk assessment should be reviewed and updated as appropriate, and particular attention should be given to minimizing or eliminating the use of hazardous chemicals in the workplace. Finally, if the risk assessment reveals health risk(s) for the workers, a health surveillance, conducted by a doctor specializing in occupational medicine, should be performed.[57]

Enforcement and Penalties

Chemical product labels must contain at a minimum, according to Annex I-1, these elements: the hazard pictogram, signal words, a hazard statement, precautionary statements and precautionary pictograms, a product identifier, supplier information, main first aid measures, and an expiration date, if applicable.[58] The label must be firmly adhered to the container and remain with it, as well as being readable "in transport vehicle, at storage or any use conditions."[59] Additionally, Annex I,

---

54. Parlemento del Uruguay, http://www.parlamento.gub.uy/palacio3/index1280.asp?e=0&w=1600.
55. Ministry of Labor & Soc. Welfare, Presidential Decree 307/009 (July 3, 2009), http://archivo.presidencia.gub.uy/_web/decretos/2009/07/T1397%20.pdf.
56. Id.
57. Id.
58. Id.
59. Id.

section I-3.1 mandates the label material (paper or PVC) and weight (at least 80 g/m² or at least 135 g/m², respectively), and table I-1 mandates the minimum sizes of the labels, based on the container size.

Uruguay adopted GHS July 3, 2009, under Decree No. 307/009, and the system has been fully operational in the country since the close of 2010.[60] "The previous system can still be used for substances up to December 31, 2012, and for mixtures up to December 31, 2017. After the transition periods, only GHS can be used."[61]

Annex I of Decree No. 307/009 addresses GHS, declaring that "The labeling of chemicals will be mandatory [sic] carried out according to the Directives of the Globally Harmonized System . . ." and Annex II of the decree directs the format and content of the SDS.[62] The SDSs, required to be in Spanish, should be submitted with the first shipment of the product and after each/any modifications.[63] According to part 3 of table 1.5.2 of the decree, "In the information about the components, the regulations of the competent authority on confidential commercial information prevail over the regulations related to the identification of the product."[64]

5.6
Peru

The Ley General de Salud No. 26842 (General Health Law, Law 26842, July 9, 1997) defines the regulation of "dangerous substances and products." It sets forth the requirements for the management of such products, and the Ministry of Health is required to establish norms for their classification, as well as packaging, transportation,

---

60. Id.
61. GHS Legislation, GHS Implementation in South America, http://www.ghslegislation.com/south-america/ghs-implementation-in-south-america/.
62. Presidential Decree 307/009 at Annex I.
63. Id. at Annex II.
64. Id. at tbl.1.5.2.

and labeling requirements.⁶⁵ In addition, regulations are published in *El Peruano, Diario Oficial (Peruvian Official Daily)*.⁶⁶

Imports and Exports

Labeling requirements for industrial products that were manufactured outside of Peru may be found in Supreme Decree No. 1-84-ITI/IND (January 18, 1984).⁶⁷ As per article I of the decree, such products must be labeled with, at a minimum, the brand name (in the case of consumer products), the country of manufacture for certain categories (e.g., medicines), and the expiration date. Failure to adhere to the decree could result in rejection of the good(s) by customs.

Key Business Issues

SDSs in Peru are governed by Technical Standard No. G 050. Although this is a construction industry standard, section 19.1 lays out the requirements for an SDS for chemical substances and directs that storage and project personnel must be trained on SDS usage.

5.7
Paraguay

Paraguay, a Mercosur member, maintains a series of obligatory and voluntary standards, much the same as multiple other countries. In general, the Instituto Nacional de Tecnología, Normalización y Metrologia (INTN, www.intn.gov.py) is responsible for developing Paraguayan standards.⁶⁸ Official decrees in Paraguay are published on the website of the Presidencia de la República del Paraguay

---

65. Ley General de Salud No. 26842, *available at* http://www.wipo.int/wipolex/en/details.jsp?id=8244.
66. *El Peruano, Diario Oficial*, http://www.elperuano.pe/Edicion/.
67. Supreme Decree 1-84-ITI/IND, *available at* http://www.glin.gov/download.action?fulltextId=51850&documentId=19063&glinID=19063.
68. Latin American Laws and Regulations, *supra* note 49.

(http://www.presidencia.gov.py/). Regulations are published in the *Gaceta Oficial de la República del Paraguay*.

REGULATION OF SUBSTANCES/PRODUCTS

As in Peru, Paraguayan regulations require the country of origin to be presented on labels for imported and domestic goods, and expiration dates on some consumer goods.[69]

Paraguay has several relatively unique regulations regarding chemical management. For example, under Decree No. 7.505/2011, sodium tripolyphosphate is effectively banned in domestic and imported cleaning products as of November 18, 2011. Manufacturers and importers must provide objective evidence that their products are free from the compound. Additionally, on August 2, 2012, the Instituto Nacional de Alimentación y Nutrición (National Food and Nutrition Institute) and the Ministerio de Salud Pública y Bienestar Social (Ministry of Public Health and Social Welfare) jointly submitted a draft resolution to the World Trade Organization (Committee on Technical Barriers to Trade) that would prohibit the manufacture, import, and marketing of baby bottles containing bisphenol A.[70]

Paraguayan Decree No. 14.390/92 (July 28, 1992), Reglamento General Técnico de Seguridad, Higiene y Medicina en el Trabajo, defines "hazardous chemicals" as those that have at least one of the following effects: narcotic; toxic, with slight, acute, or chronic effects on any organ; corrosive or irritant; allergic or sensitizing; carcinogenic, teratogenic, or mutagenic reproductive effects; flammable; explosive; combustible; or reactive.[71] Such chemicals must be clearly marked for those working with them, and they must be labeled with the name of the product; the commercial name; the appropriate danger symbol; the nature of the associated risks; safety precautions to be taken; the name, address, and telephone number of the supplier; and a statement that SDSs are available from the employer.[72]

---

69. Id.
70. Members discuss guidelines for trade-friendly regulation and STOP sign for 'junk food.' http://www.wto.org/english/news_e/news13_e/tbt_13mar13_e.htm.
71. Reglamento General Técnico de Seguridad, Higiene y Medicina en el Trabajo, *available at* http://www.leyes.com.py/resultados-busqueda.html?cx=006074493042265127504%3A6k cptmof0to&cof=FORID%3A10&ie=UTF-8&q=No.+14.390%2F92.
72. Id.

The country also has established a Comisión Nacional de Seguridad Química (National Chemical Safety Commission) through Decreto No. 21919.[73]

Trends

Although Paraguay has not implemented GHS at present, in April 2012, the secretary of the environment initiated the implementation of GHS in the country, in partnership with the National University of Asunción. No date has been set for the enactment.[74] Currently, the country sets out its SDS requirements also via Decree No. 14.390/92, directing a 16-section format.

5.8

Brazil

Fernando Tabet

The chemical industry market is flourishing in Brazil and growing rapidly. Although the market is ripe for expansion, the ability to place new chemical substances on the market is challenging due to the complexities of the registration process.

Regulation of Substances/Products

Main National Laws and Regulations on Chemicals

Brazil has no general or systemized codification of chemicals. The legislation and regulations in this area are not integrated and are oriented to specific types of chemicals, considering the respective associated uses,

---
73. Paraguay Decreto No. 21919, *available at* http://www.temasactuales.com/assets/pdf/gratis/PARchemSafComm.pdf.
74. R. Aguirre, *Paraguay to Implement Globally Harmonized System of Classification and Labeling of Chemicals*, Berkemeyer Att'ys & Counselors, Apr. 19, 2012, http://www.berke.com.py/news/news-details.aspx?p1=WHFPRlYvMDJMMFp2MHdvVWlVQ0dpTHBpTEp3U1FpWW50dk1tdTZweThrND01&tit=Paraguay+implement+Globally+Harmonized+System+Classification+Labeling+Chemicals.

such as transportation, agriculture, and hygiene.[75] Brazil has not developed a national inventory or a registration scheme for chemicals, and there are different authorities and levels of control in connection with different types of chemical products and their uses.

Combat against Traffic in Illegal Drugs (Federal Law No. 10,357/2001, Regulated by Federal Decree No. 4,262/2002)

Federal Law No. 10,357/2001 establishes monitoring and control rules on chemical products that could directly or indirectly be destined for the illegal production of narcotics, psychotropic substances, or drugs that cause physical or mental dependence. Monitoring and control actions are performed by the Federal Police Department. The full list of substances that are controlled by the Federal Police Department are described in the Ministry of Justice Administrative Edict No. 1,274/2003.

Individuals and legal entities that perform activities that are subject to control pursuant to Federal Law No. 10,357/2001 are obliged to obtain an operating license from the Federal Police Department. Such an operating license must be renewed on an annual basis. In addition, legal entities that perform the aforementioned activities must periodically provide reports on its operations to the Federal Police Department.

At the state level, in certain states such as São Paulo, a similar control is performed by the Civil Police.

Products Controlled by the Brazilian Sanitary Surveillance Agency (Federal Law No. 9,782/1999, Products Involving Risk to Public Health); Products Also Controlled by the Ministry of Agriculture, Livestock, and Food Supply; the Brazilian Institute of the Environment and Renewable Natural Resources; and State Environmental Agencies (Federal Law No. 7,802/1989, on Agrochemicals)

Pursuant to Federal Law No. 9,782/1999, the Brazilian Sanitary Surveillance Agency (the Agência Nacional de Vigilância Sanitária, or ANVISA) is authorized to regulate, monitor, and control products and services that involve risk to public health. The products that are expressly subject to ANVISA's control include, among others, the

---

75. *See* ABIQUIM (Associação Brasileira da Indústria Química), About Us, http://abiquim org.br/english/content.asp?princ=wwa&pag=whowe.

following, pursuant to the same law: human-use medicines; cosmetics, personal hygiene products, and perfumes; sanitation products used for hygiene, disinfection, or infestation combat in residential, health-care, or collective environments; and devices and reagents used in diagnostic procedures.

Federal Decree No. 3,029/1999 addresses the activities performed by ANVISA, including the registration of the products that are subject to its control.

ANVISA is also involved in the process of evaluating agrochemicals (pesticides, fungicides, and herbicides) for registration purposes before the Ministry of Agriculture, Livestock, and Food Supply (the Ministério da Agricultura, Pecuária e Abastecimento, or MAPA). Pursuant to Federal Law No. 7,802/1989, as regulated by Federal Decree No. 4,074/2002, agrochemicals, their components, and similar substances can only be manufactured, exported, imported, commercialized, and used if they are registered with the competent federal entity (MAPA, in this case), in accordance with the directives imposed by the federal agencies responsible for the sectors of health (ANVISA), environment (the Brazilian Institute of the Environment and Renewable Natural Resources, the Instituto Brasileiro do Meio Ambiente e dos Recursos Naturais Renováveis, or IBAMA), and agriculture (MAPA). Additionally, state legislation imposes registration and reporting requirements involving state environmental agencies.

Land Transportation of Hazardous Cargoes (Resolution No. 420/2004 of the Brazilian Agency of Terrestrial Transportation)
Resolution No. 420/2004 of the Brazilian Agency of Terrestrial Transportation (the Agência Nacional de Transportes Terrestres, or ANTT) addresses the land transportation (both highway and railroad) of hazardous products. The provisions of this resolution include requirements on the classification of hazardous products, labeling, packaging, segregation, identification of vehicles, and documentation, among others.

Transit of Hazardous Products in Port Facilities (Resolution No. 2,239/2011 of the Brazilian Agency of Waterway Transportation)
Resolution No. 2,239/2011 of the Brazilian Agency of Waterway Transportation (the Agência Nacional de Transportes Aquaviários, or ANTAQ) establishes procedures for the safe transit of hazardous products through port facilities, incorporating aspects related to

occupational health and safety, physical integrity of installations in port facilities, and protection of the environment, extracted from the International Maritime Dangerous Goods Code, the International Ship and Port Facility Security Code—regulations of the International Maritime Organization (IMO)—and other regulations adopted at the national level. Pursuant to the ANTAQ resolution, the entities that are responsible for hazardous products must provide the port authority with all the documents and information about the products.

Globally Harmonized System: Brazilian Standard NBR 14725:2009

Brazilian Standard NBR 14725:2009 formally introduced Brazil to the Globally Harmonized System of Classification and Labelling of Chemicals (GHS). The standard was approved by the Brazilian Association of Technical Standards (Associação Brasileira de Normas Técnicas, or ABNT), which is the organization responsible for technical standardization in Brazil. This standard made Brazil the first South American country to implement the GHS in full (the first version of this standard was published by ABNT in 2001 and entered into force on January 28, 2002).

The standard is divided into four parts: part 1, Terminology (a lexicon of terms common in hazard communication); part 2, Hazard Classification System (classification criteria for hazard attribution in different GHS classes and categories); part 3, Labelling (relationship between hazards and label elements); and part 4, Safety Data Sheet (format and content of a 16-section SDS). Pursuant to Federal Decree No. 2,657/1998, the Decree on which the International Labor Organization Convention No. 170 (safety in the use of chemicals at work) and Regulation No. 229/2011 of the Ministry of Labor and Employment are based, every chemical product that is considered hazardous based on the GHS must have a SDS (the manufacturer and the importer are responsible for making the SDS available to their clients or to the users of the product).

BIOTECHNOLOGY AND THE ACCESS TO GENETIC RESOURCES— PROVISIONAL MEASURE NO. 2,186-16/2001

Biotechnology activities in Brazil involving the molecular manipulation of genetic resources originated from native species are subject to stringent regulations. Access to national genetic resources is subject to the requirements imposed by Provisional Measure No. 2,186-16/2001 and numerous regulations enacted by the Genetic Heritage Management Council (Conselho de Gestão do Patrimônio Genético, or CGEN). These requirements include the prior obtainment of an authorization from CGEN to access national genetic resources and associated traditional knowledge (information or individual or collective practice adopted by a local or indigenous community, with potential or effective value, associated with genetic resources) for the performance of research, bioprospection, defined as exploratory activity that identifies a genetic resource component and information about traditional knowledge, with potential commercial use, or technological development activities. Once economical perspective in the use of a certain genetic resource is identified, benefit sharing with the provider of the genetic resource must be addressed pursuant to a mandatory Genetic Resource Use and Benefit Sharing Agreement. The legislation on the access to genetic resources in Brazil is currently being reviewed and the approval of a new law on this matter is expected, seeking less complexity and more effectiveness of the applicable legal regime.

Enforcement and Penalties

In Brazil, noncompliance with laws and regulations on chemicals may result in liabilities on three different levels at the same time: civil, administrative, and criminal. Civil liability can involve the payment of indemnities (for both material and moral damages) and/or the obligation to do or not to do something. If an environmental damage is involved, Brazil adopts a strict and joint liability regime, addressing both the direct and indirect polluter.

As for administrative liability, penalties (such as warnings, fines, and interdiction of activities) can be imposed against both individuals and legal entities by the controlling authority (environmental agencies, for instance).

Criminal liability may occur depending on the type of product and the consequences of the conduct, based on culpability. When toxic, hazardous, or noxious substances are involved, not complying with regulations on trading, packaging, storage, transportation, and use, among other activities, regardless of the existence of actual damage, may characterize an environmental crime, which is punishable with fines (against both individuals and legal entities) and imprisonment (against individuals) and other sanctions.

Trends

Brazil's growth and the prospects for its continuation present great potential while at the same time imposing significant challenges. There are challenging tax issues that require careful analysis before companies decide on conducting business in Brazil. Industry and the government are working to ensure the safety and well-being of Brazil's citizens and environment while also supporting growth of the chemical business sector. It would not be surprising to see Brazil begin to reevaluate fundamentally its chemical management program in the near term.

6

Europe

6.1

REACH

Leslie S. MacDougall and Ruth C. Downes-Norriss

Executive Summary

Overview

The Registration, Evaluation, Authorization, and Restriction of Chemicals (REACH) scheme, EC No. 1907/2006,[1] which took effect June 1, 2007, is making fundamental and complex changes in the way chemicals are regulated and used throughout Europe.[2] REACH replaces some 40 legal instruments with a single regulation, the product of a consultation and negotiation process that took years to culminate. All Member States within the European Union (EU) and European Economic Area (EEA) are included in REACH, creating a single chemical regulatory regime across Europe.

Both new and existing chemicals produced in, or imported into, EU countries in amounts of one metric ton or more per year must be registered. An 11-year phase-in period, now underway until 2018,

---

1. The most recent consolidated version as of this writing is available at http://eur-lex.europa.eu/LexUriServ/LexUriServ.do?uri=CONSLEG:2006R1907:20121009:EN:PDF.
2. REACH replaced the prior scheme for notifying chemical substances to be placed upon the (European Union (EU) market, the Notification of New Substances (NONs) scheme (EC Directive 67/548/EEC).

places priority on high volume and chemicals that present the greatest hazards to human health and/or the environment. All registered substances are subject to evaluation screening, and those deemed the most dangerous are further subject to authorization, with possible restrictions on use.[3] The registration and authorization process is overseen by the Helsinki-based European Chemicals Agency (ECHA), which was established pursuant to REACH Article 75 and is composed of regulatory specialists and scientific experts from throughout the EU.

REACH transfers the burden of proof for testing and evaluating chemical risks from the regulatory authorities to industry. In support of this process, and to avoid duplication of (especially) vertebrate animal testing, REACH provides for data sharing and data compensation, principles that also are incorporated in somewhat different form in the Federal Insecticide, Fungicide, and Rodenticide Act (FIFRA) statute governing pesticide registration and use in the United States and to a much lesser extent under the Toxic Substances Control Act (TSCA) (although data compensation provisions have seldom, if ever, been invoked under TSCA). It also includes obligations of duty of care for the industry and of communication to the public about dangerous substances in products.

Basic Provisions

REACH is an ambitious chemical control initiative, imposing significant new duties on industry to manage effectively the risks from chemical manufacture, import, and use and to provide greater safety information on the substances. The varied objectives sought to be achieved are set out in Article 1(1) as follows:

> The purpose of this Regulation is to ensure a high level of protection of human health and the environment, including the promotion of alternative methods for assessment of hazards of substances, as well as the free circulation of substances on the internal market while enhancing competitiveness and innovation.

---

3. While all substances are candidates for review, only approximately 5 percent of substances will be reviewed and candidates are selected at random with emphasis placed on those substances with the greatest hazards.

Key provisions and elements include the following:

- *Registration:* Central to REACH, the registration requirement applies to all chemicals, unless considered to be registered already or otherwise exempt, that are manufactured or imported into the EU at or above the one metric ton per year threshold. Registration entails, in part, generating or citing to substance-specific health and safety study data, the results of which are presented in a technical dossier for review, along with a Chemical Safety Report (CSR) and additional information intended to communicate other vital information about the substance and facilitate its safe use. For "phase-in" chemicals, a term that refers to existing substances, registration is a phased process. By preregistering these substances between June 1 and December 1, 2008, manufacturers and importers (and other entities in certain cases) obtained the benefit of extended registration deadlines, ahead of which they are entitled to continue manufacture or import of the chemical substance involved. Article 10 and Annexes IV to VIII of the regulation specify the information needed for registration. The general technical, commercial, and administrative information needed for all registrations for the technical dossier is specified in Annex IV.
- *Burden of proof:* Prospective registrants have the burden of establishing that the chemical substances they seek to register meet the requisite levels for safety to assure the protection of human health and the environment. This is in contrast to TSCA, which places the burden on the government to show that a substance is harmful. The REACH approach does not mean that registrants are spared the expenditures of financial and other resources necessary to underpin their registration dossiers, but it is a fundamental distinction in approach.
- *Data sharing and data compensation:* Although the process differs in certain significant respects from data sharing and data compensation under FIFRA, these principles and practices are essential to REACH's implementation. The regulation articulates a commitment to the avoidance of vertebrate animal testing insofar as possible and to the avoidance of duplicative testing in general. To encourage and facilitate efficiencies in data acquisition and (if needed) development, REACH

requires registrants of similar substances to join together in a Substance Information Exchange Forum (SIEF). A SIEF is composed of all preregistrants and late preregistrants of a substance that have an interest in the substance. Active members will typically join together to ensure that successful registration of a substance, including technical dossier preparation activities, is completed. Where participants in a SIEF or other arrangement cannot reach agreement on cost sharing, REACH provides an avenue for ECHA to become involved in the process, although this occurs infrequently. Still, this option is in contrast to FIFRA's data compensation provisions, under which the U.S. Environmental Protection Agency (EPA) is entirely outside the dispute resolution process.

- *Authorization and restriction:* REACH provides that substances that pose dangers or heightened concern due to their properties and/or risks can be listed and managed separately through procedures that impose rigorous controls, including bans where justified, on their manufacture, sale, and use.
- *Information dissemination/protection of confidential business information (CBI):* REACH reflects a commitment to the wide dissemination of information to the public, online and free of charge. The regulation specifies what types of data will be made publicly available with no possibility to claim confidentiality for them. It also specifies what types of data may be protected as a matter of commercial interest if the affected registrant has made and persuasively supported a claim of confidentiality. Certain other types of information specified in the regulation are expressly characterized as normally protected from disclosure as a matter of business confidentiality.

Scope of REACH

All chemical substances, on their own or in mixtures or articles, are subject to REACH under Article 6, unless specified exemption criteria are met or unless they are considered as already registered. REACH defines a substance as follows:

> A chemical element and its compounds in the natural state or obtained by any manufacturing process, including any additive necessary to preserve its stability and any impurity deriving from the process used, but excluding

any solvent which may be separated without affecting the stability of the substance or changing its composition.[4]

Under this definition, all substances supplied either neat or in a mixture or preparation (defined as a mixture or solution composed of two or more substances[5]) must be registered in compliance with REACH. Mixtures are not otherwise subject to REACH.

"Article," a term of art under REACH, is defined as "an object which during production is given a special shape, surface or design, which determines its function to a greater degree than does its chemical composition."[6] Articles per se do not require registration, but substances in articles must be registered if they are classified as dangerous, are intended to be released from the article, and are supplied at or above the one ton per year threshold.[7]

There are a number of categorical exclusions from REACH, and various other substances are exempted from registration requirements but not necessarily from other obligations that REACH imposes. The REACH regulation as a whole does not apply to radioactive substances, genetically modified organisms, substances in transit, or substances regulated by equivalent EU legislation (human and veterinary pharmaceuticals, food additives and flavorings, animal feed, and substances used in animal nutrition).

The many and diverse categories exempt from REACH registration include the specific substances known to be safe and covered in Annex IV of the regulation; substances covered in Annex V, such as chemical and other degradants, products of use or from reaction with additives; by-products; hydrates (if the anhydrous form is registered); nondangerous naturally occurring substances; natural gas, crude oil, and coal; food and food ingredients; waste and certain recycled materials; minerals ores and ore concentrates; substances necessary for defense purposes; polymers; and some monomers.[8] Intermediates are

---

4. REACH art. 3(1).
5. *Id.* art. 3(2); ECHA, GUIDANCE FOR IDENTIFICATION AND NAMING OF SUBSTANCES UNDER REACH AND CLP 21 (Version 1.3, Feb. 2014), http://echa.europa.eu/documents/10162/13643/substance_id_en.pdf.
6. REACH art. 3(3).
7. A product with an intended release (such as an ink cartridge) is considered to be a substance or mixture (as applicable) within a container. Although the container may meet the definition of an article, the substance or mixture intended for release must be registered.
8. Monomers bound into polymers that are present at <2% (w/w) in the polymer, or that are supplied at <1 ton per annum, are exempt from registration. By contrast, registration is required for monomers present at 32% (w/w) in the polymer and supplied at 31 ton per annum.

treated as a special category of chemicals, some but not all of which must be registered.

Substances used only for product and process-orientated research and development (PPORD) are exempt from registration for five years (extendable for a further five years on application in exceptional circumstances or ten years for substances used exclusively to develop human or veterinary medicines). The manufacturer or importer must submit to ECHA a PPORD notification, which contains basic information including substance identity, labeling, and quantity, and a list of customers that may use the PPORD substance. The substance can be used only by those customers and it cannot be supplied to the public.

Scope of Activities Subject to REACH

Under REACH Article 1(2), the provisions "apply to the manufacture, placing on the market or use of [the subject] substances on their own, in mixtures or in articles and to the placing on the market of mixtures." For purposes of REACH, "manufacturing" means "production or extraction of substances in the natural state," and "import" means "the physical introduction into the customs territory of the Community."[9] The phrase "placing on the market" is defined broadly to mean "supplying or making available, whether in return for payment or free of charge, to a third party. Import shall be deemed to be placing on the market."[10] Thus, the manufacture (including research and development (R&D)), use, import, or otherwise placing on the market of industrial chemicals all are activities subject to REACH. Distribution activities, however, are not covered within REACH's scope. A distributor is defined under REACH as "any natural or legal person established within the Community, including a retailer who only stores and places on the market a substance, on its own or in a mixture, for third parties."[11] The exclusion of distribution activities reflects the reality that many participants have minimal knowledge of the substances intended for distribution, compared with the other entities involved.

---

9. REACH arts. 3(8), 3(10).
10. *Id.* art. 3(12).
11. *Id.* art. 3(14).

Scope of Persons/Entities Subject to REACH

It follows from the above discussion of subject activities that manufacturers and importers must shoulder many key responsibilities of complying with REACH, including but not limited to obtaining registrations timely and submitting robust evidence to ECHA to support a determination that the substance involved does not adversely affect human health or the environment. The term "person" is not defined in REACH, but "manufacturer" means "any natural or legal person established within the Community who manufactures a substance within the Community." Similarly, "importer" means "any natural or legal person established within the Community who is responsible for import."[12]

For purposes of the REACH registration process, "registrant" is defined as "the manufacturer or the importer of a substance or the producer or importer of an article submitting a registration for a substance."[13] While manufacturers and importers must be "established within the Community" for REACH purposes, non-EU entities also may elect to participate in REACH by appointing an Only Representative (OR), pursuant to Article 8, in lieu of taking the necessary steps to become "established" in the EU. The basic OR relationship as authorized under Article 8 is as follows: "A natural or legal person established outside the Community who manufactures a substance on its own, in mixtures or in articles, formulates a mixture or produces an article that is imported into the Community may by mutual agreement appoint a natural or legal person established in the Community to fulfil, as his only representative, the obligations on importers under this Title." Appointment of an OR also enables the non-EU entity to maintain the confidentiality of product compositions, volume, uses, and downstream user details (client list) from an EU importer, who in many instances may be a competitor.

In instances where the confidentiality of actors in a supply chain is critical (such as registration, data sharing, and disputes), any or all actors (manufacturers, importers, or downstream users) may appoint a third-party representative (TPR). The TPR acts upon the instruction of the appointee company and represents the company's interest

---

12. *Id.* arts. 3(9), 3(11).
13. *Id.* art. 3(7). It should be noted that generally a single registration will cover all parts of the supply chain. Downstream users of a registrant must be identified in the registration and need to obtain documentation from their supplier to support their downstream user status. Such documentation can be shared as needed with the Competent Authority representative upon inspection.

in discussions with other supply chain actors. The TPR does not adopt the obligations of the registrant as the OR does but provides a level of confidentiality in the registrant's interactions with other legal entities.

NOTIFICATION/REGISTRATION/APPROVAL REQUIREMENTS AND COMPLIANCE

Registration Requirements

The basic registration mandate is set out in REACH Article 6 as follows: "Save where this Regulation provides otherwise, any manufacturer or importer of a substance, either on its own or in one or more mixture(s), in quantities of one tonne or more per year shall submit a registration to the Agency."[14] The connection of the imperative to register chemical substances to the registrant's ability to market them is bluntly stated in Article 5, which is captioned "No data, no market": "Subject to Articles 6, 7, 21 and 23, substances on their own, in mixtures or in articles shall not be manufactured in the Community or placed on the market unless they have been registered in accordance with the relevant provisions of this Title where this is required."[15]

REACH is grounded on the principle that those who place chemical substances on the market must assure that those substances do not cause adverse effects in human beings or the environment. The burden of proof to support the substance as safe for use and safe to the environment (including classification, labeling, and hazard potential) is placed upon the registrant. This differs from the approach in TSCA, under which it is up to EPA to show that a chemical substance poses "an unreasonable risk of injury to health or the environment" as a prerequisite to imposing regulatory restrictions.[16]

Accordingly, among the key elements of registering a chemical involves generating or citing to relevant study data and other pertinent information to support the registration and presenting the data

---

14. The exclusion of quantities less than one ton per year is in contrast with TSCA, which provides for no such de minimis exclusion.
15. The referenced articles address, respectively, the general obligation to register substances on their own or in mixtures; registration/notification of substances in articles; the manufacturing and import of substances; and specific provisions for phase-in substances.
16. As discussed in chapter 2.1, TSCA also differs from REACH in that TSCA does not use a "registration" mechanism for regulating chemicals.

in a technical dossier to ECHA for review. Information included in the registration must be robust and scientifically sound.

Registration is a prerequisite for the manufacture or import of new substances when supplied at or greater than one metric ton per annum, that is, those never before marketed in Europe. An Article 26 inquiry must be submitted prior to exceedence of the one metric ton volume threshold and a registration submitted. Manufacture or import of such new substances may begin three weeks after submission of a registration, unless ECHA informs the registrant to the contrary.[17] New substances also are referred to "non-phase-in substances," in contrast to "phase-in" substances, which are eligible for a different registration pathway.

"Phase-in substances" include—and largely are composed of—those listed on the European Inventory of Existing Commercial Chemical Substances (EINECS).[18] Phase-in substances also include those that were manufactured in EU countries but not placed on the market within 15 years before REACH took effect, and substances that were "notified" under the directive that REACH replaced.[19] REACH provides for a phase-in period of some 11 years from its June 1, 2007, effective date, with priority placed on high volume and chemicals that present a hazard to human health and/or the environment.[20] Depending on the annual volume and the hazard associated with a substance, the applicable "phase-in" registration deadline could be either 3½ (December 31, 2010), 6 (May 31, 2013), or 11 years (May 31, 2018) from June 1, 2007.

REACH established a six-month "preregistration" period that ran from June 1, 2008, until December 1, 2008.[21] Timely preregistration, which is considered a "duty" under Article 28, enabled the manufacture or import of the substance until the extended registration deadline.[22] Phase-in substances that were not timely preregistered,

---

17. REACH art. 21(1).
18. Id. art. 3(20)(a). The EINECS list contains more than 100,000 substances and is available on the website of the European Commission's (EC) Joint Research Centre (JRC) Institute for Health and Consumer Protection (IHCP), http://esis.jrc.ec.europa.eu/.
19. See REACH art. 3(20)(b), (c).
20. Priority substances are considered those with total volume >1,000 tons per annum throughout the EU as a whole and those substances that are either hazardous to human health—those classified as Category 1 or 2 carcinogens, mutagens, or toxins for reproduction (CMR)—or hazardous to the environment, that is, persistent, bioaccumulative, and toxic (PBT), or very persistent and very bioaccumulative (vPvB). Endocrine disruptors not covered by these criteria are to be added to the list of very high concern substances on an ad hoc basis.
21. REACH art. 28(2).
22. Preregistration does not represent a commitment to register the chemical substance.

however, are not afforded the benefit of an extended registration period but are treated, instead, like new (non-phase-in) substances for registration purposes.

ECHA guidance notes that potential registrants who for the first time manufacture or import a phase-in substance in a quantity of one ton per year after the December 1, 2008, close of the main preregistration period, may be eligible for "late preregistration." To do so, the potential registrant must submit a limited preregistration notification to ECHA within six months of first manufacturing or importing the substance and no later than 12 months before the deadline prescribed for the corresponding tonnage band. Otherwise, the substance will be subject to the requirements for "non-phase-in" substances.[23] ECHA does not require fees be paid to conduct a preregistration or late preregistration.

The Registration Dossier and Chemical Safety Report

REACH Article 10 sets out the mandatory information for general registration purposes. The two basic elements for submission are the technical dossier (Article 10(a)) and the Chemical Safety Assessment (CSA)/CSR (Article 10(b)). Annexes IV–VIII further specify the information for submission, beginning with the general technical, commercial, and administrative information needed for the technical dossier, as provided in Annex IV, while Annexes V–VIII are tonnage band-dependent.

The technical dossier must contain various elements, including robust summaries of study data; proposed classification and labeling information; guidance for safe use; and any proposal for further testing. The dossier is prepared using the International Uniform Chemical Information Database (IUCLID), a well-established database format for communicating and storing information on chemicals, and submitted to ECHA electronically via REACH-IT.[24]

Significantly, under Article 11(a), REACH directs multiple registrants of the same substance to submit the information specified in Article 10(a)(iv), (vi), (vii) and (ix), and any relevant indication

---

23. ECHA, ECHA-13-B-04-EN, GUIDANCE IN A NUTSHELL ON REGISTRATION 8 (Sept. 2013), http://echa.europa.eu/documents/10162/13632/nutshell_guidance_registration_en.pdf.
24. ECHA has issued a REACH-IT user manual for industry. ECHA, REACH-IT INDUSTRY USER MANUAL, PART 05-PRESIEF (July 2012), http://echa.europa.eu/documents/10162/13654/reachit_presief_en.pdf.

under Article 10(a)(viii) through a single registrant (the "Lead Registrant) on behalf of them all and with their consent.[25] Those include classification and labeling information; study summaries; robust study summaries (if required); and testing proposals (as needed). This approach is consistent with the intent to streamline the process and avoid duplicative effort, including, certainly, duplicative testing. Each of multiple registrants then is to submit separately the relevant information corresponding to the remaining portions of Article 10(a).

A CSA, documented in a CSR, is required for substances registered at or greater than ten tons per annum and for persistent, bioaccumulative, and toxic (PBT), very persistent and very bioaccumulative (vPvB), and carcinogenic, mutagenic, or toxic to reproduction (CMR) substances registered at or greater than one ton per annum.[26] The CSA is a risk assessment, the first step in developing which is identifying the substance's hazard profile by evaluating the available data against standardized hazard categorization criteria. Following the general provisions of REACH Annex I, the CSA includes a human health hazard assessment; a physicochemical hazard assessment; an environmental health assessment; and a PBT and a vPvB assessment.[27] Where, based on these various assessments, the registrant concludes that the substance is either "dangerous" (in accordance with a referenced directive) or is either a PBT or a vPvB, the CSA must include both an exposure assessment and a risk characterization.[28] Registrants also are charged with the responsibility of identifying and applying control measures appropriate to the risks identified in the CSA and, where suitable, to communicate them in the safety data sheet (SDS) to be supplied in accordance with Article 31.[29] The ultimate goal of the CSA/CSR is to demonstrate that the substance being registered can be and is used in a manner that is safe to human health and the environment.

---

25. As discussed later in this chapter, these registrants already will have joined together as participants in a SIEF, as required under Article 29(1).
26. REACH art. 14(1).
27. *Id.* art. 14(3). The PBT and vPvB assessments are to be done in accordance with Annex XII.
28. *Id.* art. 14(4).
29. *Id.* art. 14(6). The CSA/SDS relationship also is referenced in Article 31(2).

ECHA Fees and Review

When a registration dossier for a substance is submitted, it is accompanied by a fee payable to ECHA. In 2013, the Commission recognized the burden of the fees being placed on small- and medium-sized enterprises (SME) and offered a reduction in the standard fees to further assist SMEs. This fee is in addition to the cost for obtaining a letter of access (LOA) for use of scientific and technical data to support the registration.[30] The fees are delineated by the volume of substance placed on the market and the company size of the registrant (or the legal entity that an Only Representative is acting on behalf of). Company size of the registrant is determined via the type of legal entity, revenue, and number of employees (headcount).[31]

Once the fees are paid upon submission, ECHA will conduct a completeness check and assign a registration number if deemed complete. Under Article 41(5), to ensure compliance, ECHA is directed to undertake "compliance checking"—that is, detailed review—of at least 5 percent of the total submitted dossiers for each tonnage band, although this floor and the criteria for priority review may be adjusted by the EC.[32] The presence of low-percentage floor written into the regulation creates a very different reviewing situation than under TSCA, even though prospective REACH registrants have the burden of establishing safety. ECHA has been criticized, particularly by nongovernmental organizations (NGO), for going no further than the 5 percent minimum for detailed review prescribed in the regulation. This very modest percentage of dossiers reviewed by ECHA means that many registrants opt to take their chances and provide a lower standard of document support. ECHA has indicated that it intends to increase its review floor to a higher percentage of submissions.

Despite its detailed review of only 5 percent of the registration dossiers, ECHA does undertake detailed checks on substance sameness, based upon spectral and analytical data, as standard on all registrations. These quality checks on substance sameness

---

30. COMMISSION IMPLEMENTING REGULATION (EU) No 254/2013 of 20 March 2013 amending Regulation (EC) No 340/2008 on the fees and charges payable to the European Chemicals Agency pursuant to Regulation (EC) No 1907/2006 of the European Parliament and of the Council on the Registration, Evaluation, Authorisation and Restriction of Chemicals (REACH).
31. COMMISSION RECOMMENDATION of 6 May 2003 concerning the definition of micro, small and medium-sized enterprises (notified under document number C(2003) 1422) (Text with EEA relevance) (2003/361/EC).
32. REACH, art. 41(7).

provide confirmation that registrants are supplying and registering the substance(s) that they claim. There has been much discussion on the topic of substance sameness (which is not defined under REACH), both as a result of Article 26 inquiries[33] and the analytical data being generated and then evaluated for comparison against the Substance Identification Profile (SIP) developed for purposes of addressing Annex VI. In many cases, a potential registrant may discover that a substance it manufactures or imports falls within a different and more appropriate substance identifier than originally contemplated. In other cases, detailed discussions with ECHA scientific experts are needed to explain processes and identifiers to support a substance's identity. Ensuring correct substance identity early in the process is critical to preventing delays in registration.

REGULATION OF SUBSTANCES

Authorization/Substances of Very High Concern

Substances of very high concern (SVHC) are what their name cautions: chemicals singled out for use restrictions or other special controls due to the inherent risks that they pose. To enable the identification of particular SVHC, the hazardous properties warranting such heightened concern are set out in REACH Article 57 and include the following:

- "CMR" Substances—Those meeting the criteria for classification as carcinogenic, mutagenic, or toxic for reproduction category 1A or 1B in accordance with Commission Regulation EC No. 1272/2008;
- Substances that are PBT or vPvB under REACH Annex XIII; and
- Substances identified on a case-by-case basis for which there is scientific evidence of probable serious effects that cause a level of concern equivalent to that for PBT or vPvB substances.[34]

---

33. As discussed below in connection with data sharing, this term refers to an inquiry dossier submitted by a legal entity under Article 26, informing ECHA that the submitter intends to place a substance on the EU market. This inquiry is usually a short dossier, stating the name and contact details of the legal entity and the description of the substance with spectral and analytical characterization, along with request to reference existing data, if any, in support of the necessary registration.
34. The relationship between SVHC and the authorization procedure is summarized usefully at ECHA, Authorisation, http://echa.europa.eu/regulations/reach/authorisation.

An SVHC substance may be included in Annex XIV, List of Substances Subject to Authorisation (Authorization List), through a regulatory process set out in Article 58, which involves consultation with the Member State Committee, public notice and an opportunity for public comment, and a recommendation (if ECHA determines to go forward) to the Commission. The first step in the authorization process is the identification of a substance as an SVHC and its inclusion in the Candidate List of SVHC for Authorization. The current Candidate List is maintained on the ECHA website.[35]

Once a substance is subject to authorization, it cannot be placed on the EU market or used after a given date unless an authorization is granted for a specific use or the use is exempt from authorization.[36] Manufacturers, importers, or downstream users of a listed substance may apply for authorization.

To apply for an authorization under REACH, an authorization dossier must be submitted to ECHA pursuant to Article 62 that includes, among other elements, the specific use(s) covered by the application, a detailed CSR[37] (unless already submitted with the substance registration), and an investigation of potential alternative substances and/or technologies. The authorization dossier is accompanied by a base fee of €50,000 for a normal size business per application with additional fees assessed per use and per substance. Joint applications for authorization are allowed, in which fees are reduced per applicant. As with registration, ECHA assesses reduced fees for SMEs.[38] This dossier is the starting point and often the fulcrum issue for authorizations. Because of the time and cost required to progress through the authorization process, entities may opt to make a business decision to cease supply of certain substances in favor of other, less hazardous substances.

---

35. See ECHA, Candidate List of Substances of Very High Concern for Authorisation, http://echa.europa.eu/web/guest/candidate-list-table.
36. REACH art. 56; see also ECHA, Authorisation, http://echa.europa.eu/regulations/reach/authorisation.
37. REACH art. 62(4)(d). As the CSR must show that the use of the substance is safe and that all risks are suitably contained, the underlying risk assessment may be more detailed and encompassing than what usually is submitted with a registration.
38. COMMISSION REGULATION (EC) No 340/2008 of 16 April 2008 on the fees and charges payable to the European Chemicals Agency pursuant to Regulation (EC) No. 1907/2006 of the European Parliament and of the Council on the Registration, Evaluation, Authorisation and Restriction of Chemicals (REACH).

Under Article 60(1), it is for the EC to decide whether to grant an authorization and under what conditions.[39] The authorization is to be granted if the applicant can show that the risks posed by the substance are adequately controlled under that use or, in the event that this showing is not made, only if the socioeconomic benefits of using the substance outweigh the risks and no alternative substances or technologies are available.[40]

Once granted, authorization holders must comply with the conditions of the authorization, as must their downstream users.[41] Authorizations are limited to uses within the scope of REACH. Uses outside its scope (such as biocidal products or use in medical or veterinary medicines), cannot be authorized through this process.

Restrictions

Separate and distinct from the authorization process for SVHC substances is the restriction process, intended to address the situation in which there is an "unacceptable risk" to human health or the environment posed by the manufacture, use, or placing on the EU market of certain dangerous substances (or mixtures or articles), and where the risk needs to be addressed on a Community-wide basis.[42] REACH provides for the amendment of Annex VXII to identify the implicated substance and to adopt corresponding restrictions (or amend those that may already be in effect), pursuant to the procedure set out in Articles 69–73.

Restrictions may limit or even ban the manufacture, placing on the market, or use of a substance.[43] They may apply to imports and to substances for which a REACH registration is not required.[44] Any such restriction decision is to take into effect the socioeconomic impact, including the availability of alternatives.[45] Once restricted, the substance may not be manufactured, used, or placed on the market

---

39. Article 60(2) provides that the Commission is to take into account the opinion of ECHA's Committee for Risk Assessment. *See also id.* art. 64(1), (4)(a).
40. REACH art. 60(2), 60(4). The socioeconomic analysis is generally considered as the last stage in the authorization process because of the high cost and amount of time required to compile the analysis.
41. REACH art. 65.
42. REACH art. 68(1).
43. *See* ECHA, Restriction, http://echa.europa.eu/regulations/reach/restriction.
44. *See id.*
45. REACH art. 68(1).

except in accordance with the restriction, assuming that the restriction allows for any such latitude.[46] The restriction does not apply to a substance in scientific R&D.[47]

Any Member State[48] or ECHA, at the Commission's request, may propose a restriction, so long as the risks posed by the substance should be addressed Community-wide. Anyone may comment on a proposed restriction, including persons from outside the EU. Although most of the work is undertaken by governmental bodies, the opportunity for input from interested parties can be an important part of the process. Information regarding the substance, use practices, release and exposure data, and potential socioeconomic analysis all are potentially valuable data that can be submitted to ECHA through public consultation, as is feedback on initial opinions and conclusions. While this type of information gathering can be expensive (both in time and financially) to industry and other third-party stakeholders, stakeholders' input can assist in providing a more complete and accurate picture of the use and life cycle of a restricted substance.

Test Data Development

Achieving the health and environmental safety objectives stated in Article 1 of REACH relies in significant part on ensuring that the prescribed data requirements are satisfied, as addressed in detail in Annexes VII to X. A potential competing concern is achieving the articulated goal of avoiding testing on vertebrate animals, except "as a last resort," as stated in Article 25(1). The same provision also directs that measures be taken to avoid the duplication of other tests. Article 25 also encourages the "sharing and joint submission of information in accordance with this Regulation" as "concern[s] technical data and in particular information related to the intrinsic properties of substances."[49] Article 25(3) further authorizes that "[a]ny study sum-

---

46. *Id.* art. 67(1).
47. *Id.*
48. Restrictions may be proposed to an EU Member State by any interested party based within that Member State. Interested parties may be any legal entity based within the Member State that has a stake in the substance market. These parties may include individual companies, industry groups, consumer organizations, nongovernmental organizations, and governmental agencies not directly responsible for REACH.
49. REACH art. 25(2). This provision also cautions registrants from exchanging information about their market behavior.

maries or robust summaries of studies submitted in the framework of a registration under this Regulation at least 12 years previously can be used for the purposes of registration by another manufacturer or importer."

To minimize animal testing and duplicative testing generally, REACH provides for the use of data already available (literature, historic, published data) and from other sources, such as quantitative structural-activity relationships (QSAR) and other mathematical and qualitative models. REACH encourages the use of such information, where reliable, to fulfill data requirements.[50] Additionally, the relevant annexes are designed so that each builds upon the information required in the previous one; thus, only basic information is required for substances placed upon the EU market at low tonnages, whereas more information is needed to address the potential hazards posed by higher-volume substances. Annex VII, which specifies the information requirements for the lowest tonnage level, allows that in some instances, the required information may be "omitted, replaced by other information, provided at a different stage, or adapted in another way."[51] It also sets out specific instances in which otherwise required studies need not be performed. While each time a new, higher tonnage band is reached, the corresponding Annex adds information requirements, but these "shall be considered in conjunction with Annex XI, which allows variation from the standard approach, where it can be justified."[52] Annex XI sets out the general rules for adaptation of the standard testing regime—also known as test plan waivers—which must be acceptable to ECHA.

ECHA's examination of testing proposals is governed by Article 40. According to Article 40(1), priority of review is given to substances that have or may have PBT, vPvB, sensitizing and/or CMR properties, or substances classified as "dangerous" substances at ≥100 tons per year with uses resulting in diffuse and widespread exposure.

---

50. *Id.* art. 13.
51. REACH, Annex VII.
52. Annex VI, Guidance Note on Fulfilling the Requirements of Annexes VI to XI.

ECHA recommends that all data submitted for REACH registration purposes are scored under the Klimisch system to ensure that the registration is based only on reliable and relevant data.[53]

Data Sharing and Compensation

To further the objectives of information sharing, avoidance of duplicative testing, and facilitating agreement on substance classification and labeling, REACH requires most registrants, potential registrants, downstream users, and third-party data submitters to be participants in a SIEF for the substance involved.[54] When a substance is preregistered, the potential registrant is assigned to a SIEF along with other preregistrants of the same substance. SIEF participants are to "provide other participants with existing studies, react to requests by other participants for information, collectively identify needs for further studies for [registration] purposes . . . and arrange for such studies to be carried out."[55] Each SIEF is to remain in operation until May 31, 2018, the eleventh year, and final registration deadline.[56]

While SIEF membership is mandatory, parties engaged in registration support activities, such as technical dossier preparation, may opt simultaneously to join in other, voluntary arrangements for cooperation for data sharing, data compensation, and registration preparation. Consortia are a more formalized mechanism to accomplish these tasks, typically pursuant to a written agreement among the members, although REACH requires no such agreement.[57] Other, looser arrangements also may be formed for similar purposes.[58] A consortium may incorporate different SIEFs for similar substances, whereas each SIEF addresses a single substance only. One party to a consortium frequently will act as Lead Registrant to promote efficiency and streamline various interactions.

---

53. ECHA, ECHA-10-B-05-EN, Practical Guide 2, How to Report Weight of Evidence § 3.1(2010), http://echa.europa.eu/practical-guides. "Klimisch" refers to H.J. Klimisch, E. Andreae & U. Tillmann, *A Systematic Approach for Evaluating the Quality of Experimental and Ecotoxicological Data*, 25 Reg. Toxicol. Pharmacol. 1–5 (1997).
54. REACH art. 29(1), (2).
55. *Id.* art. 29(3). ECHA has issued comprehensive and detailed guidance on data-sharing and related issues. ECHA, 2012-G-1-EN, Guidance on Data Sharing (Version 2.0, Apr. 2012), http://echa.europa.eu/documents/10162/13631/guidance_on_data_sharing_en.pdf.
56. REACH art. 29(3).
57. Guidance on Data Sharing 132 (§ 8.1).
58. *See id.*

For purposes of data sharing under REACH, a data owner—often called a data holder or third party—need not be a prospective registrant of the chemical substance nor have any interest in placing the substance on the market in the EU. ECHA encourages such data holders to share the data they own with participants in the relevant SIEF to support their registration dossiers, in exchange for compensation. Data owners who also are SIEF members and prospective registrants, of course, also are entitled to compensation for sharing relevant data with the other members.[59]

For purposes of data sharing, typically a data owner will grant permission to refer to the relevant full study report to the SIEF members and/or to the Lead Registrant on behalf of a consortium. A LOA is the usual instrument by which this right is granted. The data owner infrequently will provide a hard copy of all the study data but more often will provide a robust study summary, along with the right conveyed by the LOA to reference and rely on the full study report in exchange for compensation. A data owner also may elect to grant SIEF or consortium members the right to sublicense data citation rights to other legal entities in the future; variations on this theme may be negotiated depending what needs are envisioned.

Data sharing and compensation procedures under REACH differ for non-phase-in substances and phase-in substances that were not preregistered, on the one hand, and for phase-in substances, on the other. As to the former, pursuant to Article 26, every potential registrant of a non-phase-in substance or of a phase-in substance that was not preregistered must inquire of ECHA whether a registration has been submitted for the same substance and must provide information about substance identity and the anticipated data requirements. As noted above, this has come to be known as an "Article 26 inquiry." If the substance has been registered less than 12 years previously, ECHA will provide the contact information for the previous registrants as well as a listing of available robust study summaries submitted within that period so that they may be shared with the prospective registrant pursuant to Article 27. The latter directs that the prospective registrant *shall* request such information from the previous registrants where the information involves tests on vertebrate animals and *may* request it where the information does not involve such testing.[60] Article 27 also directs the potential registrant and

---

59. *See* GUIDANCE ON DATA SHARING 16 (§ 1.2.6).
60. REACH art. 27(1)(a), (b).

the previous registrants to "make every effort" to reach an agreement on information sharing and to ensure that the costs of sharing are determined in a "fair, transparent and non-discriminatory way."[61]

Data sharing and compensation for phase-in substances are governed by Article 30. Before undertaking any testing to address data requirements, SIEF participants must determine whether any study data are available within the SIEF from testing involving vertebrate animals; if so, that information must be requested before any further testing proceeds.[62] Study data from testing involving nonvertebrate animals may (but not must) be requested also. The data owner is to provide proof of its cost within one month after it is requested.[63] The data owner is obligated to make the data available to the other participants, subject to sharing of the cost—which, again, must be determined in a fair, transparent, and nondiscriminatory way. REACH does not prescribe a specific method for cost sharing, and a participant group has flexibility in arriving at a mutually acceptable approach; where no agreement can reached, REACH directs that the cost shall be shared equally.

Article 30 also addresses the situation in which SIEF participants determine that data development—new testing—is needed to complete the registration dossier.[64] It further addresses problematic scenarios, as where the data owner refuses to provide proof of cost or the study itself to other participants, with specific reference to vertebrate or invertebrate testing.[65]

Legal and other professionals who have worked with FIFRA will notice familiar principles in REACH's data-sharing and data compensation provisions but also some conspicuous differences. Most notably, perhaps—although this does not occur routinely in practice—ECHA is authorized to take a hands-on role in resolving data disputes, whereas EPA stands aside in these instances under FIFRA, which provides for arbitration when the parties to such disputes cannot reach agreement. Unfortunately, to date, ECHA has seldom engaged and although ECHA is tasked with resolving disputes between potential and prior registrants over compensation for studies supporting the registration of non-phase-in substances, this is not happening. In the context of phase-in substances, it is tasked with resolving disputes in

---

61. REACH arts. 27(2), 27(3).
62. Id. art. 30(1).
63. Id. art. 30(1).
64. Id. art. 30(2).
65. Id. arts. 30(3), 30(4).

which SIEF participants believe that a data owner is impeding efforts to register their products by failing to share supporting data or to provide proof of its cost. This fallback authority given to ECHA does not preclude the parties to a dispute from choosing to resolve it by another means, such as by binding third-party mediator. Additionally REACH provides avenues for appeal from ECHA's decisions in these circumstances. Still, expectations formed by U.S. practitioners after decades of arbitrations under FIFRA will not transfer seamlessly to the resolution of data-sharing and data compensation disputes within the REACH framework

Registration Software

REACH Article 111 provides that for registration purposes, the format for the Article 10(a) technical dossier shall be the IUCLID system. IUCLID has a long history as a key, and perhaps the most versatile, database designed to assist jurisdictions with the collection and sharing of relevant hazard data in chemical control and similar programs. Nonetheless, a complete restructuring was needed to meet the requirements of REACH while also considering multiuse functionality by other disciplines, namely pesticides/biocides. IUCLID is free to download and use and thus adds nothing to the fees and costs associated with REACH registration and compliance. As of this writing, the most current version is 5.6.0.[66]

IUCLID version 5 is closely linked with ECHA's submission tool REACH-IT, the mandatory online portal for registration. Within REACH-IT, the registration dossier is appended and submitted to ECHA electronically. When IUCLID is updated, REACH-IT is modified to correspond. As a result, REACH-IT accepts the only registration dossiers created in the most up-to-date version of IUCLID 5 for submission to ECHA. Therefore it is essential that all entities that have submitted, or will submit, a REACH registration keep current with version changes to IUCLID, as well as to understand how to use it.

The IUCLID 6 program is now under development with a targeted release of sometime in 2014. As of this writing, the impacts of the upgrade on REACH registration procedures are unknown. The expectation is that significant modifications will be made to allow for better overall functionality.

---

66. *See* IUCLID 5, http://iuclid.eu/.

If multiple registrants are identified within the SIEF, according to Article 11, one registrant, or the Lead Registrant, can submit a Full Registration dossier while each SIEF member performing a registration can subsequently submit its registration dossier with reduced requirements, or as a Joint Registrant. The Joint Registrant can refer to the robust study summary submitted by the Lead Registrant without having to duplicate the relevant sections of the dossier. Upon submission, the Joint Registrant will need to supply a joint registration token provided by the Lead Registrant and complete the information requirements within the dossier, as set out in Article 10(a)(i)–(iii).

ECHA suggests and strongly supports the "One Substance, One Registration" principle, which means that only one Full Registration dossier is submitted per substance by a Lead Registrant, but in practicality this has become problematic for a minority of registrants for various reasons. If a legal entity wishes to submit further information on its substance separate from the Lead Registrant, it is possible to do so in accordance with Article 11(3), as long as there is no duplication of animal testing.[67] There are other circumstances to consider, such as when the self-appointment of a Lead Registrant occurs and only structure analogue data are submitted to support required endpoints or when there are significant discrepancies in reaching agreement on data compensation. For the vast majority of registrants, the above ECHA principle is true, but there are circumstances that can make this principle challenging and not achievable.

Duty to Inform ECHA of Current Information

Article 22 requires that a registrant, on its own initiative and "without undue delay," inform ECHA of updates affecting its registration or of other relevant new information. The various types of information specified in the regulation that must be updated are wide-ranging and include (but are not limited to) any change in the composition of the substance; any change in classification or labeling; new identified uses or uses advised against; changes in the annual quantities of the substance manufactured or imported, or in the quantities of substance in the article manufactured or imported if these result in a change in tonnage band; changes in the registrants' status (including cessation

---

67. The registrant must submit evidence why additional information has been submitted separately to the Lead Registration dossier, such as disproportionate costs or confidential data.

of manufacture of import) or identity; a new knowledge of risks to health or the environment that lead to changes in the SDS or CSR; and any other update or change in the CSR.[68]

IMPORTS AND EXPORTS

While each Member State in the EU regulates its own borders as to imports[69] and exports, "free movement" of REACH-compliant substances is among the principles set out in the regulation. Under Article 128(1), "Member States shall not prohibit, restrict or impede the manufacturing, import, placing on the market or use of a substance, on its own, in a mixture or in an article" if that substance falls within the scope of REACH and complies with it. This broad directive is subject to the qualification, set out in Article 128(2), that Member States nonetheless may maintain national rules to protect workers, human health, or the environment in cases where REACH does not harmonize the requirements for manufacture, placing on the market, or use.

CONFIDENTIAL AND TRADE SECRET INFORMATION

REACH is oriented in favor of the wide public dissemination of information on all registered substances, as well as classification and labeling information that it maintains pursuant to Articles 114 and 115, which, under Article 77(2)(e), ECHA is directed to make publicly available via the Internet, free of charge, except where a manufacturer or importer requests that information submitted for registration purposes under Article 10 should not be made available and can justify why publication would be harmful to its own or any other party's commercial

---

68. REACH art. 22.
69. It should be noted that for purposes of REACH, "import" is defined as "the physical introduction into the customs territory of the Community" and that, as discussed in this chapter, "imports" are subject to REACH.

interests.[70] Each such request should be justified robustly, and ECHA has indicated that justifications should include as much supporting information as possible on the reasons for requesting confidentiality.[71] Each confidentiality claim is assessed a fee. Confidentiality fees range from €113–€4,500 dependent upon the item/field in which confidentiality is sought and whether the submission is individual or joint.[72]

Article 119 specifies the many types of information that ECHA is to make publicly available online, free of charge, while Article 118(2) specifies the types of information that "shall normally be deemed to undermine the protection of the commercial interests of the concerned person."[73]

The types of information about substances, on their own or in mixtures or articles, which ECHA is to make publicly available online under Article 119(1) include the name, in International Union of Pure and Applied Chemistry (IUPAC) nomenclature, for "dangerous substances"; the name of a substance given in EINECS, if applicable; the classification and labeling of a substance; physicochemical data on the substance and on pathways and environmental fate; the results of each toxicological and eco-toxicological study; any derived no-effect level or predicted no-effect concentration; safe use guidance provided under Annex VI; and, if properly requested, analytical methods that make it possible to detect a dangerous substance discharged into the environment and to detect the direct exposure of humans.

Article 119(2) sets out a second category of information about substances, on their own or in mixtures or articles, which ECHA is to make publicly available online under Article 77(2)(e), *except* where

---

70. See REACH art. 10(a)(xi). There is an additional service fee charged under EC regulation 340/2008 (known as the "ECHA fees regulation") for a request in a registration submission that certain information be kept confidential. *See, e.g.*, EC, Fees and Charges, http://ec.europa.eu/enterprise/sectors/chemicals/reach/fees/index_en.htm. Reduced fees are available for small and medium-sized enterprises (SME). *See generally* REACH art. 74, which addresses fees. The exhibits at the end of this chapter depict fees for confidentiality requests in general and for SME.
71. *See* ECHA, DATA SUBMISSION MANUAL: PART 15— DISSEMINATION 27 (Version 2.0, July 2012), http://echa.europa.eu/documents/10162/13653/dsm_15_dissemination_manual_en.pdf.
72. COMMISSION REGULATION (EC) No 340/2008 of 16 April 2008 on the fees and charges payable to the European Chemicals Agency pursuant to Regulation (EC) No 1907/2006 of the European Parliament and of the Council on the Registration, Evaluation, Authorisation and Restriction of Chemicals (REACH).
73. Under the same provision, ECHA is authorized to disclose such information nonetheless in cases where urgent action is needed to protect human health or the environment.

a party submitting the information can justify why such publication would be commercially harmful. These types of information include, if essential to classification and labeling, the purity of a substance and the identity of impurities and/or additives known to be dangerous; the tonnage band within which a particular substance is registered; study summaries or robust study summaries for the physicochemical, toxicological, and eco-toxicological data referenced in Article 119(1); information contained in the SDS, other than that referenced in Article 119(1); the trade name of a substance; the name in IUPAC nomenclature for non-phase-in substances that are "dangerous" for a period of six years; and the name in IUPAC nomenclature for substances that are "dangerous" and used only (1) as an intermediate, (2) in scientific R&D, and/or (3) in PPORD.

The types of commercially sensitive information normally considered protected under Article 118(2) includes details of the full composition of a mixture; the precise use, function, or application of a substance or mixture, including precise information about its use as an intermediate; the precise tonnage of the substance or mixture manufactured or placed on the market; and the links between a manufacturer or importer and its distributors or downstream users.

As a practical matter, in cases where it is critical to keep the identity of a registrant confidential, it may be useful for the registrant to appoint a TPR. Likewise, those non-EU entities that opt to appoint an OR also benefit, as a result, from the masking of the legal entity name. Under these options, only the TPR or OR names are used in all communication and official documentation, and the registrant's name is not disclosed.

Enforcement and Penalties

REACH delegates enforcement obligations to each EU Member State. Below is a brief summary overview of key enforcement provisions.

In accordance with Article 126, the EU Member States are responsible for REACH enforcement, and thus must take the appropriate measures to ensure that REACH is enforced within its jurisdiction; penalties must be "effective, proportionate and dissuasive." Delegation to the Member States, in practice, means that the levels of enforcement between them may vary widely. For example, 15 Member States, such as the United Kingdom Member State Authority, have adopted sanctions provisions relating to the specific offense, under which a violation could result in prosecution, leading to a fixed penalty fine or several years' imprisonment, depending upon the type of violation and the level of offense.[74] Other Member States, such as Denmark, have adopted more severe, and less flexible, sanctions provisions under which all violations are treated similarly, irrespective of the severity.[75] The inevitable variations and inconsistencies among Member State provisions for violations may prove challenging for entities placing substances on the EU market because most registrants (particularly SMEs) may not have the resources to maintain on-staff regulatory experts or outside expert counsel who are aware of the nuances of each Member State.

Unlawful Activities

It is an offense for any person to violate a REACH provision or to cause or permit another person to do so. Enforcement regulations in various EU Member States specify additional criminal offenses, including obstruction of inspectors, providing false statements, and failing to comply with enforcement notices.

Civil and Criminal Penalties

Prosecution for REACH violations can take place summarily or by indictment, and a maximum fine and/or imprisonment can be given for each offense. Penalties are set under the European Communities Act of 1972 and are shown in table 1. Sentences may include both the maximum fine and the maximum imprisonment, or any combination or omission of the two.[76]

---

74. MILIEU LTD., REPORT ON PENALTIES APPLICABLE FOR INFRINGEMENT OF THE PROVISIONS OF THE REACH REGULATION IN THE MEMBER STATES (Final Report, 2010). http://ec.europa.eu/enterprise/sectors/chemicals/files/reach/docs/studies/penalties-report_en.pdf.
75. Id.
76. Health & Safety Exec., The U.K. Enforcement Regime for REACH, http://www.hse.gov.uk/reach/regime.htm

Table 1. Penalties for REACH Violations

MANNER OF TREATMENT	MAXIMUM FINE	MAXIMUM IMPRISONMENT
Summarily (per offence)	£5,000	3 months
Indictment (per offence)	No maximum limit	2 years
Source: Health & Safety Executive, The U.K. Enforcement Regime for REACH, http://www.hse.gov.uk/reach/regime.htm.		

KEY BUSINESS ISSUES

Substance Identity

A key issue encountered under REACH is the challenge of substance identification.[77] Although a substance may have been historically identified globally as a particular substance, based upon substance name or Chemical Abstracts Service (CAS) identifiers, REACH registrants must establish that the substance being registered meets the Lead Registrant's SIP. ECHA requires that each registrant provide the required analytical identification within the registration to document substance sameness. Differences in a substance's identification may result in a SIEF further dividing into subcategories to ensure that substance identification is complete, with the potential for inefficiency, confusion, and delay.

For example, one CAS number may cover a substance that is manufactured in different ways, resulting in a chemical substance that can take multiple forms generating multiple CAS numbers, including those that are potentially Unknown, of Variable Composition, or of Biological Origin (UVCB). In non-REACH jurisdictions, this single substance could be registered under multiple forms. Under REACH, the substance would be identified as two or more separate substances and would require separate lead registrations for each. The lack of identity consistency can lead to severe commercial confusion as well as REACH classification disarray.

---

77. *See* ECHA, First Substance Evaluation Results—Further Information Needed on 32 Substances, Mar. 6, 2013, http://echa.europa.eu/view-article/-/journal_content/title/first-substance-evaluation-results-further-information-needed-on-32-substances.

The issue of substance identity is further complicated for entities that must perform an Article 26 inquiry. Each Article 26 inquiry is subject to review by ECHA. As a result, ECHA will offer its opinion on substance identity issues before being placed in contact with the Lead Registrant. It is not unusual for naming conventions or substance identity disagreements to cause delays with registration activities. The effect of realignment of substance identification is not confined to the EU because substance identification issues under REACH affect entities placing substances on other markets. For a variety of reasons, companies prefer to use the same descriptor when placing substances on markets globally. Changes in the substance identification can invite adverse commercial impacts on companies that prefer consistent chemical descriptors.

In addition, a change in substance identity can affect the commercial identity of the same substance when listed on existing public inventories, as is often the case. If a new substance identity is required, then it is possible that the new substance identity may not be listed on an existing inventory and, hence, the product—tagged incorrectly as a "new" substance—cannot be marketed. The implications of such changes are broad and generally adverse, and specific to each jurisdiction in which an identification change is applicable; it may involve, for example, correction notices, potential removal from market, and even penalties until the "new" substance is either notified or registered. The potential effects are so broad and significant that the Organization for Economic Co-operation and Development (OECD) is assisting countries in helping to identify problematic scenarios and develop solutions.

Preregistrations and SIEF Issues

ECHA has previously asked industry stakeholders to limit the number of preregistrations performed.[78] Many companies have elected to preregister more substances than they intended to register or place on the EU market, either as a precautionary measure to accommodate future market expansion, or as a marketing and networking exercise. Not all entities that preregistered substances, however, legitimately

---

78. ECHA, Press Release, ECHA Re-emphasises Its Approach to Pre-Registrations, Oct. 6, 2008, http://apps.echa.europa.eu/legacy/doc/press/pr_08_32_pre_reg_followup_20081006.pdf.

represented companies placing substances on the EU market. Many of these entities, instead, have appointed themselves as data holders based on QSAR model results or as SIEF Formation Facilitators (SFF),[79] and, in rare cases, appointed themselves as a Lead Registrant. These "protective" preregistrations and questionable role appointments make it difficult to conduct SIEF business for legitimate registrants. In extreme cases, legitimate registrants have had to contact ECHA or the Member State Competent Authority in an attempt to obtain assistance in resolving the internal SIEF issues created as a result of these actions. SIEFs are to be self-supporting and, as such, involvement from ECHA or a Competent Authority is regarded as a last resort. Nonetheless, ECHA and the Competent Authority need to be made aware of significant issues that threaten to impede preregistrants from conducting a fair registration process, unburdened by actors who seek to game the system to their own advantage.

Guidance and Interpretation Shortfalls

REACH registrants have been hampered by a lack of clarity as to what certain REACH provisions mean. By the 2013 six-year registration deadline, ECHA had released 30 guidance documents, including five "Guidance in a Nutshell" documents, ten Guidance Fact Sheets, 14 Practical Guides, and a list of 137 Frequently Asked Questions on REACH, REACH-IT, and IUCLID.[80] While these are useful resources insofar as they go, a registrant with further questions for resolution has little additional resource. ECHA's view is that the interpretation of REACH is the obligation of the registrant, leaving it to the national REACH "helpdesks" appointed by Member States under Article 124 to address those issues not covered by ECHA guidance. Because ECHA places the responsibility of interpretation squarely on the registrant, difficulties arise when interpretations conflict. In some cases, ECHA guidance documents may imply one interpretation, a registrant may have come to a different conclusion, and then a Member State may offer yet a third interpretation. In cases where the registrant is an SME, or lacks an adequate working knowledge of REACH—or where the issue is manifestly a complex one under a still-evolving regulatory regime—

---

79. This term is not defined in REACH but was created in REACH-IT. *See* ECHA REACH-IT INDUSTRY USER MANUAL, *supra* note 24, at 6.
80. All ECHA guidance documentation is available online at http://echa.europa.eu/support.

this confusion can cause delays in registration, or even withdrawal from the market to avoid long, complicated and unrewarding communications with the regulatory authorities.

Data-Sharing and Cost-Sharing Issues

Data sharing and cost sharing can be problematic, even though they are essential to REACH's functioning effectively. Difficulties are prone to arise especially in cases where data holders are late preregistrants, which can occur for various reasons. When a data holder joins a SIEF as a late preregistrant, the SFF or Lead Registrant may already have been appointed and activities to support a REACH registration may already have been undertaken. This situation, admittedly, is highly variable by SIEF because activities typically begin one to two years before the applicable registration date. For specialty substances, first-time registrations began in 2013 and are anticipated to increase leading up to the 2018 registration deadline. As part of this activity, a data availability survey, test plan preparation, and placement of new testing may have occurred. As a result, it is possible that a late preregistrant may hold data relevant to studies that already may be coordinated or where new studies have been contracted for in what appeared to be the absence of sufficient data.

The SIEF participants will need to agree upon how to address this type of situation, particularly if vertebrate data are available from the late preregistrant. Formulas for compensation vary by SIEF, and in some cases, the lead members of a SIEF may determine to pay compensation for only one study, even if additional data are available. Other SIEF members may press the late-joining data owner to share its data regardless of compensation. The data owner, however, may elect not to share its data for inclusion in the registration without compensation because this would undercut the perceived value and reliability of those data. Under a different scenario, the existence of duplicative data may result in SIEF members needlessly paying compensation for all data included in the Lead Registrant's dossier and thus incurring unanticipated additional costs.

Data-sharing issues can also arise from the Lead Registrant failing to share the cost breakdown of the LOA from the data owner in a clear and transparent way with the other SIEF members. It is the duty of the data owner and the Lead Registrant to agree upon the most appropriate data cost, which can become difficult because of confidentiality issues and competition infringement concerns. Where data sharing and cost issues cannot be resolved, ECHA may need to become involved, although it will step in only on a demonstration, based on appropriate documentary evidence that efforts were made to come to an agreement but have failed. If the cost-sharing issue is left for ECHA to decide, that means it is no longer in the hands of the registrants to negotiate a result. In such instance, ECHA bases its decision on several factors, including assessment of the parties' respective efforts to reach an agreement, any sharing of data, and whether costs are communicated and sought to be shared in a fair, transparent, and nondiscriminatory way.

Managing Adverse SVHC Inferences

Once a chemical substance has been proposed and identified as a SVHC, there are global inferences, all of them adverse to the substances and products containing it. The SVHC "label" attaches almost immediately, leading the affected products to be deselected by customers and retailers, and subject to criticism by NGOs, who emphasize the adverse aspects of the chemical and products containing the chemical. This treatment puts chemical products and users of these substances at a competitive disadvantage that may prove difficult to overcome, especially if reasonably efficacious alternatives are available on the market.

Practical Tips on Supply Chain Communication

Supply chain communication is a significant portion of REACH and the EU's Classification, Labeling, and Packaging (CLP) regulation, which aligns EU legislation with the United Nations' Globally Harmonized System of Classification and Labelling of Chemicals (GHS).[81] Supply chain communication is instrumental in the following commonly encountered stages of REACH registration and supply:

- Identification of known uses;
- Identification of known hazards and risk for substances and mixtures;

---

81. EC Regulation 1272/2008, *available at* http://echa.europa.eu/web/guest/regulations/clp/legislation.

- Communication of risk assessment when appropriate;
- Confirmation of use under strictly controlled conditions;
- Collection of any existing endpoint data;
- Communication of registration/notification numbers or identifiers;
- Continued communication of new hazard information up and down the supply chain;
- Communication of volume information; and
- Deciding key roles of each supply chain actor.

Supply chain communication for REACH registration purposes can include EU and non-EU entities, that is, EU companies; non-EU manufacturers and reformulators; and downstream users (including end users). EU entities recognized under REACH may undertake their own preregistration and/or registration activities. The regulation also allows for EU entities to appoint an independent third party as a consultant or to function in the role of the TPR. As discussed, the latter often occurs when there is a need to protect CBI in the supply chain. Non-EU entities may participate in REACH preregistration and registration activities via an OR, but in their own right may be data owners. Many data owners are identified as such within a SIEF or may be known and identified by another potential registrant. In general, data owners are more actively involved in registration activities and therefore often are privy to the analysis of the data and the ultimate effect on a substance's hazard classification. As a result, the data owner is positioned to share these relevant details with downstream users through hazard communication.

The principal instrument for hazard communication under REACH, as well as most global programs, is the SDS. In the EU, the format of the SDS is set out in REACH Annex II, which was amended by EC Regulation 453/2010, Annexes I and II.[82] Annex I of EC Regulation 453/2010 specifies the format of an SDS and generally applies to substances from December 1, 2010, forward and preparations or mixtures from June 1, 2015, forward. The SDS for mixtures placed

---

82. Commission Regulation (EU) No 453/2010, *available at* http://eur-lex.europa.eu/JOHtml.do?uri=OJ%3AL%3A2010%3A133%3ASOM%3AEN%3AHTML.

upon the EU market before December 1, 2010, must have complied with Annex I of EC Regulation 453/2010 by November 30, 2012.[83]

Many downstream users are end users, often buying products through distributors, and therefore may be unaware of who within the supply chain provides them with the proper documentation. Downstream users, regardless of whether an end use is involved, typically will require their supplier, or an entity further up the supply chain, to ensure that appropriate registrations/notifications are performed. This approach makes practical sense because upstream entities are better positioned to provide the information and invest the resources needed for registration/notification. But even if it is reasonable for a downstream user to expect that these requirements will be addressed by an upstream entity, it is essential for the downstream user to confirm that this is the case. Communication throughout the supply chain about whether these basic requirements, and others, have been satisfied is vital to avoid any breach of the regulations through ignorance or lack of awareness. Downstream users must ensure that their supply of substances and preparations is legal, in that all relevant substances subject to registration/notification have been or will be registered/notified before the applicable registration deadline, or within the time frame given.

In the EU, the REACH registration number or preregistration number must be stated in Section 1 of the SDS.[84] This is one of, if not the primary, communication methods for sharing with the supply chain that a substance has been registered or preregistered under REACH. REACH "conformity letters" (letters demonstrating compliance with REACH) are also typically provided to downstream users to show a supplier's intent to comply with REACH and document the supplier's agreement to address the downstream user in its registration.

Trends

REACH is a game-changer. Many countries are adopting chemical regulatory programs similar to it. China's Ministry of Environmental Protection (MEP) Order Number 7, South Korea's K-REACH

---

83. EC Regulation 453/2010 art. 2(7) states:

    Without prejudice to Article 31(9) of Regulation (EC) No 1907/2006, safety data sheets for mixtures provided to any recipient at least once before 1 December 2010 may continue to be used and need not comply with Annex I to this Regulation until 1 November 2012.

84. EC Regulation 453/2010, Annex I § 1.1.

(scheduled to go into effect on January 1, 2015), and aspects of Canada's and Turkey's current programs emulate elements of REACH, particularly inventory inclusion, volume triggers, and increased data requirements for higher-volume chemicals.

The EC recognizes the need to monitor REACH implementation, and as such, the EC conducted an evaluation of REACH in early 2013. The EC published its review of REACH in February 2013,[85] which accounts for the five years since REACH was implemented within the EU. From this report, the EC highlighted the following key findings and recommendations:

- Recommendations to improve REACH implementation;
- Recommendations to reduce the financial and administrative burden on SMEs;
- REACH does not impede upon other EU legislation;
- Alternatives to animal testing are progressing well;
- Improved enforcement is needed; and
- Insurance is needed for EU businesses to ensure stability throughout the market.

In 2012, ECHA reviewed 427 dossiers for compliance, of which 93 were carried over from 2011 and 334 were newly initiated compliance checks.[86] ECHA issued 364 draft decisions to registrants based on its evaluation, and the remaining open decisions were overlapped into 2013.[87] In addition, ECHA adopted 171 decisions on testing proposals out of 586 submitted.[88]

ECHA has a targeted evaluation process. By screening submissions for substances of high concern, for example, SVHCs, CMR substances, and PBT chemicals, and by detecting submissions containing incomplete sections, for example, missing CSR and missing exposure scenarios, ECHA ensures that priority substances and substances with

---

85. EC, *Report from the Commission to the European Parliament, the Council, the European Economic and Social Committee and the Committee of the Regions in accordance with Article 117(4) of REACH and Article 46(2) of CLP, and a review of certain elements of REACH in line with Articles 75(2), 138(2), 138(3) and 138(6) of REACH* (Feb. 5, 2013), http://eur-lex.europa.eu/LexUriServ/LexUriServ.do?uri=COM:2013:0049:FIN:EN:PDF.
86. ECHA, EVALUATION UNDER REACH, PROGRESS REPORT 2012, at 23 (Feb. 27, 2013), http://echa.europa.eu/documents/10162/13628/evaluation_report_2012_en.pdf.
87. *Id.* at 8.
88. *Id.* at 20, tbl. 4.

incomplete datasets (and therefore potentially unknown hazards) are more likely to be evaluated. Based upon the findings from compliance checks, it appears that at least one or more significant deficiencies were identified in the submissions.

Competition issues have been raised by trade representatives from numerous countries, including the United States, under the framework of the World Trade Organization (WTO). To date, no modifications have been made to the REACH legislation or its guidance to resolve such matters.[89]

REACH, in its current form, will be present in the EU until at least 2018. Although aspects of REACH are challenging and could affect a company's ability to continue to place certain substances on the market, industry is becoming increasingly aware of these regulatory imperatives and considers them to be aspects of its business planning and decision-making process. Non-EU countries looking to establish or modify chemical management programs have a unique opportunity to look closely at all aspects of REACH as it stands, some six-plus years (as of this writing) into its implementation, with registration more than midway to its June 2018 milestone deadline and with its programs and supporting tools still evolving. The present day as well as the near future will yield lessons learned, successes, and ongoing challenges. All of these will merit study by regulators in other jurisdictions as well as by members of the regulated community and other stakeholders within the EU.

Resources

The European Chemicals Agency (ECHA): The ECHA website provides useful information on REACH as well as practical guidance and links to other governmental organizations involved in REACH. The ECHA website also provides a REACH helpdesk for inquiries regarding REACH obligations and technical problems (http://echa.europa.eu/web/guest).

---

89. Duygu Yaygir, The Compatibility of REACH Regulation with WTO TBT Agreement (master's thesis, Tufts University, May 2012), *available at* http://www.academia.edu/1591105/The_Compatibility_of_REACH_Regulation_with_WTO_TBT_Agreement; CEFIC Legal Guidance for REACH Compliance (May 2009), http://www.cefic.org/Documents/IndustrySupport/Cefic%20Legal%20Guidance-for-REACH-Compliance-WTOrules.pdf.

EU Member State Competent Authorities: Each EU Member State has appointed a Competent Authority, which enforces REACH within its borders and assists industry in fulfilling its REACH obligations. Correspondence with each Competent Authority is usually in one of the official languages of the Member State.

IUCLID 5: The most current IUCLID 5 program can be downloaded from the official IUCLID 5 website. The website also provides a user guide for the program, which assists with the technical aspects of setting up and managing the IUCLID 5 system (http://iuclid.eu/).

REACH-IT: REACH-IT is the official submission system for REACH. Legal entities must register within REACH-IT and obtain a Unique Universal Identifier (UUID) to participate in REACH registration activities (https://reach-it.echa.europa.eu/reach/public/welcome.faces).

6.2

RoHS and WEEE

Lynn L. Bergeson and Leslie S. MacDougall

EXECUTIVE SUMMARY

For over a decade, European Union (EU) legislation has been in effect to restrict the use of hazardous substances in electrical and electronic equipment (the RoHS 2 Directive[1] or RoHS 2) and to promote the collection and recycling of this type of equipment (the WEEE 2 Directive[2] or WEEE 2). Any company considering a medium- to long-term marketing and supply strategy for marketing electrical/electronic equipment in the EU should understand the implications of these two far-reaching pieces of legislation.[3]

Because these legislative instruments are directives, they do not instantly apply uniformly across all EU Member States, but they

---

1. Directive 2011/65/EU of the European Parliament and of the Council of 8 June 2011 on the restriction of the use of certain hazardous substances in electrical and electronic equipment, 2011 O.J. (L 174), 88–110 [hereinafter RoHS 2]. "RoHS" is short for "restriction of the use of certain hazardous substances in electrical and electronic equipment."
2. Directive 2012/19/EU of the European Parliament and of the Council of 4 July 2012 on waste electrical and electronic equipment, 2012 O.J. (L 197), 38–71 [hereinafter WEEE 2]. "WEEE" stands for "waste from electric and electronic equipment."
3. Each of these directives was "recast," in 2011 and 2012, respectively, to update them in response to evolving circumstances.

must instead be written into law by each Member State separately. This process can, and typically does, result in minor differences in the adopting legislation among Member States. As a result, prior to placing electrical or electronic equipment or components upon the EU market, each Member State's legislation adopting the RoHS and WEEE should be reviewed carefully.

Overview

The RoHS 2 Directive and its predecessor establish a regulatory framework for EU Member States to harmonize broadly their requirements for safety standards for electrical and electronic equipment (EEE). This framework is complemented by the WEEE 2 Directive, which seeks to harmonize the safe treatment of waste EEE and more stringently control cross-border trade in waste EEE. The "recasts" of these directives were adopted, and are being implemented, with similar deadlines, which allow for a gradual phaseout of noncompliant EEE on the EU market, as well as for recouping the costs of Member States' treatment of hazardous waste EEE from industry.

Basic Provisions

As recasts or consolidated modifications of previous directives,[4] both RoHS 2 and WEEE 2 seek to improve upon implementation and enforcement challenges across the EU, close loopholes, and reduce the administrative burdens on Member States.

The original RoHS Directive effectively banned the use of the dangerous substances in EEE listed in table 1. This ban was later modified to grant, instead, a maximum allowable tolerance for the listed substances. These tolerances allow for a small fraction of the hazardous substance(s) to be present in homogenous materials within the EEE article.[5]

---

4. The earlier versions, respectively, are Directive 2002/95/EC on the restriction of the use of certain hazardous substances in electrical and electronic equipment and Directive 2002/96/EC on waste electrical and electronic equipment.
5. U.K. NAT'L MEASUREMENT OFFICE, RoHS GUIDANCE, PRODUCER SUPPORT BOOKLET (2010), http://webarchive.nationalarchives.gov.uk/20121212135622/http://www.bis.gov.uk/assets/nmo/docs/rohs/support-literature/producer-support-booklet.pdf.

Table 1. Hazardous Substances and Maximum Allowable Tolerances

SUBSTANCE	MAXIMUM ALLOWABLE TOLERANCE (% (PPM))
Lead	0.1 (1,000)
Mercury	0.1 (1,000)
Cadmium	0.01 (100)
Hexavalent chromium	0.1 (1,000)
Polybrominated biphenyls	0.1 (1,000)
Polybrominated diphenyl ethers	0.1 (1,000)

Source: 2011 O.J. (L 174), Annex II, Restricted substances; referred to in Article 4(1) and maximum concentration values tolerated by weight in homogeneous materials.

"Homogenous material" is a term of art as defined by RoHS 2 to mean "one material of uniform composition throughout or a material, consisting of a combination of materials, that cannot by disjointed or separated into different materials by mechanical actions such as unscrewing, cutting, crushing, grinding and abrasive processes."[6] An example of a homogenous material in an EEE article is the plastic casing for a laptop computer. The tolerances for dangerous substances in homogenous materials listed in Annex II of RoHS 2 apply to the weight of the homogenous material alone, not the weight of the entire EEE article.

Exemptions

Exemptions from the outright ban and tolerable thresholds were created through amendments to the original text and were carried through in the recast. They apply to specific industries and to EEE articles and materials/formulations. As of July 1, 2006, RoHS prohibits the placing on the EU market of EEE containing the listed hazardous substances, with the following exceptions for certain equipment specified in WEEE:[7]

- Arms, munitions, and war materials intended for specifically military purposes;
- Equipment designed and installed as part of another type of equipment that is excluded from the WEEE Directive; and
- Filament bulbs.

---

6. RoHS 2, *supra* note 1, at 92.
7. WEEE 2, *supra* note 2, at 42–43.

In addition, the following EEE will be excluded from the WEEE 2 Directive beginning August 15, 2018:[8]

- Equipment designed to be sent into space;
- Large-scale stationary tools;
- Large-scale fixed installations (excluding equipment not specifically part of those installations);
- Means of transport for people or goods (excluding electric two-wheeled vehicles that are not type-approved);
- Nonroad mobile machinery available only for professional use;
- EEE solely for research and development made available on a business-to-business basis;
- Medical devices and in vitro diagnostic medical devices that are expected to be ineffective before the end of life; and
- Active implantable medical devices.

The original RoHS Directive is complemented by the original WEEE Directive. Where waste EEE (e-waste) in articles used in private households is involved, the WEEE Directive opts to place the obligation to deal with this waste on an actor better positioned than the residents of the home typically would be. Accordingly, the responsibility for the recovery, treatment, and/or disposal of articles containing e-waste substances in their pure forms from private households is on the supplier of new or replacement equipment on a like-for-like basis. Under a "polluter pays"[9] approach, the producer of the EEE is obligated to fund the appropriate collection, recovery, treatment, and or disposal of e-waste placed on the EU market after August 13, 2005.

SCOPE AND IMPLEMENTATION DEADLINES OF RoHS 2 AND WEEE 2

As with the original laws, RoHS 2 and WEEE 2 complement each other and set a unified standard for EEE waste in the EU. The scope of both regulations has been expanded to include all EEE placed upon the EU market with the exception of those types listed in article 4(2)

---

8. *Id.* at 43.
9. *Id.* at 48.

of RoHS 2[10] and article 2(3) and (4) of WEEE 2.[11] Waste from EEE already covered by the original RoHS became subject to the requirements of RoHS 2 as of January 2, 2013. New categories of EEE were identified in Annex I of RoHS 2, along with the associated implementation deadlines set out in table 2.

Table 2. RoHS 2 Implementation Dates for Each EEE Category

CATEGORY NUMBER	DESCRIPTION	IMPLEMENTATION DATE
1	Large household appliances	January 2, 2013
2	Small household appliances	
3	IT and telecommunications equipment	
4	Consumer equipment	
5	Lighting equipment	
6	Electrical and electronic tools	
7	Toys and leisure and sports equipment	
8	Medical devices	July 22, 2014
8	In vitro diagnostic medical devices	July 22, 2016
9	Monitoring and control instruments	July 22, 2014
9	Industrial monitoring and control instruments	July 22, 2017
10	Automatic dispensers	January 2, 2013
11	Other EEE not covered by any of the categories above	July 22, 2019

Source: 2011 O.J. (L 174), Articles 2.2 and 4.3 and Annex 1; also cited in http://ita.doc.gov/td/standards/Markets/Western%20Europe/European%20Union/Webpage%20RoHS%20II%20FAQ%20final.pdf.

These dates are legal deadlines that apply to EEE placed on the EU market. There is no grace period or other escape clause that authorizes the use of existing stockpiles in the supply chain or at the retail/commercial level. Participants in the EEE industry should allow for a

---
10. RoHS 2, *supra* note 1, at 91.
11. WEEE 2, *supra* note 2, at 42–43.

sufficient period of planning and preparation before each deadline to ensure that all EEE within the supply chain timely complies with the applicable RoHS 2 requirements.[12]

WEEE 2 specifies a transitional period for most categories of EEE (table 3). This period allows for older types of EEE to be replaced gradually in the household and consumer markets over a period of years.

Table 3. WEEE 2 Transitional Periods and Implementation Dates for Each EEE Category

CATEGORY NUMBER	DESCRIPTION	IMPLEMENTATION DATE
1	Large household appliances	August 15, 2018 (transitional period: August 13, 2012, to August 14, 2018)
2	Small household appliances	
3	IT and telecommunications equipment	
4	Consumer equipment and photovoltaic panels	
5	Lighting equipment	
6	Electrical and electronic tools (with the exception of large-scale stationary industrial tools)	
7	Toys and leisure and sports equipment	
8	Medical devices (with the exception of all implanted and infected products)	
9	Monitoring and control instruments	
10	Automatic dispensers	
11	Other EEE not covered by any of the categories above	August 15, 2018

Source: 2012 O.J. (L197) Article2.1

---

12. It should be noted that waste EEE exported outside the EU must comply with EC Regulations 1013/2006 and 1418/2007, as well as WEEE 2. Commission Regulation (EC) No 1418/2007 of 29 November 2007 concerning the export for recovery of certain waste listed in Annex III or IIIA to Regulation (EC) No 1013/2006 of the European Parliament and of the Council to certain countries to which the OECD Decision on the control of trans-boundary movements of wastes does not apply, 2007, O.J. (L 316), 6–52.

Notification/Registration/Approval Requirements and Compliance

Manufacturer Obligations

EEE placed upon the EU market does not need to be registered and is presumed by EU Member States to comply with all obligations. This includes the materials and components that EEE are made from.

Under RoHS 2, all EEE articles placed upon the EU market must conform and must be certified "Conformité Européenne" (CE marking), in line with Decision 768/2008/EC.[13] Manufacturers must prepare an "EU declaration of conformity" to accompany the CE marking. The elements of the declaration are set out in figure 1.

The format for the declaration must follow that set out in Annex VI of RoHS 2 and must contain the information specified in that Annex and also in article 13, as follows:

Figure 1. Format and Content of EU Declaration of Conformity

EU DECLARATION OF CONFORMITY

1. No. _____ (unique identification of the EEE).
2. Name and address of the manufacturer or his authorized representative.
3. This declaration of conformity is issued under the sole responsibility of the manufacturer (or installer).
4. Object of the declaration (identification of EEE allowing traceability. It may include a photograph, where appropriate).
5. The object of the declaration described above is in conformity with Directive 2011/65/EU of the European Parliament and of the Council of 8 June 2011 on the restriction of the use of certain hazardous substances in electrical and electronic equipment.
6. Where applicable, reference to the relevant harmonized standards used or references to the technical specifications in relation to which conformity is declared.
7. Additional information.

Signed for and on behalf of: _____
(place and date of issue): _____
(name, function) (signature): _____

---

13. RoHS 2, *supra* note 1, at 94.

These documents must be kept on file for a period of ten years after placing the EEE on the market.[14] EEE must bear identification such as type, batch, or serial number to allow traceability.[15]

Series production must include procedures to ensure that EEE remain in conformity in the face of changes in such variables as product design or characteristics, applicable harmonized standards, or technical specifications.[16] Nonconforming EEE and product recalls must be documented, including information on distributors notified of the recall. As with RoHS legislation worldwide, recordkeeping is essential to document such recalls and noncompliant products, as well as changes in product design and technical specifications.[17]

To place EEE on the market of a Member State in which it does not have a legal presence, RoHS 2 permits a manufacturer to appoint by written mandate an authorized representative who will assume all of its legal obligations within the Member State.[18] At a minimum, these obligations should include the following:

- Keeping records of the EU declaration of conformity and the technical documentation for national surveillance for ten years after placing the EEE on the market;
- Providing the Member State Competent Authority with all requested information; and
- Cooperating with the Member State Competent Authority during inspection and surveillance activities.

EEE Waste Compliance

Under WEEE 2, all EEE manufacturers and importers must be registered in either the Member State where they are based or, where the manufacturer is outside of the EU, in the Member State they

---

14. Id.
15. Id. at 95. Where the size or nature of the product does not allow for identification to be placed directly on it, the identifying information may be provided on the packaging or in accompanying documentation.
16. Id.
17. Id.
18. Id.

supply. A non-EU manufacturer, by written mandate, may appoint an authorized representative to register on its behalf and to assume legal responsibility for fulfilling the manufacturer's obligation. The appointment applies only in the Member State in which the representative is established.[19]

Regulation of Substances/Products under WEEE

Manufacturers and importers may be required by Member States to provide, at the time of sale, a best estimate of costs of environmentally sound collection, treatment, and disposal to the purchaser.[20]

Suppliers of EEE to private households must provide the following information to householders about the EEE and WEEE:[21]

- Information on correct collection and disposal of EEE and notice of the requirement not to dispose of EEE as unsorted municipal waste;
- Information that describes the various available systems of collection and collection points and encourages householders to participate (irrespective of the manufacturer or operator who established and operates the collection);
- Information on the roles of reuse, recycling, and recovery for the environmentally sound disposal of EEE;
- Information on potential environmental and human health effects as a result of improper disposal of EEE; and
- The meaning of the symbol stated in Annex IX of WEEE 2 (shown in figure 2).

---

19. WEEE 2, *supra* note 2, at 50.
20. Id. at 48.
21. Id. at 48–49.

Figure 2. Symbol for the Marking of EEE

This symbol must appear on all EEE (preferably in accordance with European Standard EN 50419) placed upon the EU market.[22] Where the size, shape, or function of the EEE does not allow for the symbol to appear directly on the item, it shall be reproduced on the packaging, instructions for use, and warranty of the EEE.[23]

Member States may require that the above information be provided to householders by the supplier in a variety of different ways, including public awareness campaigns, point of sale information, and instructions for use.[24]

Importer Obligations

Importers may place compliant EEE onto the EU market. Each importer must ensure that the manufacturer has complied with the obligations described in the chapter before placing its own EEE or incorporating any imported EEE into its own product, before placing the product on the EU market.

For traceability purposes, the importer must indicate the manufacturer's name, registered trademark or registered trade name, and contact address on the EEE or, if this is not possible, must provide the information on the packaging or in the accompanying documentation.

---

22. CENELEC (2006) Marking of electrical and electronic equipment in accordance with Article 11(2) of Directive 2002/96/EC (WEEE), EN 50419:2006, European Committee for Electrotechnical Standardization.
23. WEEE 2, *supra* note 2, at 49.
24. *Id.*

Importers must also keep records of recalls and noncompliant EEE and keep distributors informed of such developments.

Importers must retain the EU declaration of conformity and technical documentation for a period of ten years after placing the EEE on the market. This documentation may be required by Member States during surveillance activities. An importer who discovers noncompliant EEE or believes it may be noncompliant must take immediate steps to bring the EEE involved into conformity. Alternatively, the importer must withdraw or recall it (as appropriate) and must inform the Competent Authority of the Member State where the EEE was manufactured.

Distributor Obligations

A distributor of EEE must ensure that (1) it bears the CE marking; (2) it is accompanied by the required documentation in language that can be understood by consumers in the Member State in which it is placed on the market; and (3) the manufacturer and/or importer have complied with these obligations. If the distributor believes that the EEE fail to meet these, it must not place the EEE on the market until full compliance has been achieved. As with importers, if a distributor believes that EEE it has placed on the market is noncompliant, the distributor either must ensure that the necessary corrective measures are implemented by the relevant actors in the supply chain or must withdraw or recall the EEE, as appropriate. Like other such actors in the supply chain, distributors are subject to Member State surveillance and inspection activities and must comply fully with Member State Competent Authority requests for information and corrective action.

Enforcement and Due Diligence

Enforcement of RoHS 2 and WEEE 2 is the responsibility of each Member State, which generally nominates a single government agency to conduct enforcement activities within its territories. By way of example, the United Kingdom enforcement agency for RoHS 2 is the National Measurement Office (NMO), and the enforcement agency for WEEE 2 is the Vehicle Certification Agency, an executive agency of the U.K. Department of Transport. Addressing RoHS, the NMO's website states that "[m]anufacturers, authorised representatives, importers and distributors need to understand the requirements placed on them to ensure compliance." The website provides information about the regulations and guidance and addresses related topics such as due diligence.[25]

In accordance with RoHS 2, article 23, Member States must establish rules on penalties for infringements of the national provisions adopted pursuant to the directive and such penalties must be "effective, proportionate and dissuasive." This reflects a goal of both recast directives, to standardize enforcement across the EU. Punitive measures are expected to be set at similar levels across the Eurozone, with equivalent costs for non-Eurozone, EU, or European Economic Area Member States that have adopted the directives.

CE Markings

The "CE" marking requirements, as an indicator of quality assurance and compliance with RoHS 2, are a key aspect of the directive.[26] The marking requirements mean that manufacturers must conduct (or arrange to have conducted on their behalf) and then document the internal production control procedure detailed by Decision 768/2008/EC, prior to placing EEE products on the EU market. Importers and distributors must ascertain from the particular manufacturer that the EEE it supplies is in compliance with the RoHS 2 Directive and is properly marked "CE."[27]

As this chapter describes, in most cases CE marking, quality assurance, and compliance with the maximum tolerance levels of hazardous substances regulated by RoHS 2 are responsibilities of the manufacturer. Where an importer or distributor places the EEE on the EU market under its own trademark or name, however, or where the importer or distributor modifies the EEE to such an extent that compliance with the CE marking requirements of RoHS 2 may be affected, the responsibility for compliance then lies with that importer or distributor.[28]

---

25. *See* U.K. Nat'l Measurement Office, RoHS: Compliance and Guidance, http://www.bis.gov.uk/nmo/enforcement/rohs-home
26. *See* Decision No. 768/2008/EC of the European Parliament and of the Council of 9 July 2008 on a common framework for the marketing of products, and repealing Council Decision 93/465/EEC, 2008 O.J. (L 218), 82–128.
27. RoHS 2, *supra* note 1, at 95–96.
28. *Id.* at 96.

Waste EEE Containing Hazardous Substances

WEEE 2 clearly distinguishes between waste EEE and used EEE that is placed upon the EU market as functional. The obligations of the WEEE Directive concerning international trade of waste EEE were previously circumvented by making the claim that waste EEE was instead "secondhand" EEE.[29] This loophole essentially has been closed in WEEE 2 by placing specific, and potentially costly, obligations on EU companies trading in used EEE internationally.[30] Although these obligations are unlikely to affect the trade in used EEE, they are designed to curb the international trade in waste EEE, and particularly in reducing illegal shipments of waste EEE disguised as used EEE.

As presented earlier in this chapter, the requirements for waste EEE to be compliant with WEEE 2 mirror those of RoHS 2, allowing for affect entities to make life-cycle decisions about their subject products and providing sufficient time for phasing in new technologies, formulations, and designs.

Resources

Detailed information on enforcement is published by the Competent Authority in each Member State, usually in the official language of that Member State. The U.K. Competent Authorities have published detailed information on enforcement of RoHS and WEEE and are expected to update this information to comply with the adoption of RoHS 2 and WEEE 2.

U.K. RoHS guidance: U.K. National Measurement Office, Legislation, http://www.bis.gov.uk/nmo/enforcement/rohs-home/legislation
WEEE guidance: U.K. Department of Transport, Vehicle Certification Agency, Waste Electrical and Electronic Equipment Directive (WEEE), http://www.dft.gov.uk/vca/enforcement/weee-enforcement.asp
Full text of all RoHS and WEEE legislation, as well as the full text of other associated legislation, can be found on the European Commission website:

- EC RoHS information: European Commission, Environment, Recast of the RoHS Directive, http://ec.europa.eu/environment/waste/rohs_eee/
- EC WEEE information: European Commission, Environment, Recast of the WEEE Directive, http://ec.europa.eu/environment/waste/weee/index_en.htm

---

29. European Comm'n, Recast of the WEEE Directive, http://ec.europa.eu/environment/waste/weee/index_en.htm
30. WEEE 2, *supra* note 2, at 61.

6.3

Biocides

Lisa R. Burchi

Executive Summary

On September 1, 2013, the Biocidal Products Regulation (BPR) entered into force.[1] The BPR repeals and replaces the Biocidal Products Directive (BPD).[2] The BPR builds upon the BPD and seeks to harmonize the regulation of active substances and biocidal products across the European Union (EU) area. The BPR includes several new regulatory tools from its predecessor (the BPD), including authorization of biocidal products at the EU level, rather than simply at the Member State level, and tighter control of treated articles.

---

1. Biocidal Products Regulation (BPR), Regulation (EU) No 528/2012 of the European Parliament and of the Council of 22 May 2012 concerning the making available on the market and use of biocidal products. Subsequent amendments can be found at http://echa.europa.eu/regulations/biocidal-products-regulation/legislation.
2. Biocidal Products Directive (BPD), Directive 98/8/EC of the European Parliament and of the Council of 16 February 1998 concerning the placing of biocidal products on the market.

Overview of BPR

The BPR regulates active substance and biocidal products placed on the market in the EU. All substances and products meeting the definitions for the active substances and biocidal products must be approved, in the case of active substances, or authorized, in the case of biocidal products, in accordance with this regulation.

Basic Provisions

The BPR is meant to improve the EU marketplace by ensuring the harmonization of the rules for placing active substances and biocidal products on the market, and the use of these products, while ensuring a high level of protection to human health and the environment. As with most EU regulations that use assessment techniques, the provisions are based on the precautionary principle approach, "the aim of which is to safeguard the health of humans, the health of animals and the environment," which places the burden of proof on the applicant. It is therefore the applicant that must demonstrate the active substance or biocidal product is not harmful in the use for which it is intended.[3]

The BPR builds upon the BPD in the following key areas:

- Biocidal products (including treated articles) should neither be placed upon the market nor used unless they are authorized for that use.[4]
- All active substances contained in biocidal products must be approved for that use.[5]
- Similarly, all active substances contained within or that are used to treat treated articles must be approved before the treated articles can be made available.

SCOPE OF BPR

Scope of Substances/Products Subject to and/or Exempt from BPR

The purpose of the BPR is to improve the functioning of the EU biocides market through promoting greater harmonization between

---

3. BPR art. 1(1).
4. Id. art. 1(2).
5. Id. art. 95.

EU Member States and their Competent Authorities (CAs), while ensuring a high level of human health and environmental protection.[6] Article 95 of the BPR requires that all legal entities placing active substances, either on their own or in a biocidal product, on the market after September 1, 2013, do the following:

- Submit a dossier for approval of the active substance; or
- Provide a letter of access (LoA) to a dossier for approval of the active substance; or
- Provide a reference to a dossier for the approval of the active substance and for which all data protection periods have expired.

This provision sets a level playing field for all legal entities making active substances available on the EU market.

Active Substances

An active substance is the actual chemical substance that either solely or in combination with other substances exerts an action against an organism:

> "[A]ctive substance" means a substance or a micro-organism that has an action on or against harmful organisms.[7]

Active substances are further divided into two subcategories:

> "[E]xisting active substance" means a substance which was on the market on 14 May 2000 as an active substance of a biocidal product for purposes other than scientific or product and process-orientated research and development;[8]

> and

> "[N]ew active substance" means a substance which was not on the market on 14 May 2000 as an active substance of a biocidal product for purposes other than scientific or product and process-orientated research and development.[9]

---

6. *Id.* art. 1.
7. *Id.* art. 3(1)(c).
8. *Id.* art. 3(1)(d).
9. *Id.* art. 3(1)(e).

Biocidal Products

Biocidal products are mixtures or articles containing active substances for the purposes of exerting a controlling effect on an organism. "Biocidal product" is defined as:

> —any substance or mixture, in the form in which it is supplied to the user, consisting of, containing or generating one or more active substances, with the intention of destroying, deterring, rendering harmless, preventing the action of, or otherwise exerting a controlling effect on, any harmful organism by any means other than mere physical or mechanical action,
> —any substance or mixture, generated from substances or mixtures which do not themselves fall under the [above definition], to be used with the intention of destroying, deterring, rendering harmless, preventing the action of, or otherwise exerting a controlling effect on any harmful organism by any means other than mere physical or mechanical action.
> A treated article that has a primary biocidal function shall be considered a biocidal product.[10]

Treated Articles

The BPR introduces the new term "treated articles" into the biocides regulation, a term well defined under U.S. pesticide law. Treated article "means any substance, mixture or article which has been treated with, or intentionally incorporates, one or more biocidal products."[11]

Treated articles (articles not meeting the definition of a biocidal product) were previously exempt from authorization in the EU. Under the BPR, treated articles are regulated, but do not need to be authorized. All active substances contained in treated articles must be approved prior to the treated article being made available on the market.

Research and Development

Unauthorized biocidal products and unapproved active substances can be used for research and development purposes, provided that the following criteria are met:

1. Records must be created and maintained by the persons carrying out the experiment or test, including the following information at a minimum:

---

10. *Id.* art. 3(1)(a).
11. *Id.* art. 3(1)(l).

- The identity of the biocidal product or active substance;
- Labeling data;
- Quantities supplied;
- Names and addresses of persons receiving the biocidal product or active substance; and
- A dossier containing all available information on possible effects on human health, animal health, and the environment.[12]

This information must be retained and made available to the relevant Member State CA upon request.

2. Where a person intends to carry out an experiment or test that would result in release of the biocidal product to the environment, the persons must notify the Member State CA of the above information. The Member State CA must inform the persons within 45 days of their opinion. If no opinion is received within 45 days, the experiment or test may take place.[13]

If the experiment or test may have harmful or adverse effects on humans, animals, or the environment, the Member State CA may prohibit or impose conditions for the experiment or test. The opinion of the Member State CA shall be passed to the European Commission (EC) and other Member State CAs without delay.[14]

Nanomaterials

The EU has a recommended definition of a nanomaterial.[15] In accordance with this recommended definition, the BPR defines a nanomaterial as:

> a natural or manufactured active substance or non-active substance containing particles, in an unbound state or as an aggregate or as an agglomerate and where, for 50% or more of the particles in the number size distribution, one or more external dimensions is in the size range 1–100 nm.[16]

Fullerenes, graphene flakes, and single-wall carbon nanotubes with one or more external dimensions below 1 nm shall be considered nanomaterials.

---

12. *Id.* art. 56(1).
13. *Id.* art. 56(2).
14. *Id.* art. 56(3).
15. 2011 O.J. (L 275) 38–40.
16. BPR art. 3(1)(z).

Exemptions

Biocidal products and treated articles meeting the following definitions are exempt from the BPR, but are otherwise addressed in the noted EC directive or regulation:[17]

- Medicated feedingstuffs (under the definition of EC Directive 90/167/EEC);
- Active implantable medical devices (EC Directive 90/385/EEC);
- Medical devices (EC Directive 93/42/EEC);
- In vitro diagnostic medical devices (EC Directive 98/79/EC);
- Veterinary medicinal products (EC Directive 2001/82/EC and EC Regulation 726/2004);
- Medicinal products for human use (EC Directive 2001/83/EC and EC Regulation 726/2004);
- Additives for use in animal nutrition (EC Regulation 1831/2003);
- Hygiene of foodstuffs (EC Regulation 852/2004);
- Hygiene of food of animal origin (EC Regulation 853/2004);
- Food additives (EC Regulation 1333/2008);
- Flavorings and food ingredients with flavoring properties (EC Regulation 1334/2008);
- Feed (EC Regulation 767/2009);
- Plant Protection Products (1107/2009);
- Cosmetic products (EC Regulation 1223/2009); and
- Toys (EC Directive 2009/48/EC).

Biocidal products with final approval under the International Convention for the Control and Management of Ships' Ballast Water and Sediments are considered to be authorized under the BPR.[18] Member States can elect to exempt biocidal products in the interest of defense.[19]

Substances that are used during the manufacture of biocidal products, but are not intentionally included in the finished biocidal product, are considered "processing aids" and are thus exempt from approval under the BPR.[20]

---

17. *Id.* art. 2(2).
18. *Id.* art. 2(6).
19. *Id.* art. 2(8).
20. *Id.* art. 2(5)(b).

Processing aids are defined by the BPR as those substances meeting the criteria for processing aids contained in EC Regulation 1333/2008 on food additives or EC Regulation 1831/2003 on additives for use in animal nutrition. These criteria are as follows:

"processing aid" shall mean any substance which:
(i) is not consumed as a food by itself;
(ii) is intentionally used in the processing of raw materials, foods or their ingredients, to fulfill a certain technological purpose during treatment or processing; and
(iii) may result in the unintentional but technically unavoidable presence in the final product of residues of the substance or its derivatives provided they do not present any health risk and do not have any technological effect on the final product;[21]

and

"processing aids" means any substance not consumed as a feedingstuff by itself, intentionally used in the processing of feedingstuffs or feed materials to fulfill a technological purpose during treatment or processing which may result in the unintentional but technologically unavoidable presence of residues of the substance or its derivatives in the final product, provided that these residues do not have an adverse effect on animal health, human health or the environment and do not have any technological effects on the finished feed.[22]

Scope of Activities Subject to BPR

The BPR establishes responsibilities for manufacturers of active substances, producers of biocidal products, and importers of active substances and/or biocidal products. Where it is necessary for a manufacturer of the active substance, the producer of the biocidal product, or the importer to obtain an approval or authorization, the legal entity is referred to as the "applicant." In general, it is the responsibility of the EU manufacturer to obtain approval of the active substance and of the biocidal product producer to obtain the authorization of the biocidal product. In the case of non-EU manufacturers, the importer would be responsible for approval of the active substance. Where the

---

21. EC Regulation 1333/2008 art. 3(2)(b).
22. EC Regulation 1831/2003 art. 2(2)(h).

active substance is imported as a component of a biocidal product, the importer also is responsible for obtaining authorization.

All manufacturers and importers of active substances must be able to support that they have the proper approval to place an active substance on the market by one of the following means: submitting a dossier to the European Chemicals Agency (ECHA), providing a LoA to a dossier already approved, or providing a reference to data contained within the dossier for which the data protection period has expired.[23] Similarly, all biocidal products must be authorized by the relevant legal entity prior to "making them available on the market," and all legal entities must meet one of the above requirements for all active substances contained within the biocidal product. Because distribution is covered under the definition of "making available on the market,"[24] it is expected that all active substances will be approved and all biocidal products authorized by their respective manufacturers or importers. Distributors are therefore not expected to play a role in the regulatory aspects of the BPR.

Scope of Persons Subject to BPR

A key aspect of the BPR is to ensure that all manufacturers and importers of active substances and biocidal products are treated similarly by regulatory authorities and participate in sharing the regulatory burden for approval and authorization of active substances and biocidal products that they make available on the market.[25] Any legal entity making available on the market a biocidal substance or product is subject to the provisions of the BPR. The BPR defines "making available on the market" as follows:

> "making available on the market" means any supply of a biocidal product or of a treated article for distribution or use in the course of a commercial activity, whether in return for payment or free of charge.[26]

To ensure that all legal entities making active substances or biocidal products available on the market have an equal and level

---

23. BPR art. 95.
24. *Id.* art. 3(1)(i).
25. *Id.* art. 63(1), (2).
26. *Id.* art. 3(1)(i).

marketplace and that the regulatory burden is not placed solely upon a single company or group, all legal entities are required to hold a dossier or have a LoA to a dossier or its relevant data for both the active substances and the biocidal products they make available.[27]

NOTIFICATION/REGISTRATION/APPROVAL REQUIREMENTS AND COMPLIANCE

Summary of Authority Governing the Requirements for Notification/Registration/Approval

Which agency governs depends upon the type of approval (for active substances) or authorization (for biocidal products) being sought. Each type of approval or authorization also dictates the process that will apply.

Approval of Active Substances

Application dossiers are made to ECHA, which administers the approval process. The evaluation of the application dossier is made by a Member State CA. The applicant must make contact with the CA prior to submission to ensure that the CA agrees to evaluate the application once it is submitted. The agreement from the CA must be in writing and must accompany the submitted application.

Active substances are approved for specific uses, which are characterized by the 22 product-types listed in Annex V of BPR and shown in table 1 below. An application for active substance approval must contain information on at least one product-type and related biocidal product. Multiple product-types can be grouped together into a frame. A frame describes a set of related product-types that share similar properties, uses, and efficacy data. Frames do not usually cover more than a single main use group; however, in some circumstances product-types from multiple main groups can be framed together provided that the hazard, use, and efficacy data support this framing.

---

27. *Id.* art. 95.

Table 1. List of Product-Types

MAIN GROUP 1: DISINFECTANTS
Product-Type 1: Human hygiene
Product-Type 2: Disinfectants
Product-Type 3: Veterinary hygiene
Product-Type 4: Food and feed areas
Product-Type 5: Drinking water
MAIN GROUP 2: PRESERVATIVES
Product-Type 6: Preservatives for products during storage
Product-Type 7: Film preservatives
Product-Type 8: Wood preservatives
Product-Type 9: Fiber, leather, rubber, and polymerized materials preservatives
Product-Type 10: Construction material preservatives
Product-Type 11: Preservatives for liquid-cooling and processing systems
Product-Type 12: Slimicides
Product-Type 13: Working or cutting fluid preservatives
MAIN GROUP 3: PEST CONTROL
Product-Type 14: Rodenticides
Product-Type 15: Avicides
Product-Type 16: Molluscicides, vermicides, and products to control other invertebrates
Product-Type 17: Piscicides
Product-Type 18: Insecticides, acaricides, and products to control other arthropods
Product-Type 19: Repellants and attractants
Product-Type 20: Control of other vertebrates
MAIN GROUP 4: OTHER BIOCIDAL PRODUCTS
Product-Type 21: Antifouling products
Product-Type 22: Embalming and taxidermist fluids
Source: Biocidal Products Regulation (BPR), Regulation (EU) No 528/2012, Annex V.

Simplified Authorization for Biocidal Products

In cases where all active substances in a product are listed in Annex I of the BPR and the product does not contain any substance of concern or any nanomaterials, the product is sufficiently effective, and the handling and use of the product do not require personal protective equipment, a simplified authorization may be applied for.

National Authorization of Biocidal Products

Where an applicant wishes to place a biocidal product on the market of a single EU country, a national authorization may be the most effective option. This option allows for the marketing of the product in those countries for which application has been made and authorized.

Applications must include all information required in Annex III. It is possible to submit several dossiers in a single application, and it is also possible to group products together provided that they are suitably similar to reduce the total number of dossier submitted.

Union Authorization of Biocidal Products

Unlike a national authorization, a "Union authorization" allows the biocidal product to be placed upon the market across the EU area. Although the requirements of the application are the same as those for the national authorization, submission is made to ECHA, accompanied by written agreement from a Member State CA to undertake the evaluation duties.

Mutual Recognition

Mutual recognition is the process by which a biocidal product already evaluated and authorized for use in an EU Member State can obtain authorization in other Member States. This process can occur concurrently with the initial authorization (mutual recognition in parallel) or can be initiated once the authorization is granted (mutual recognition in sequence). There is no limit to the number of Member States that can grant a mutual recognition.

Standard of Review

Historically, the level of review has varied among EU CAs. The BPR seeks to harmonize the standard of the review across the EU area. Evaluation of applications for approval is based upon the criteria stated in articles 4 and 5 on the BPR conditions for approval and exclusion criteria. These criteria are set out fully in the sections below. Variation may occur in additional information requested by EU CAs because some CAs are traditionally more experienced with the review of certain chemical substances.

Burden of Proof

The burden of proof is on the applicant to prove efficacy of the active substance for the use applied for, and for the efficacy of the product to meet label claims. This requirement means that efficacy studies must be conducted, both on the active substance to prove that it is effective in the use for which application is made and for additional efficacy studies conducted upon the biocidal product to prove label claims.

Inventories or Other Listing Procedures

Information on active substances approved for use will be disseminated and electronically available to the public. A database of information including the following details will be available via the ECHA website:[28]

- International Organization for Standardization (ISO) name and International Union of Pure and Applied Chemistry (IUPAC) name;
- European Inventory of Existing Commercial Chemical Substances (EINECS) name;
- Classification and labeling information;
- Physicochemical endpoints and data on pathways and environmental fate and behavior;
- Results of toxicological and ecotoxicological studies;
- Acceptable exposure level or predicted no-effect concentration;
- Guidance on safe use;
- Analytical methods for identification;
- Terms and conditions of the authorization (for biocidal products);
- Summary of the biocidal product characteristics (for biocidal products); and
- Analytical methods of detection (for biocidal products).

ECHA and Respective Member State Competent Authorities Review and Response

ECHA and respective Member State CAs are engaged in BPR authorization and approval activities. Review and response periods vary among approval and the different types of authorization. A brief overview of the process follows.

Active Substance Approval

Applicants must submit the completed application dossier to ECHA and specify which EU Member State CA will perform the assessment. Agreement of the assessment duties must be provided in writing from the CA to ECHA in the application dossier. ECHA shall inform the

---

28. Id. art. 67.

applicant of the fees payable; if the applicant fails to pay these fees within 30 days, ECHA will inform the CA and the assessment process may be halted. The application shall only be accepted upon payment of the fees. Once the application has been accepted, both the applicant and the CA will receive the unique identification code from ECHA for administration of the approval process.

The CA has 30 days from the date of acceptance of the application to validate the application and provide feedback to the applicant and ECHA whether regarding missing information, data waiving, or substances meeting the criteria for exclusion (see below). The CA will also inform the applicant of the fees relevant to the assessment of the application, and shall reject the application if these fees are not paid within 30 days of notice.[29]

The CA will inform the applicant when it considers the information to be incomplete and will set a reasonable deadline for submission, usually not exceeding 90 days. Within 30 days of receipt of all additional information, the CA will validate the application and confirm to the applicant whether any more information is necessary. If the applicant fails to provide the requested additional information within the deadline, the CA will reject the application.[30]

Evaluation of the dossier will be performed by the CA within one year from the date of validation. If any additional information is found to be required, the CA will set an appropriate deadline for submission by the applicant. During this time, the evaluation period is frozen; this will not usually last more than 180 days, unless additional time is required to perform testing.[31]

Once the evaluation is complete, the CA will provide its conclusions to ECHA, which will prepare and submit an opinion on the approval of the active substance to the EC within 270 days.[32]

In practical terms, this means that the time frame for approval of a new active substance or for approving a new use of an active substance can be as long as two and a half years upon submission. This time frame does not include the data generation phase, which is necessary to complete the dossier that accompanies the application. If additional long-term or multigenerational studies are required, this time frame may be longer.

---

29. *Id.* art. 7.
30. *Id.*
31. *Id.* art. 8.
32. *Id.*

Simplified Authorization for Biocidal Products

As with applications for approval of active substances, application dossiers must be submitted to ECHA, which administers the authorization process. An EU CA must agree to evaluate the application dossier in writing, and this agreement must accompany the application upon submission to ECHA. The estimated time frame for evaluation and issuance of the authorization is seven months from application submission (30 days for payment of fees, 90 days for validation, and 90 days for assessment), assuming that no additional information is required.[33]

National Authorization of Biocidal Products

In the case of performing a national authorization, the application is made directly to the EU CA concerned. Additional countries can be added to this authorization either at the time of submission or once the authorization has been granted through the process of mutual recognition (see below).

The validation and evaluation period for the authorization of the biocidal product is 30 days for payment of fees, 90 days for validation, 30 days for revalidation, 365 days for evaluation, 180 days suspension for submission of additional information, and 30 days for comment period, and broadly follows that of approval for an active substance. This means that the estimated time frame after application for the authorization of a biocidal product may be more than two years. This time frame does not include data collection/generation or dossier preparation activities.

Union Authorization of Biocidal Products

Union authorization follows the same general submission process as active substance approval and simplified authorization of biocidal products. The application is made to ECHA, which administers the process, whereas the evaluation is handled by an EU CA. Again, the EU CA must agree to this prior to submission, and the written agreement must be provided to ECHA at the time of submission.

The estimated time frame for Union authorization is a year and three months; the work flow broadly follows that of an active substance

---

33. *Id.* ch. V.

approval (30 days for payment of fees, 30 days for validation, 365 days for evaluation, and 30 days for development of opinion).[34]

Mutual Recognition

Under mutual recognition, the EU CA granting the original authorization (the reference Member State) can provide their assessment to one or more Member States (the Member States concerned) to speed up the evaluation process and reduce the financial cost to the applicant. For mutual recognition in parallel, the application for authorization is sent to a single Member State (the reference Member State), who takes all evaluation obligations. At the same time, applications for mutual recognition are made to all other Member States in which the applicant wishes to market the substance. Once the reference Member State has evaluated the application and reached its decision, it makes contact with all other Member States to provide them with the evaluation materials and reach a joint conclusion. Authorizations are then provided in all Member States that agree to authorize the request.[35]

For mutual recognition in sequence, authorization may take up to five months (30 days for payment of fees, 30 days for validation, 90 days for Member State CAs to agree on the summary of the biocidal product and record their agreement, and 30 days to authorize the biocidal product).[36] Mutual recognition in parallel may take up to four months from the time of authorization of the reference Member State.[37]

REGULATION OF SUBSTANCES/PRODUCTS

The BPR builds upon the foundation of the BPD and its amendments. Active substances continue to be regulated at the Union level; that is, active substances are approved to be made available on the market across the whole EU area.[38] Biocidal products have previously only been authorized at the Member State level. Under the BPR, biocidal products may now be authorized at Member State or Union level.[39]

---

34. *Id.* arts. 43 and 44.
35. *Id.* art. 32.
36. *Id.* art. 33.
37. *Id.* art. 34.
38. *Id.* ch. II.
39. *Id.* chs. VI and VIII.

Listed Substances or Products

Active substances must be approved to be made available upon the EU market.[40] Approval involves the applicant making a submission of a scientific dossier meeting the information requirements of Annex II of the BPR.[41] Approvals for active substances must be made for specific product-types; these types are listed in Annex V of the BPR (see table 1, above).[42] The approval process involves both scientific testing to determine the hazard potential of the active substance and also efficacy testing to determine that the active substance is effective when used in the product-types applied for.[43]

Active substances are approved for an initial period of ten years.[44] After this period of approval has elapsed, active substances can be renewed for an additional 15 years and applies to all product-types previously approved, provided that they meet the relevant conditions.[45]

Biocidal products are authorized for a set period (maximum ten years) specified by the authorizing Member State CA in the authorization. Authorizations can be renewed for a term decided by the evaluating Member State CA. Renewals of Member State authorized biocidal products must be received by the evaluating CA at least 550 days before the original authorization expires.[46]

Prioritization and/or Risk Assessments

Risk assessment is specified but is not detailed by the BPR. Historically, risk assessments for biocide applications are complex, requiring expert input and close ties with the supply chain to obtain the necessary use and release information. It is likely that new guidance documents will be created to provide additional insight into the standard of information required by ECHA.

Use Restrictions

The evaluating Member State CA can specify conditions of use for an active substance within their approval opinion. This information

---

40. Id. art. 1(2).
41. Id. art. 6.
42. Id. art. 4.
43. Id. art. 6.
44. Id. art. 4.
45. Id. art. 12.
46. Id. art. 31.

is then communicated through ECHA to the European Commission, which will specify the conditions of use within the approval regulation. Use of the active substance must be within the conditions of use for all approved product-types.

Chemical or Product Bans

The BPR instigates bans on active substances, either through exclusion or substitution.

In addition to these Union-level measures, Member States can initiate additional bans of substances. Neither the BPR nor an approval or authorization (as appropriate) precludes a Member State from restricting or banning the use of biocidal products in the public supply of drinking water.[47]

Exclusion

Active substances meeting the following definition are subject to exclusion measures:[48]

- Carcinogen category 1A or 1B (in accordance with EC Regulation 1272/2008);[49]
- Mutagen category 1A or 1B (in accordance with EC Regulation 1272/2008);
- Toxic for reproduction category 1A or 1B (in accordance with EC Regulation 1272/2008);
- Endocrine disruptors;[50] or
- Persistent, bioaccumulative, and toxic (PBT) or very persistent, very bioaccumulative (vPvB) (in accordance with Annex XIII of REACH (Registration, Evaluation, Authorization, and Restriction of Chemicals)).

---

47. *Id.* art. 2(7).
48. *Id.* art. 5(1).
49. 2008 O.J. (L 353), 1–1355.
50. BPR art. 5(3). The BPR provides that no later than December 31, 2013, the EC shall adopt official scientific criteria for the classification of endocrine disruptors. These criteria have not yet been established as of the date of this writing. Until such time that official scientific criteria are adopted, the following criteria shall be used to classify an active substance as an endocrine disruptor:

    1. Classified as Carcinogen Category 2 and Toxic for Reproduction Category 2 (in accordance with EC Regulation 1272/2008), or
    2. Classified as Toxic for Reproduction Category 2 and have toxic effects on endocrine organs (hypothalamus, pineal gland, pituitary gland, and thyroid).

Active substances meeting the exclusion criteria may only be approved if it is shown that:[51]

- The risk of exposure to humans, animals, and the environment is negligible under realistic worst-case conditions of use (in particular where the product is used in closed systems or under controlled conditions that aim to exclude contact with humans and release into the environment);
- It is shown by evidence that the active substance is essential to prevent or control a serious danger to human health, animal health, and the environment; or
- Not approving the substance would have disproportionate socioeconomic effects, when compared to the risk to human health, animal health, and the environment from the proposed use.

The availability of suitable and sufficient alternative substances or technologies is important when deciding whether to approve an active substance that meets the exclusion criteria. The use of any biocidal product that contains an active substance approved in accordance with the exemption considerations must be restricted to the Member States where the above conditions are met (for example, the Member States in which the socioeconomic analysis was carried out), and are subject to risk management measures to ensure that exposure to humans, animals, and the environment is minimalized.[52]

Substitution

Substitution is the process by which the more hazardous active substances are identified and gradually replaced by less hazardous substances that can be used for the same purpose with a comparable level of efficacy.

Active substances that meet any of the following criteria are considered a candidate for substitution:[53]

- Active substances that meet the criteria for exclusion and that may be granted approval under the conditions stated above;

---
51. *Id.* art. 2.
52. *Id.* art. 5.
53. *Id.* art. 10.

- Active substances classified as respiratory sensitizers (in accordance with EC Regulation 1272/2008);
- The acceptable daily intake, acute reference dose, or acceptable operator exposure level is significantly lower than the majority of active substances for the same product-type and use scenario;
- The active substance meets two of the criteria for consideration as PBT (in accordance with Annex XIII of REACH);
- There exist reasons for concern linked to the nature of the critical effects of the active substance, which in combination with the use patterns amount to use that could still cause concern (for example, high risk to groundwater, even with very restrictive risk management measures); or
- The active substance contains a significant proportion of nonactive isomers or impurities.

Prior to submitting its opinion on the approval or renewal of any active substance to the EC, ECHA posts information on its website regarding the potential alternatives to an active substance. This information shall be open to public consultation for 60 days, during which any interested third parties may make submission of additional information. ECHA will consider all information when finalizing its opinion for the EC.

Active substances that are candidates for submission shall be identified as such in the regulation that adopts their approval. Such approvals are only valid for a maximum period of seven years.[54]

Test Data Development

Chemical Substances and Products That Must Be Tested

All active substances and biocidal products must meet the information requirements listed in Annex II or III of the BPR. This may mean that additional testing may be required above what is already available to the applicant.

Information on specific endpoints can be waived if the test is not applicable to the substance or if specific considerations listed in Annex II and III are met.[55]

---
54. *Id.* art. 10(4).
55. *Id.*, Annex II tbl., col. 3, and Annex III tbl., col. 3.

Testing Triggers

The testing requirements for all active substances and biocidal products are generally the same. Additional information may be required under the criteria set out in Annexes II and III based upon the intrinsic properties of the active substance or biocidal product, the product-type, and/or to refine the initial risk assessment.

The approval or authorization application must address all the information requirements stated in the appropriate annex and must also be sufficient to support the risk assessment.[56] In practical terms, this means that, although it may be possible to waive multiple testing requirements, sufficient information must be available to undertake the risk assessment on the proposed use of the active substance or biocidal product in a robust manner.

Implementing Test Requirements

Any applicant wishing to undertake testing on vertebrate animals for the purposes of the BPR must submit a written request to ECHA to determine whether the information and results of such tests have already been submitted by another applicant and are therefore available for data sharing.[57] Tests not conducted upon vertebrate animals (for example, physicochemical tests and environmental fate tests) do not require the applicant to make this submission to ECHA.[58]

Data Sharing

The BPR seeks the dual goals of reducing the amount of animal testing needed to meet the endpoint requirements of Annexes II and III and avoiding duplicate animal testing.[59] To facilitate achieving these goals, the BPR sets out data-sharing procedures for information previously submitted by an applicant and data protection for data shared with other applicants and (in part) disseminated to the public.[60]

---

56. *Id.* at art. 8.
57. *Id.* art. 62(2).
58. *Id.* In the context of a company seeking to be listed on the Active Substance Suppliers List under article 95, mandatory data sharing applies to tests involving vertebrate animals and to all toxicological and ecotoxicological studies not involving tests on vertebrates. This extension of data sharing applies only for active substances in the review program.
59. *Id.* art. 62(1).
60. *Id.* art. 62.

Following a data-sharing inquiry, ECHA will issue a confirmation of existing data to the applicant. It is the obligation of the applicant to contact the data owner and arrange either for the sharing of data or for the right to refer to that data for the purposes of the BPR.[61] In lieu of obtaining a full share of the data, the applicant may be able to purchase a LoA that grants the right to refer to the data already submitted for the purposes of the BPR.[62]

The content of the LoA is prescribed in Article 61 of the BPR.[63] These requirements are as follows:

- The name and contact details of the data owner and the beneficiary;
- The name of the active substance or biocidal product for which access to the data is authorized;
- The date on which the LoA takes effect; and
- A list of the submitted data to which the LoA grants citation rights.

Article 61 also states:

> Revocation of a letter of access shall not affect the validity of the authorisation issued on the basis of the letter of access in question.[64]

This statement indicates that even where a LoA is revoked because of misuse or violation of contract, a submitted authorization or approval using the data in question would still be valid.

Data Protection Periods

Data protection periods should not be confused with intellectual property. The data protection period is the duration during which the data owner can charge new applicants a fee for relying on those data for purposes of processing their registrations. Although the data will always be the property of the owning legal entity (or entities), once the data protection period elapses, ECHA and EU Member State CAs may consider the information already submitted by an applicant to be in support of a new application for approval or authorization

---

61. *Id.* art. 62(2).
62. *Id.* art. 61.
63. *Id.*
64. *Id.*

without a LoA or data-sharing agreement in place. Data protection periods are set out in table 2.

Table 2. Data Protection Periods

PRODUCT	DATA PROTECTION PERIOD
Existing active substances or biocidal products	10 years from the first day of the month following the decision of approval or authorization
New active substances or biocidal products	15 years from the first day of the month following the decision of approval or authorization
Renewal or amendment of approval or authorization	5 years from the first day of the month following the decision concerning renewal or amendment
Source: Biocidal Products Regulation (BPR), Regulation (EU) No 528/2012, Article 60.	

Where a biocidal product contains both new and existing active substances, the data protection period lasts 15 years and applies from the first day of the month following the decision concerning the authorization.

Compensation for Data Sharing

Compensation for data sharing must be conducted in a fair, transparent, and nondiscriminatory manner.[65] Guidance produced by ECHA for data sharing under REACH shall be applicable to the BPR.[66]

The level of compensation is at the discretion of the data holder and the applicants relying on their data, using the above information to guide them. The reality, however, is that the data holder largely controls the compensation cost. The agreement can be replaced by the referral of the matter to an arbitration body, and may require the commitment of both sides to accept and abide by the decision of the arbitration body.[67]

If the parties cannot reach an agreement (either by themselves or through third-party arbitration), the prospective applicant may refer the matter to ECHA. The parties must make every effort to reach agreement. The matter can only be referred to ECHA at the earliest one month after the prospective applicant receives the contact

---

65. Id. art. 63(4).
66. Id.
67. Id. art. 63(1).

details of the data owner from ECHA.[68] Provided that the prospective applicant can demonstrate that it has made every effort to come to an agreement, ECHA must provide permission to refer to the requested information on vertebrate animals within 60 days of being informed.[69] Pursuant to this intervention by ECHA, the data holder cannot refuse an offer of reimbursement.[70] Acceptance of such an offer does not prejudice the rights of the data holder to have the proportionate share of the costs determined by a national court.[71]

An appeal may be brought against ECHA in regard to data sharing. Where the potential applicant or data holder has a concern regarding the treatment and judgment of ECHA, such an appeal can be submitted to the ECHA Board of Appeal.[72]

Reporting and Recordkeeping Requirements

Although the BPR does not require reporting of tonnage information on active substances and biocidal products, records must be kept in case of inspection by a Member State CA.

Manufacturers of biocidal products must maintain records of the manufacturing process (either paper or electronic) relevant to the quality and safety of biocidal products. These records must include at least the following information:

- A Safety Data Sheet (SDS) and specification of each active substance and other ingredients used for manufacturing the biocidal product;
- The manufacturing process itemized into each operation;
- Results of internal quality control; and
- Identification of production batches.

In addition to this information, biocidal product manufacturers must also retain production batch samples under appropriate storage conditions.

---

68. Id. art. 63(3).
69. Id.
70. Id.
71. Id.
72. Id. art. 63(5).

IMPORTS AND EXPORTS

The authority to regulate and inspect imports and exports remains with the EU Member State to which import is made or from which material is exported. The BPR does not affect customs procedures and inspection duties.

CONFIDENTIAL AND TRADE SECRET INFORMATION

Authority Relating to Confidential Business Information

Public access to documents submitted by the applicant and kept on file by ECHA is generally regulated under the provisions of EC Regulation 1049/2001 regarding public access to European Parliament, Council, and Commission documents.[73]

Information considered by the applicants to be detrimental to their business interests or considered to pose a risk to privacy or safety of the persons concerned can be kept confidential. The following information is considered to fulfill these criteria and so is not normally disclosed by ECHA. Where urgent action is essential to protect human health, animal health, safety, or the environment, or for other reasons of overriding public interest, ECHA may disclose this information, either in whole or in part:[74]

- Details of the full composition of a biocidal product;
- The precise tonnage of the active substance or biocidal product manufactured or made available on the market;
- Links between a manufacturer of an active substance and the person responsible for the placing of a biocidal product on the market or between the person responsible for the placing of a biocidal product on the market and the distributors of the product; and
- Names and addresses of persons involved in testing on vertebrates.

---

73. Id. art. 66.
74. Id. art. 66(2).

Once an authorization of a biocidal product has been granted, the following information shall be made available to the public and cannot be kept confidential:[75]

- The name and address of the authorization holder;
- The name and address of the biocidal product manufacturer;
- The name and address of the active substance manufacturer;
- The content of the active substance or substances in the biocidal product and the name of the biocidal product;
- Physical and chemical data concerning the biocidal product;
- Any methods for rendering the active substance or biocidal product harmless;
- A summary of the results of the efficacy tests and effects on humans, animals, and the environment (where appropriate, this summary shall include data on the product's ability to promote resistance);
- Recommended methods and precautions to reduce dangers from handling, transport, and use, as well as from fire or other hazards;
- SDSs;
- Methods of analysis;
- Methods of disposal of the product and its packaging;
- Procedures to be followed and measures to be taken in the case of spillage or leakage; and
- First aid and medical advice to be given in the case of injury to persons.

Provisions Relating to Confidential Business Information

The following information can be claimed as confidential by the applicant. Any such claim must be accompanied with a robust justification:[76]

- The degree of purity of the substance and the identity of impurities and/or additives of active substances that are known to be hazardous (only where essential to classification and labeling);

---

75. Id. art. 66(3).
76. Id. art. 67(3).

- Study summaries and robust study summaries submitted in support of the approval of the active substance and/or biocidal product;
- Information contained within the SDS that is not subject to the general obligations of dissemination;
- The trade name(s) of the active substance; and
- The assessment report.

Enforcement and Penalties

In accordance with Article 87 of the BPR, individual Member States must lay down provisions for penalties for violation of the BPR within their borders. Such penalties must be effective, proportionate, and dissuasive. As of this writing, although enforcement duties and penalties were expected to be written into the legislation of each Member State by September 1, 2013, many Member States have not yet implemented the BPR with respect to enforcement and penalties.

As one example, based upon the enforcement provisions for the BPD, enforcement of the BPR will likely be split between a number of U.K.-based agencies, depending upon the type of incident or biocidal product in question. Table 3 breaks down the areas of authority.

Table 3. BPR Enforcement Authorities in the U.K.

PESTICIDE PRODUCTS	
TYPE OF INCIDENT	RELEVANT AUTHORITY
Issues around advertising or retail sale	Local Trading Standards Office
Incident involving people in the workplace	Health and Safety Executive (HSE) or Local Authority
Environment	Environmental Agency
Wildlife, livestock or pets/companion animals	Wildlife Incident Investigation Scheme

BIOCIDAL PRODUCTS	
TYPE OF INCIDENT	RELEVANT AUTHORITY
Issues around advertising or retail sale	Local Trading Standards Office
All issues relating to biocidal products	HSE or Local Authority

Source: http://www.hse.gov.uk/biocides/enforcement.htm

Penalties vary considerably based upon the damage caused and the level of the offense.

COMMONLY ENCOUNTERED ISSUES

Issues commonly encountered under the BPR are likely to mirror those seen under REACH because of the similarities between the two regulations. Those issues that are most commonly encountered under REACH are therefore explored below in regard to the provisions of the BPR.

Substance Sameness and Technical Equivalence

The BPR introduces the new concept of technical equivalence, which is similar to substance sameness under REACH.[77] Technical equivalence is used where it is necessary to establish that active substances produced by different manufacturers or through different processes are the same substance and can therefore benefit from data sharing.

Because of the challenges of ensuring substance sameness under REACH, ECHA has sought to reassure industry by releasing information on how technical equivalence will be assessed before the full implementation of the BPR.[78] Technical equivalence is assessed in a two-tier system as follows:

Tier 1: A Tier 1 assessment is based upon analytical and spectral information. These data are composed of a combination of the following data, which are considered both to be appropriate to the substance and also to provide adequate characterization of the substance:

- Gas chromatography (GC);
- High-performance liquid chromatography (HPLC);
- Mass spectroscopy (MS) (usually performed alongside chromatography);
- Ultraviolet spectroscopy (UV/Vis);
- Nuclear magnetic resonance (NMR); and
- Infrared spectroscopy (IR).

Where it is difficult to characterize a substance using the above methods, titrations and other less commonly used analytical tests can be useful.

For those familiar with REACH, the requirements in Tier 1 are similar to those identified for performing an Article 26 Inquiry.

---

77. *Id.* ch. XI.
78. ECHA, Technical Equivalence, http://echa.europa.eu/regulations/biocidal-products-regulation/technical-equivalence.

Tier 2: Where it is not possible to characterize the active substance using the spectral and analytical methods identified in Tier 1, available physicochemical and nonvertebrate toxicological data will be reviewed and compared on the active substances by ECHA to determine if indeed the substances are the same substance. This method has been pioneered by CONCAWE (Conservation of Clean Air and Water in Europe), a division of the European Petroleum Refiners Association, for substance sameness of Unknown, of Variable Composition, or of Biological Origin (UVCB) petroleum substances under REACH.[79] Testing on vertebrate animals may not be duplicated solely for the purposes of assessing technical equivalence.[80] If vertebrate animal data are available, however, there is nothing that precludes this data from being submitted and considered. It is important that for comparison purposes, the testing guideline, species, and conditions be fully presented. There is the opportunity for nuances between test methods, or even impurities in the active substance, to affect an analysis of technical equivalence.

REACH requires that a single registration be submitted for each substance placed on the EU market. Similarly, BPR requires that ECHA confirm the technical equivalence (substance sameness) of active substances sourced from different manufacturers, or where the manufacturing process is changed (including where starting materials are changed). Under REACH, this has been an intricate process, using spectral and analytical data to characterize and compare the composition and purity of substances from different manufacturers and suppliers. ECHA has previously indicated that technical equivalence will take place in a tiered manner, using spectral and analytical data at an initial level and physicochemical and toxicological assessment to derive the hazard characterization at the second.[81]

Data and Cost Sharing

Data and cost sharing have caused issues for many legal entities under REACH[82] and can be expected as well in BPR. Companies have found it difficult to interpret what information to which they have a legal

---

79. CONCAWE, GUIDANCE ON ANALYTICAL REQUIREMENTS FOR PETROLEUM UVCB SUBSTANCES—COMPLETION OF REACH REGISTRATION DOSSIERS (Version 1.5, Dec. 1, 2010).
80. BPR art. 62.
81. ECHA, Technical Equivalence, http://echa.europa.eu/regulations/biocidal-products-regulation/technical-equivalence.
82. *ECHA Steps Up Support to Tackle REACH Data-Sharing Disputes*, CHEM. WATCH, Aug. 2, 2010, http://chemicalwatch.com/4683/echa-steps-up-support-to-tackle-reach-data-sharing-disputes.

right to refer. In extreme cases, REACH registrants have submitted information published by the U.S. Environmental Protection Agency (EPA) High Production Volume (HPV) program, or even information obtained through freedom of information requests.[83] Data-sharing disputes may now be referred to a third-party arbitration body, provided that both sides accept their judgment. This may lessen the pressure upon ECHA to provide judgments in this area.

Data Protection

Data protection periods are more heavily specified under the BPR than under REACH. Initial and renewal periods for data protection are set out in the BPR, providing a concrete regulatory framework for information referral by other potential registrants.

Trends

Ongoing trends in biocide regulation are generally characterized by increases in safety assessments for known hazards in previously approved active substances.

Review Program

The review program was launched in 2000 with the adoption of EC Regulation 1896/2000, which specified that manufacturers and formulators had to identify or notify the active substances they produced/used before March 28, 2002.[84] The EC regulation set out the deadline for those substances identified but not notified as of January 31, 2003.[85] This initial phase of the review program was performed with the aim of cataloging active substances present in biocidal products on the EU market and categorizing them based upon their uses. From this initial phase, and throughout the following phases, the list of biocidal product-types was created and diversified, and the amount of information on the hazardous properties of each substance was increased.

This process was costly to industry because the burden for testing and submission in a standardized format was placed almost entirely

---

83. ECHA Makes Substantial Changes to Guidance on Data-Sharing, CHEM. WATCH, Apr 4, 2012, http://chemicalwatch.com/10693/echa-makes-substantial-changes-to-guidance-on-data-sharing.
84. 2000 O.J. (L 228) 6–17; see European Comm'n, Environment, Regulations, http://ec.europa.eu/environment/biocides/regulation.htm
85. 2002 O.J. (L 258) 15; see European Comm'n, Environment, Regulations, http://ec.europa.eu/environment/biocides/regulation.htm

upon the private sector. For this reason, industry has been sluggish to develop new active substances for use in biocidal products.

Ecolabel

The EU Ecolabel program is a voluntary labeling program for consumer products that has experienced an increase in popularity in recent years. The scope of the BPR covers consumer products such as soaps, cleaners, detergents, textiles, footwear, coatings, and in-can preservatives. Ecolabel criteria are set by experts, including industry and consumer group stakeholders, and cover the entire life cycle of the product. Ecolabel products must be both good quality (effective) and environmentally friendly.

The adoption of the Ecolabel program by industry has increased the need for less hazardous but effective alternatives to traditional active substances. This has also fueled the need for reliable information on active substances used in consumer products.

New Active Substance Development

The demand for new active substances that are less hazardous and can be used as alternatives for existing active substances without the need for extensive or costly alterations to biocidal products or manufacturing techniques is likely to increase under the BPR and in line with the uptake of the Ecolabel program.

While industry has been slow to develop new active substances, the need for alternatives and the potential for the phasing out of certain existing active substances may well accelerate the development of new technologies to produce new, greener active substances.

Resources

European Commission: The EC is a valuable source of information on the development of EU biocidal regulations, as well as a repository for information on active substances and the current status of the review program. See http://ec.europa.eu/environment/chemicals/biocides/regulation/regulation_en.htm.

European Chemicals Agency: On September 1, 2013, ECHA assumed the responsibility for administration of the Union-level activities of the BPR. While ECHA has not released practical information on the electronic submission, communication, and inventory system to be used under the BPR, this information is expected to be available on the ECHA website, http://echa.europa.eu/regulations/biocidal-products-regulation.

Regulation for Biocidal Products (R4BP): The legacy system for information on biocidal products and active substances is expected to be kept online and available for stakeholders. See https://webgate.ec.europa.eu/env/r4bp2/.

7

Asia

EXECUTIVE SUMMARY

Asian economies offer cost-effective alternatives to the domestic manufacture of chemical substances in Western Hemisphere countries. Asia is thus of significant and growing interest to multinational companies, both as a base for manufacture and a source of chemicals for import. Because of the rapid growth in these markets—and, in some cases, because past practices or unfortunate chemical accidents have underscored the benefits of reducing and managing pollution—Asian governments are placing increased emphasis on chemical control programs and greatly strengthening and expanding these programs. Similar to their U.S. and European counterparts, Asian chemical control programs reflect an approach that imposes the greatest scrutiny and data demands on new or existing chemical registrations of chemical substances manufactured or imported in the greatest volumes with an eye, as well, to hazardous properties and potential. Because these markets are a critical part of the world economy, it is important to recognize the key elements of the chemical control programs and the challenges U.S.-based and other entities face in doing business in Asia. Chapters 7.1 and 7.2 describe the regulatory landscapes in China and South Korea, respectively, and point up specific challenges that come with the territory for foreign manufacturers and importers who are, or seek to be, active in these jurisdictions.

7.1

China

Leslie S. MacDougall, Andrew G. Burgess, and Lynn L. Bergeson

OVERVIEW

Regulation of chemical substances in China is complex. Three primary laws regulate industrial chemicals similar to the way the Toxic Substances Control Act (TSCA) regulates industrial chemicals in the United States:

- State Council Decree No. 591, Regulations on the Safe Management of Hazardous Chemicals in China,[1] issued on March 2, 2011;
- State Administration of Work Safety (SAWS) Order No. 53, Measures for the Administration of Hazardous Chemicals, issued on July 11, 2012,[2] which regulate "dangerous" or "hazardous" chemicals; and

---

1. Available in Chinese at http://aqbzh.chinasafety.gov.cn:8080/wss/abxxfbAction1!list XxfbMore.action?xxlx=XXFB_FLFG.
2. Available in Chinese at http://www.chinasafety.gov.cn/newpage/Contents/Channel_5330/2012/0711/173367/content_173367.htm.

- Ministry of Environmental Protection (MEP) Order No. 7, Environmental Management of New Chemical Substances in China (Order No. 7),[3] which regulates "new" chemical substances and was issued on January 19, 2010.

Chemicals management in China and the Asian region in general is in a state of tremendous flux and understanding these authorities is challenging for a variety of cultural reasons. Language barriers exacerbate these challenges as official translations of laws and regulations are not available from the government. While there are similarities between the Chinese chemical management spectrum and TSCA and the European Union (EU) Registration, Evaluation, Authorization, and Restriction of Chemicals (REACH), the programs are quite different, and, as noted, very much evolving.

Governance Framework

Before delving into the specifics of chemical regulation in China, we describe generally the governance framework for the management of chemicals. The framework consists of three core elements. The first element consists of laws issued by the Standing Committee of the National People's Congress (NPC). The NPC is the highest level of state power in China. The Standing Committee is the permanent organ of the NPC.

The second element of the framework consists of administrative regulations issued by the various ministries included within the State Council of the People's Republic of China, specifically the Central People's Government, which is the highest executive level of state power. The State Council consists of a premier, vice premiers, state councils, ministers in charge of ministries and commissions, the auditor-general, and the secretary-general. The MEP and the Ministry of Transportation (MOT) are two of many ministries under the State Council. The MEP is responsible for the supervision and administration of the disposal of hazardous waste, the assessment of chemical hazards and the risks chemicals pose, and the registration of chemical substances. In mid-2013, the Competent Authority for new chemical notification

---

3. Available in Chinese at http://www.crc-mep.org.cn/news/NEWS_DP.aspx?TitID=225&T3=01&LanguageType=CH&Sub=12.

activities shifted to the MEP – Solid Waste and Chemical Management Center (MEP-SCC), formerly referenced as the MEP – Chemical Registration Center (MEP-CRC). The MEP-CRC website and references are currently being redirected and/or replaced to reference MEP-SCC. The MOT is responsible for the regulating the rail, road, air, and water transport of chemicals.

The third element consists of departmental rules issued by SAWS, the authority responsible for the national work safety management and various technical standards, including national standards and industrial standards.

Together, these ministries and departments have issued various environmental and worker protection laws, laws intended to address emergency response, and laws regulating the rail, land, and air transport of chemicals.

Cultural Framework

As important as the governance framework in China is its cultural framework. An often repeated statement in describing Chinese practices is that legislation is developed through planning and consensus building. The NPC takes the long view, and progress is anchored by the decisions of regulatory predecessors. The core goal of the central government is to set targets and leave to the regulators the task of designing a regulatory system to achieve these targets. Regulations are intended to serve as incremental stepping stones to achieving long-term goals, and are not viewed as ends in themselves.

Two other points are important. First, regulations and guidance documents in China are typically quite short on detail and granularity. As important as the written word is what is not explicitly stated in an order, regulation, or guidance document, which makes compliance an ongoing challenge. While some have suggested that the absence of detail is a de facto trade barrier, the reality is that regulations are specifically designed to be flexible and are adapted and refined as necessary by experience and evolving science.

Second, challenges posed by language barriers are real and significant. Official documents are issued in Chinese and there are no "official" translations. It is thus critically important to confirm translations from the Chinese with knowledgeable and native speakers who are familiar with the regulations.

SCOPE OF THE THREE PRIMARY LAWS

Decree No. 591

State Council Decree No. 591 replaced Decree No. 344, which had been in effect since 2002. Decree No. 591 sets forth the regulations addressing the manufacture, use, operation, transportation, and registration of hazardous or dangerous chemicals. It entered into effect on December 1, 2011.[4]

Substances Subject to Decree No. 591

Decree No. 591 consists of eight chapters and 102 "articles." Decree No. 591 is challenging from a compliance perspective as the legislation is silent on the availability of exemptions, and the Catalog of Hazardous Chemicals was only recently updated, the prior version dating back to 2002. The decree serves as the overarching regulation for regulations expected to be published by various government ministries, including the U.N. Globally Harmonized System of Classification and Labelling of Chemicals (GHS) implemented in China.

Decree No. 591 revised the definition of hazardous or dangerous chemicals to include chemicals that are toxic or that reflect certain properties, including corrosiveness, flammability, explosiveness, and related properties that have the potential to cause harm to human health and the environment.[5] The Catalog of Hazardous Chemicals, which was updated and released for comment in September 2013, identifies chemicals satisfying the criteria for hazardous/dangerous. The revised draft version included approximately 3,000 substances, less than the almost 4,000 substances included in the current Catalog of Hazardous Chemicals. Once the draft Catalog is formally adopted, it will replace the current Catalog of Hazardous Chemicals. Decree No. 591 contemplates the issuance of a consolidated list of dangerous chemicals to be jointly issued by other government agencies under the State Council in accordance with the rollout of the GHS.

Part of the inherent complexity of Decree No. 591 relates to the fact that ten government agencies participate in the regulation of dangerous chemicals. These include SAWS, MEP, MOT, the Ministry of Agriculture, and others. Each agency is tasked with issuing its

---

4. See http://www.miit.gov.cn/n11293472/n11505629/n11506364/n11513631/n11513880/n11927781/15122828.html.
5. Decree No 591 art. 3.

own implementing regulations. The scheme for regulating dangerous chemicals is therefore highly fragmented and difficult to understand and track.

Permits and Licenses

The backbone of Decree No. 591's regulation of dangerous chemicals is a complex system of "permits" or "licenses." Virtually any activity involving dangerous or hazardous chemicals cannot occur lawfully in China without first procuring one or more of these licenses from the local or higher government agency. Under the laws in China, it is not unusual for a company to have multiple licenses, one for production, one for import, and one for use. Figure 7.1 summarizes these permits and licenses. Knowing what activity is associated with what permit or license and from which agency it must be procured is half the battle in navigating Decree No. 591 compliance.

Figure 7.1

	Authority and Regulatory Obligations	License	Information Resources	Notes Comments	Chemical Inventory
Manufacture	SAWS: Site Manufacturing Hazardous Chemicals	Safe Production License	http://www.chinasafety.gov.cn/newpage/	GHS compliant SDS and Label requirements apply	Hazardous Chemicals Inventory
	NRCC – SAWS: Registration for each Hazard Chemical manufactured into China	Hazardous Chemicals Certificate	http://register.nrcc.com.cn/		http://www.chinasafety.gov.cn/whpcx.htm
Import	Local Work Safety Department: Site Importing Hazardous Chemicals	Safe Operations License	Provincial – examples Beijing: http://www.bjsafety.gov.cn/publish/portal0/ Shanghai: http://www.shsafety.gov.cn/index.htm	GHS compliant SDS and Label requirements apply AQSIQ and CIQ address storage provisions	Hazardous Chemicals Inventory
	NRCC – SAWS: Registration for each Hazard Chemical imported into China	Hazardous Chemicals Certificate	http://register.nrcc.com.cn/		http://www.chinasafety.gov.cn/whpcx.htm
Distribution	Local Work Safety Department: Site Distributing Hazardous Chemicals (storage and warehouse)	Safe Operations License	Local Municipal and County Work Safety Department	Interested in business license for hazardous chemicals, inventory of dangerous, hazardous, toxic, explosive chemicals, etc., at the site	
Transportation	Ministry of Transport	Dangerous Goods Transportation License	http://www.mot.gov.cn/	List of Dangerous goods: GB 12268-2005. Classification and Code of Dangerous goods: GB 6944-2005	Included along with Hazardous Chemicals on http://www.chinasafety.gov.cn/whpcx.htm
	County Public Security Bureau	Toxic Chemicals Transportation License			
End User	Local Work Safety Department: Site Using Hazardous Chemicals	End Use License	Local Municipal Work Safety Department	Emergency Response program, etc.	

Manufacture, Distribution, Registration, and Use of Dangerous Chemicals

Companies producing or importing chemicals deemed dangerous or hazardous under Decree No. 591 must register these substances with the National Registration Center of Chemicals (NRCC), managed under SAWS.[6] As noted in figure 7.1, in addition to or in lieu of registering with the NRCC, some activities may require notification to obtain licensing and permitting at the local level. The registration of dangerous or hazardous chemicals in China is a cornerstone of the chemicals management system. Both the site where chemicals are manufactured and each dangerous chemical manufactured or imported into China must be registered.[7] Information that must be submitted includes a description of the chemical and its uses, hazard class, and emergency response information.[8]

Under Decree No. 591, manufacturers of regulated dangerous or hazardous chemicals are required to provide a Material Safety Data Sheet (MSDS) and affix precautionary labeling on each package containing the substance.[9] If a new dangerous characteristic is identified in a chemical, the manufacturer must notify the public and update the MSDS as appropriate.

Decree No. 591 contains separate provisions for users of "dangerous" chemicals. Users are required to obtain safety use licenses from the local authority if the volume of a chemical exceeds a prescribed amount, and the entity uses the dangerous chemical in its production process.[10] SAWS implements the requirements pertinent to this permit system through Order No. 57, Implementation Measures on the Permit for Safe Use of Hazardous Chemicals.[11]

Any company that sells, distributes, or provides storage services for dangerous substances must apply for a Safe Operation License of Dangerous Chemicals from the local authority.[12] Requirements apply to "specialized staff" and an emergency response plan must be developed and emergency rescue equipment must be obtained as a condition of operation. Entities are not permitted to purchase dangerous chemicals from unlicensed manufacturers.

Chapter 5 of Decree No. 591 addresses the requirements pertinent to the transportation of dangerous chemicals. The transport of dangerous chemicals by any means, including land, rail, water, or air,

---

6. Id. art. 67.
7. Id. arts. 66–67.
8. Id. art. 67; Order No. 53 art. 12.
9. Id. art. 15.
10. Id. art. 29.
11. Order No. 57, available in Chinese at http://www.chinasafety.gov.cn/newpage/Contents/Channel_5330/2012/1127/187011/content_187011.htm.
12. Decree No. 591 art. 32.

requires a license. The decree contains detailed requirements pertinent to the transport of dangerous chemicals.[13]

SAWS Order No. 53

Scope of SAWS Order No. 53

SAWS Order No. 53, Measures for the Administration of Hazardous Chemicals, specifies the scope, application, review, and issuance of registration certificates. Order No. 53 was issued in final in July 2012, and implements the registration provisions of State Council Decree No. 591 that came into force on December 1, 2012.

Under Order No. 53, manufacturers and importers of dangerous or hazardous chemicals must register with NRCC prior to any manufacturing or import activity.[14] Company registrations entail preparing and completing an application via an online system, which then is reviewed by the local registration office within 20 days of submission.[15] A subsequent review is conducted by the NRCC and, if successful, a registration certificate is issued to the applicant within 15 days. A registration certificate is valid for three years, and must be renewed within three months of expiration.[16]

Prioritization

Hazardous chemicals are chosen for "priority" assessment from the Catalog of Hazardous Chemicals based on their hazardous properties and risks. Companies using chemicals listed in the Catalog must obtain a Safe Use Permit at the local level when the quantity threshold is exceeded.[17] Priority chemicals will also be subject to a pollutant release and transfer registry (PRTR) similar to the U.S. Environmental Protection Agency's (EPA) Toxics Release Inventory (TRI) reporting program under the Emergency Planning and Community Right-to-Know Act (EPCRA) promulgated in 1986. While both Canada, under its national Pollutant Release Inventory program, and the EU, under its Pollutant Emissions Register program, maintain such systems and have for years, the PRTR is new in China.

---

13. *Id.* arts. 43–65.
14. Order No. 53 arts. 10–11.
15. *Id.* art. 13.
16. *Id.* art. 16.
17. *See* http://www.chinasafety.gov.cn/newpage/newfiles/hxpml2013.pdf.

MEP Order No. 7

Order No. 7 is the cornerstone of "new" chemical substance notification and registration in China. Issued by the China MEP on January 19, 2010, the order came into force on October 15, 2010. It replaces Order No. 17, titled Measures on the Environmental Management of New Chemical Substances, which was issued in 2003. Sometimes referred to as the "China REACH," Order No. 7 is similar to TSCA and REACH in that a "new" chemical is any chemical subject to Order No. 7 and that is not currently among the approximately 46,000 substances listed on the Inventory of Existing Chemical Substances Produced or Imported in China (IECSC) or otherwise exempt from notification. Under the regulation, companies are required to submit a new chemical substance notification to the MEP – SCC for substances considered within the scope of the regulation, regardless of annual tonnage amount. The notification applies to the new substance, the substance in preparations (mixtures), and in articles intended to be released. The chemical notification also covers the use of a new substance when it is used as an ingredient in pharmaceuticals, pesticides, veterinary drugs, cosmetics, food additives, and feed additives.

Notification/Registration/Approval Requirements
Inventory of Existing Substances

Chemical substances considered "existing" substances are those listed on the IECSC published by the MEP-SCC, formerly the CRC. "New" chemical substances—those not listed on the IECSC—must be notified (registered) prior to manufacture in or import into China if they are within the scope of Order No. 7. As of late 2013, there are approximately 46,000 substances reportedly listed on the IECSC, which was last updated in January 2013. According to a statement issued by CRC on December 31, 2013, the IECSC was revised to be consistent with China's updated customs regulations as well as the Rotterdam Convention on the Prior Informed Consent Procedure for Certain Hazardous Chemicals and Pesticides in International Trade.[18] The IECSC is available for downloading.[19]

---

18. The 2014 Catalog of Hazardous Chemicals Restricted for Import and Export is available in Chinese at http://www.mep.gov.cn/gkml/hbb/bgg/201312/t20131231_265886.htm.
19. *See* http://www.mep.gov.cn/gkml/hbb/bgg/201301/t20130131_245810.htm (in Chinese).

7.1. China

Inventory—Confidential Lists

There is also a "confidential" IECSC, similar to the confidential TSCA Inventory EPA maintains. As of February 2013, the confidential inventory reportedly contains approximately 3,270 substances. Chemicals are listed by category names, but no Chemical Abstracts Service (CAS) Registry Number or molecular identity information is disclosed. Similar to the TSCA bona fide submission, entities can ask MEP-SCC to search the confidential inventory by CAS number or other identifier. MEP-SCC searches the confidential inventory, for a fee (approximately 600 renminbi (RMB), or $100 USD). A reply from MEP-SCC generally takes less than two weeks.

Chemical Substance Notifications

Article 9 of Order No. 7 identifies the three types of new chemical substance notifications: scientific research record (SRR), simplified, and regular. This is an important and somewhat challenging aspect of the order and is briefly described in figure 7.2.

Figure 7.2

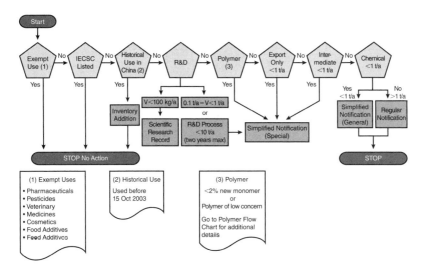

Scientific Research Record (SRR) Notification

An SRR notification[20] is intended to accommodate the use of new chemicals for research and development (R&D) purposes only at a

---

20. Order No. 7 art. 14.

volume of <100 kg/year. Unlike TSCA, where new chemicals used in R&D applications are self-implementing and exempt from new chemical notification, R&D new chemical applications in China are subject to notification obligations. Upon submission of the SRR, there is no review or waiting period and R&D activities may commence immediately. This category includes the shipment of chemical samples to Chinese laboratories that perform testing. An SRR notification must be completed for every such shipment, and the total volume of all samples manufactured in or imported into China must be addressed. Careful attention must be made to ensure that the total volume of an SRR substance remains below the 100 kg/year threshold. If amounts in excess are expected, entities are required to submit a simplified (special) notification (discussed below), which must be approved, before the threshold amount is exceeded.

Simplified Notifications

These notifications replaced what had been known as notification exemptions. There are two types of simplified notifications: simplified (special) (SNS) and simplified (general) (SNG).

Simplified notification (special):[21] A SNS applies to new substances that meet the following criteria: meets the definition of a polymer;[22] is used as an intermediate or only for export at <1 ton/year; is used for scientific research at > 0.1 but <1 ton/year; or is used for R&D for processes and products at <10 tons/annum (two years). No chemical testing is required for a simplified (special) notification except in unusual cases.

Simplified notification (general):[23] A SNG applies when the production (including manufacture and import) volume of a new substance is <1 ton/year and the substance does not qualify for a SNS or SRR. Order No. 7 requires that for a SNG, certain studies are required to support a registration or notification, including an acute fish or earthworm study and a "ready biodegradation" study. These studies must be conducted at an approved Chinese laboratory. Other study data are required, but the relevant studies need not be conducted in China. Activities relating to the new substance may not commence until the notification is reviewed by an MEP Expert Committee and a certificate of approval issued by MEP-SCC.

---

21. *Id.* arts. 12–13.
22. Polymers meeting the criteria as <2 percent new monomer or polymers of low concern.
23. Order No. 7 art. 12.

Regular Notification

A regular notification (RN)[24] applies to new chemical substances that do not meet the eligibility requirements for an SRR or a simplified notification. The regular notification category is further subdivided into four tonnage bands or Levels (Level 1: 1– ≤10 tons/year, Level 2: 10– ≤100 tons/year, Level 3: 100– ≤1,000 tons/year, and Level 4: >1,000 tons/year), with each level corresponding to a volume-based tonnage band, from lowest to highest.[25] Similar to REACH notification and registration requirements, the higher the tonnage band volume, the more substantial the applicable data requirements.

A RN must be supported by the preparation and submission of a dossier, various study reports, a risk assessment report, laboratory qualifications, classification, and labeling based on the GHS and MSDS.[26] Activities involving the new substance may not lawfully commence until a review is conducted by the Expert Committee and a registration number (certificate) is issued.

MEP-SCC assesses a nominal fee for chemical notifications, approximately 200 RMB. Generating the data necessary to support a notification entails expense, however, which can be substantial. In China, relevant studies must be submitted to support a registration. Unlike REACH, where a letter of access to cited studies is sufficient to register a new chemical, under Order No. 7 the actual study must be submitted. This point is critically important because many data owners often decline to share study reports, necessitating the development of new data, some of which must be coordinated with Chinese-based laboratories. Consulting fees for knowledgeable and reliable consultants vary.

Registered substances are typically added to the IECSC five years after the date of the first manufacture or import is submitted to the MEP-SCC, formerly the CRC.[27] Substances that were in use in China prior to October 15, 2003, may be added to the IECSC through an application filed with the MEP-SCC, formerly the CRC, and accompanied by supporting documentation or proof of use during the relevant time period.

---

24. *Id.* art. 11.
25. *Id.*
26. *Id.* art. 10.
27. *Id.* art. 41.

Under Order No. 7, article 16 requires that applicants, or their agents, who submit these notifications or filings must be registered entities in China.[28]

MEP-SCC Review of New Chemical Notifications

Upon submission of the notification, the MEP-SCC conducts a comprehensive review, beginning with a completeness check that occurs within five working days of submission of the notification. The Expert Committee reviews all regular and simplified notifications and provides a technical review of a regular notification within 60 days of receipt and simplified notification within 30 days.[29] It is not uncommon for the Expert Committee to request clarification and/or ask additional questions pertaining to the notification submission. The MEP grants a registration certificate where it determines that appropriate risk control measures are in place and that the notification meets all other applicable requirements. In cases where a notification is rejected, the MEP sends a written notice explaining the basis for the rejection.[30]

Postchemical Notification Obligations

Registration certificate holders need to fulfill different postnotification obligations depending on the management category of the substance. Chemical substances are categorized either as "general" new chemical substances or "hazardous" new chemical substances. "Hazardous" new chemical substances possessing persistent or bioaccumulative properties or considered harmful to the environment or human health will be further classified as "priority hazardous" new chemical substances.

General new substance obligations include sharing the MSDS with downstream users; implementing certain risk management measures; submitting a first-activity report; retaining documents for a ten-year period; and submitting updates based on new information, as appropriate.[31]

Hazardous new chemical substance obligations include the above noted requirements and submitting an annual report (on the preceding calendar year activities); and satisfying the requirements of SAWS Order No. 53.

Priority hazardous new chemical substance obligations include all the above and registrants must submit a disposal report, a substance

---

28. Id. art. 16.
29. Order No. 7 art. 24.
30. Id. art. 21.
31. Id. art. 36.

flow chart, and an annual plan that addresses the next year's activities regarding the substance.

Simplified notifications have the fewest obligations. Registrants must submit an annual report and retain documents for a ten-year period.[32]

Exemptions from Notifications

Under Article 2 of Order No. 7, chemicals not listed on the IECSC are considered "new" and subject to notification. There are four categories of exemptions from notifications: naturally occurring substances, chemicals subject to other laws and regulations (pesticides, drugs, and related regulated substances), unintentionally or incidentally produced substances of a noncommercial nature, and certain "special categories." These include substances similar to the TSCA exemption for statutory mixtures (frits, glass, Portland cement), nonisolated intermediates, reaction product substances that occur when an article is used as intended, and certain alloys. Unlike TSCA and REACH, polymers are not exempt from notification. See the polymer decision logic notification presented in figure 7.3.

Figure 7.3

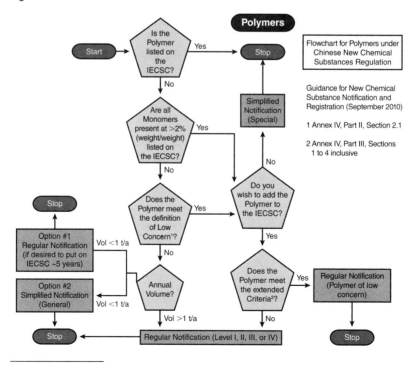

32. *Id.* arts. 29–42.

Risk Assessment

Risk Assessment Overview

Each RN must be accompanied by a risk assessment report that considers human health and environmental hazard assessments and exposure assessment to determine and thus contribute to the overall risk assessment. Ideally, risk characterization ratios (RCR) should be less than 1. The elements the risk assessment must address are set forth generally in relevant guidance documents, many of which unfortunately remain in draft.[33] This inconvenient fact underscores the critical importance of being familiar with or associating with a source that is familiar with the guidance as well as with the contents of Order No. 7. As a general matter, the level of detail and format required of the risk assessment report are less demanding than what is required under the EU REACH program.

Environmental Risk Assessment

The environmental risk assessment seeks to identify potential areas of greatest impact of a new chemical on the environment. Similar to the REACH approach, risk is based on an assessment of hazard and exposure. Predicted No Effect Concentrations (PNEC) are determined and presented in the risk assessment.

Human Health Assessment

The human health assessment is similarly calculated, based on an assessment of hazard and exposure potential. Similar to TSCA and REACH, the absence of chemical-specific data leads to the imposition of worst-case assumptions that enhance considerably the likelihood of toxicity and/or exposure concerns and the imposition of commercial limitations. Derived No Effect Levels (DNEL) are determined and presented in the risk assessment. Endpoints that are explicitly considered: acute toxicity, skin irritation, eye irritation, sensitization, mutagenicity, repeated dose toxicity (subchronic and chronic), reproductive/developmental toxicity, and carcinogenicity.

---

33. *See* THE GUIDELINES FOR RISK ASSESSMENT OF CHEMICALS, available in Chinese at http://www.zhb.gov.cn/gkml/hbb/bgth/201109/W020110930530417850520.pdf, and THE GUIDELINES FOR HAZARD IDENTIFICATION OF NEW CHEMICAL SUBSTANCES, available in Chinese at http://www.zhb.gov.cn/gkml/hbb/bgth/201109/W020110930530417896977.pdf.

Exposure is low, medium, or high depending upon the respective tonnage band of <10 tons per year, 10–1,000 tons per year, and >1,000 tons per year. Risk control measures are imposed when risk is determined to be in the medium to high ranges. Data generation needs are also considered to be a risk control option and new data requirements are often outcomes of the risk assessment process.

TEST/DATA DEVELOPMENT

Testing/data needs and projected substance-to-market timelines should be carefully considered early in the process to ensure commercial operations can commence on schedule. While describing the full array of data requirements is beyond the scope of this discussion, we note a few of the key features and issues.

Data requirements for regular notifications are somewhat similar to those required under the EU REACH regulation, with some exceptions. Many of the environmental studies required to be submitted as part of the notification must be conducted by approved laboratories located in China. Other than this requirement, data from other regulatory programs and from laboratories located outside of China may be used to address the required endpoints.[34]

While the MEP-SCC is open to considering alternative approaches, that is, read-across to chemical structural analogues for addressing endpoints, data submission is the preferred option, and all data must be fully presented and well supported to survive technical review. Not surprisingly, businesses are often deterred from pursuing alternative means for addressing endpoints, daunted by the need for iterative discussions with the regulators and the potential for delays that are so much a part of the new chemical notification process. Another approach to consider, if production options allow, is to pursue new chemical notifications associated with lower tonnage thresholds as they inspire fewer data requirements and typically take less time to procure.

---

34. *Id.* art. 19.

IMPORTS AND EXPORTS

Decree No. 591 restricts the import or export of hazardous chemicals. The China Entry-Exit Inspection and Quarantine (CIQ), which is part of the general Administration of Quality Supervision, Inspection, and Quarantine (AQSIQ), enforces these import/export requirements. The list is updated regularly. In January 2014, the MEP and Chinese Customs issued a joint notice No. 85 publishing the 2014 version of the Catalog of Toxic Substances Restricted to Be Imported into or Exported from China. The new list, which went into effect on January 1, 2014, contains 162 substances, and replaces the previous list. The Catalog was revised to align with China's updated regulations and conform with the provisions of the Rotterdam Convention on the Prior Informed Consent Procedure for Certain Hazardous Chemicals and Pesticides in International Trade. The 162 chemicals listed on the Catalog, including mercury and arsenic, are subject to additional registration and permitting requirements before they can be imported into China. Companies that import or export listed substances must register with the MEP-SCC and apply for pertinent registration certificates or custom clearance notifications prior to import or export of listed chemicals.

ENFORCEMENT AND PENALTIES

Entities that produce, market, or use prohibited hazardous chemicals are subject to a variety of enforcement measures, including the cessation of the illegal activity, the assessment of fines in amounts ranging from 200,000 to 500,000 RMB ($32,000 to $80,000 USD), confiscation of illegal gains, and prosecution for criminal liability.[35] Any illegal earnings are subject to confiscation and if the circumstances are sufficiently serious, operations can be suspended and operational permits and business licenses rescinded.[36]

---

35. Decree No. 591 art. 75.
36. Id. arts. 76, 77, 78, 83.

Unlawful Activities

Under Order No. 7, the production, import, processing, or use of any new chemical substance[37] without a registration certificate is prohibited. For violations associated with Order No. 7, penalties may include the publication of the violation and violator on an official website, which ultimately is further publicized in the trade press.

Additionally, fines may be, and typically are, issued. Most importantly, the MEP may deem the entity involved to be a "bad actor," which could result in the entity being banned from seeking a new substance notification for a period of up to three years. "Bad actor" status typically is reserved for those entities that willfully and knowingly violate the requirements in Order No. 7. In the event that an entity is made aware of a potential violation, it is wise to consult with a knowledgeable individual accustomed to interactions with both the MEP-SCC and the local environmental protection bureau. It can then be determined how best to proceed with the MEP-SCC and the local bureau to resolve the matter suitably.[38]

Enforcement of Order No. 7 is primarily a responsibility of the local environmental protection bureaus. Supervision and inspection of activities relating to Order No. 7 are managed by local environmental protection bureaus. It is not uncommon, however, for the MEP-SCC to be consulted in the process as to the status of an application in connection with enforcement activities. The local environmental protection bureaus are tasked with ensuring that the prescribed outlines are being followed, that risks are controlled, and that other needed control measures are implemented. Their duties also include identifying where and why new chemical substances may have caused adverse events.

Guidance Documents

Guidance documents are an important part of the regulatory system, and essential to understanding the intent and basis of Order No. 7. The following are some of the more helpful guidance documents:

- Guidelines for the Testing of Chemicals (HJ/T 153-2004)
- Guidelines for Hazard Evaluation of New Chemical Substances (HJ/T 154-2004)

---

37. *Id.* art. 5.
38. *Id.* ch. V.

- Guidelines for the Generic Name of New Chemical Substances
- Guidance for New Chemical Substance Notification and Registration[39]

More are expected.

KEY ISSUES AND PRACTICAL TIPS

Advance planning and careful coordination are critical prior to determining timelines for the submission of documentation to support a new chemical registration. The most significant problems that prospective registrants face are lack of planning and failure to coordinate timely and effectively with the qualified Chinese laboratories. At the time of this writing, there are approximately eleven approved and qualified laboratories from which to select, one of which is now certified under Good Laboratory Practices (GLP). Given the heavy demand for testing, it is necessary to secure laboratory time and personnel well in advance of an anticipated submission date. Building a window of six months or so into project timelines is recommended for securing laboratory space, plus an additional month or two for any miscellaneous delays that can occur in preparing and issuing the final study report.

As noted above, another challenge is the absence of final guidance on many of the testing requirements and, of course, the continuing challenge posed by language barriers. Because much is left to the discretion of the MEP-SCC, it is not uncommon to spend significant time proposing and working through "new" or unexpected situations. Experience has shown the MEP-SCC, formerly the CRC, to be cooperative in these circumstances, but Western businesses must exercise patience as they learn to maneuver through requirements that are not fully described based on draft guidance documents and regulators coping with a relatively new regulatory framework. It can be effective to engage with the MEP-SCC on specific issues and to include in these discussions individuals who are skilled at complex problem-solving and who have a demonstrated good reputation for working cooperatively

---

39. *See* Chinese Chemicals Management Regulations, http://english.mep.gov.cn/special_reports/chemical_regulation/. Note that when available, an English translation is provided. In all cases of discrepancies, the Chinese version prevails.

with the MEP-SCC, formerly the CRC, to resolve the kind of questions at issue.

Resources

People's Republic of China, Ministry of Environmental Protection, Decree No. 7, Measures on Environmental Management of New Chemical Substances, original decree in Chinese (http://www.crc-mep.org.cn/news/NEWS_DP.aspx?TitID=225&T3=01&LanguageType=CH&Sub=12) and unofficial English translation.

State Administration of Work Safety in China (SAWS), Decree No. 591, Regulations on the Safe Management of Hazardous Chemicals in China, original decree in Chinese (http://aqbzh.chinasafety.gov.cn:8080/wss/abxxfbAction1!listXxfbMore.action?xxlx=XXFB_FLFG) and unofficial English translation.

State Administration of Worker Safety in China, Decree No. 591—English translation.

The MEP has stated that it will issue the following documents in the future:

- Guidelines for Risk Assessment of Chemical Substances
- Guidelines for Hazard Identification of New Chemical Substances
- Guidelines for Chemical Testing Good Laboratory Practices (revisions expected)

This list is not exhaustive, but provides a sense of the guidance that is currently available and anticipated to be issued.

7.2

South Korea

Leslie S. MacDougall and Lynn L. Bergeson

OVERVIEW

Chemical regulation in South Korea is very much in a state of transition. Chemical notification activities long regulated under the Toxic Chemicals Control Act (TCCA) of 1991—the key law addressing the basic management of industrial and hazardous chemicals—will transition, as of January 1, 2015, to South Korea's newly minted chemical management law, the Act on Registration and Evaluation, etc. of Chemical Substance[1] (K-REACH), and the Chemicals Control Act (CCA).[2] Once effective, TCCA provisions related to the registration/evaluation/assessment of new and existing chemicals will fall under the umbrella of K-REACH while the CCA activities will focus principally on the control of hazardous substances. K-REACH's adoption in 2013 is the latest example of the growing international significance of the European Union (EU) Registration, Evaluation,

---

1. Korea Presidential Decree No 11789, adopted on May 22, 2013, as per the unofficial English translation referenced by the Korea Chemicals Management Association (KCMA), which is an affiliated organization of the Ministry of Environment (MoE).
2. Korea Presidential Decree No 11862, adopted on June 4, 2013.

Authorization, and Restriction of Chemicals (REACH) chemicals management framework.

TCCA has a broad reach in terms of the array of chemicals it currently covers, as well as the activities it affects. South Korea also maintains an extensive chemical inventory and, as under other regulatory schemes, "new" chemicals—those not listed on the inventory—generally are subject to notification requirements; differing notification options are offered. Notifications must be supported by specified data elements and by both a hazard and risk assessment for new chemicals.

As its name implies, K-REACH mirrors the EU REACH program and includes many of EU REACH's core elements. There are, however, many differences between REACH and K-REACH, as adopted, that distinguish it from its European counterpart. As under EU REACH, K-REACH shifts the burden of demonstrating chemical safety to chemical manufacturers and importers. Because the United States exported more than $6.8 billion worth of chemicals to South Korea between March 2012 and February 2013, K-REACH is expected to have significant implications for the global chemical supply chain in Asia and beyond.[3] As more fully discussed below, K-REACH goes beyond EU REACH in several key respects. In addition, K-REACH imposes new labeling, reporting, and other consumer information requirements applicable to manufacturers and importers of certain products and articles containing chemicals identified by ministerial decree. These regulations apply to chemicals in consumer products in which exposure to the chemicals is possible.

7.2.1
TCCA

Overview

At the time of this writing, TCCA remains in effect but is expected to undergo an unofficial transition to K-REACH and CCA with a firm implementation date of January 1, 2015. Based on time frames

---

3. U.S. Int'l Trade Comm., Publ'n 4393, U.S.-Korea Free Trade Agreement: Effects on U.S. Small and Medium-Sized Enterprises (May 2013), http://www.usitc.gov/publications/332/pub4393.pdf.

for the review of TCCA notifications by the competent authority, the last date that notifications/registrations were accepted under the TCCA program was October 22, 2014. A significant push is underway by industry to ensure that when possible, a chemical notification is made under the TCCA, as TCCA notification requirements for new chemical substances under K-REACH are more extensive.

TCCA is intended "to enable all people to live in a healthy and comfortable environment by preventing any harm to people's health and the environment caused by chemicals."[4] There are five chapters to TCCA: chapter I, General Provisions; chapter II, Examination of Toxicity and Assessment of Harmfulness of Chemicals; chapter III, Safety Control of Toxic Chemicals; chapter IV, Supplementary Provisions; and chapter V, Penal Provisions. The MoE, the authority that implements TCCA, consists of three divisions: Environmental Health Policy Division; Safety Division; and Hazardous Chemicals Division. The MoE works closely with the National Institute of Environmental Research (NIER), which is responsible for environmental research, related administrative functions, and generally for supporting government environmental policies.

Scope of TCCA

TCCA applies generally to chemicals. Any person who wishes to manufacture or import a chemical must determine whether the chemical falls into one of several categories, including new chemicals, poisonous (toxic) substances; substances under observation; or restricted or prohibited substances.[5] Similar to the regulation of chemical substances in other countries, chemicals regulated under other regulatory regimes in South Korea fall outside TCCA's scope. These include radioactive substances; pharmaceuticals and nonpharmaceutical drugs; narcotics; cosmetics (including materials required to undergo an examination of safety under Cosmetics Act article 4(3) and materials of cosmetics that are designated and announced by the Commissioner of the Korea Food and Drug Administration);

---

4. TCCA art. 1. An official translation of TCCA is available on the MoE website at http://www.kcma.or.kr/eng/subpage/toxic.asp.
5. Id. art. 9.

technical ingredients and agrochemicals; fertilizers; foods and food additives; livestock feeds; explosives; and toxic gases.[6]

Any person who "conducts business" to manufacture, export, import, sell, keep, store, transport, or use toxic chemicals "shall take necessary measures" under TCCA "to prevent any harm to human health or the environment caused by toxic chemicals."[7]

How a substance is regulated under TCCA depends on whether the substance is considered poisonous, observational, or restricted/prohibited. Manufacturers of poisonous chemicals must submit reports including annual volume and usage and certain import/export information. Manufacturers must also register their business and label products as required by law. Similar requirements apply to the other categories of chemicals.[8] Restricted or prohibited chemicals are subject to additional requirements. Manufacturers and importers must obtain permission from the MoE to manufacture or import the substance. Businesses selling, storing, transporting, or using prohibited or restricted chemicals must obtain permission to do so from the MoE.[9]

Notification/Registration/Approval Requirements and Compliance Chemical Inventory

Similar to other chemical control programs, South Korea maintains an inventory of chemical substances considered "existing" in South Korea. The Korea Existing Chemicals Inventory (KECI) is estimated to contain over 42,000 chemicals. KECI is available online.[10]

New Chemicals Notification

Chemicals considered "new" must be notified to MoE. New chemicals are chemicals not listed on the KECI or not otherwise exempt. The NIER is responsible for new chemical notification under TCCA. The KCMA of the MoE is responsible for accepting declarations pertinent to other chemicals and applications for confirmation certificates. Chemicals listed on the KECI are considered existing, which

---

6. *Id.* art. 3.
7. *Id.* art. 5.
8. *Id.* arts. 19–26.
9. *Id.* art. 34.
10. *See* Nat'l Chems. Info. Sys., http://ncis.nier.go.kr/.

include chemical substances commercially used in Korea prior to February 2, 1991, and were notified to MoE, or subsequently notified to MoE and listed on the KECI.

In general, TCCA requires that both a hazard and a risk assessment be conducted for new chemicals. The hazard assessment categorizes the substance into appropriate groupings, including toxic, observational, and restricted. A risk assessment is then conducted and, if it is determined necessary, the substance is evaluated for possible restriction or prohibition. TCCA requires relatively less data compared to other registration programs, especially EU REACH. There are several types of chemical notification options available under TCCA.

The type of notification to be made is dictated by the type of substance and its volume. A "simplified notification" is made when the substance to be manufactured or imported has been located on two other recognized chemical inventories since the February 2, 1991, KECI cutoff date. The European Inventory of Existing Chemical Substances (EINECS) and the U.S. Toxic Substances Control Act (TSCA) Inventory are the foreign inventories most commonly used for this purpose. Once a simplified registration is qualified, data requests are significantly reduced.

A "reduced test substance notification" can be made for certain substances considered less toxic. A "full notification" applies to other new chemical substances that do not fall under the above notifications. Further requirements are delineated by the tonnage requirements below.

Exemptions from New Chemical Notification Requirements

Certain substances are exempt from notification because they are considered outside the scope of TCCA regulation. These include substances being manufactured or imported at less than 100 kilograms a year, and chemicals manufactured or imported in amounts of less than ten tons per year for export purpose only. TCCA has no polymer exemption per se. Polymers meeting one or more of the following criteria, however, are exempt for review: polymers that meet the 2 percent rule criteria under TSCA; block copolymer of which all blocks are listed on the KECI; graft polymers in which the stem and all branches are listed on the KECI; and low concern polymers (application for exemption certifications required).

Test Data Development

A core set of data must be submitted with each chemical notification. All toxicity studies must be performed in accordance with Good Laboratory Practice (GLP). The "core" data requirements include a description of general properties and applications (chemical identity, chemical name, Chemical Abstracts Service (CAS) Registry Number, uses, volumes), certain physical-chemical properties, and environmental release information (expected release amount to the environment, daily and average working day release information).[11] All notifications must contain these data elements.

Simplified notifications must include additional data elements: acute toxicity (most relevant route) and Ames assay or other relevant data. A full notification (<1 ton/yr) must contain the simplified notification data requirements plus in vitro chromosomal test and biodegradation data. A full notification for chemicals produced or imported in excess of one ton per year must include the data elements applicable to chemicals produced below one ton per year plus skin and eye irritation data, skin sensitization data, and aquatic toxicity studies. Upon review by the regulatory authorities, additional data can be requested.[12]

Resources

Toxic Chemicals Control Act
Enforcement Decree of the Toxic Chemicals Control Act

7.2.2
K-REACH

Overview

South Korea's Act on Registration and Evaluation, etc. of Chemical Substance (K-REACH) establishes a registration and assessment scheme based on a substance's manufacture or import in South Korea

---

11. TCCA art. 10.
12. Id. art. 11.

much like the EU's REACH Program. K-REACH was passed by the South Korea National Assembly Environmental and Labor Committee on April 24, 2013, and enacted by the National Assembly on April 30, 2013, and is scheduled to come into effect on January 1, 2015. The Act is still only available in Korean, though the website of the MoE provides an overview of the Act in English.[13] An unofficial English version of K-REACH is available.[14] Once K-REACH is effective, the provisions related to registration and evaluation of TCCA will fall under the umbrella of K-REACH, and CCA will focus narrowly on the control of certain hazardous substances. Guidance documents on K-REACH are not yet available, but several reportedly are under development with the first guidance documents anticipated to be made available late in 2014.

K-REACH has adopted the basic EU REACH framework. There are differences, however. K-REACH prescribes a process for chemical registration, evaluation, and assessment of hazards and risks of chemical substances. Unlike EU REACH, where all substances produced annually in quantities over one ton must be registered, under K-REACH all new chemical substances and only designated existing substances identified by the MoE will be subject to registration. Importantly also, K-REACH includes provisions not part of the EU REACH for reporting and notification pertinent to products and articles that contain chemicals identified as toxic under K-REACH and provisions requiring product safety reviews, product labeling, sales bans, and product recall provisions.

Notification/Registration/Approval Requirements and Compliance Reporting of Chemical Substances

Chemical reporting is an essential element in K-REACH. Similar to EU REACH, K-REACH applies to manufacturers and importers of chemical substances produced in amounts above one ton per year. Manufacturers, importers, and downstream users of new chemicals and chemical substances, as described below, that are manufactured, imported, or sold at equal to or greater than one ton per year are

---

13. MoE, Enactment of Act concerning Registration, Evaluation, etc. of Chemical Substances, http://eng.me.go.kr/eng/web/index.do?menuId=167&findDepth=1.
14. See http://www.lawbc.com/uploads/docs/English_translation_-_Korea_REACH__Act_on_Registration_and_Evaluation_e.pdf.

required to make annual reports to the MoE about the volume and information on chemical use.[15]

Chemicals subject to the reporting requirement are new chemicals and those listed in the KECI, which includes those chemical substances domestically distributed in commerce before February 2, 1991, or subsequently notified to MoE.[16] Based on the results of the first annual report and available toxicity data, the MoE will select certain chemicals subject to evaluation from existing chemicals for registration.

Exemptions to the reporting requirement include chemical substances imported as "already equipped in machinery"; chemical substances imported together with machines or equipment for purposes of trial operation; chemical substances in products in solid form, performing certain functions and not intended for release during normal process or use; and other chemical substances as identified under presidential decree that are manufactured or imported for use in research and related noncommercial applications.[17] Substances used for tests and nonisolated intermediates are also exempt from reporting.

Registration of Chemical Substances

Chemical substances subject to registration include new chemical substances (regardless of tonnage); existing chemical substances subject to registration (at or greater than one ton per year) identified by MoE; and certain chemical substances that are likely to cause significant damage to human health and the environment regardless of the tonnage. Registration applies to the manufacture and import of chemical substances.[18]

Manufacturers and importers of existing substances subject to registration will be allowed to manufacture and import without registration until the registration deadline. Exemptions include chemical substances imported as incorporated in machines or in products in solid form, performing certain functions and not released during normal conditions of use, and chemicals manufactured or imported in amounts of less than ten tons per year solely for purposes of export.[19]

---

15. K-REACH art. 8.
16. *See* Nat'l Chems. Info. Sys., http://ncis.nier.go.kr/main/Main.jsp.
17. K-REACH art. 8.
18. *Id.* art. 10.
19. *Id.* art. 11.

Total volume and hazard and risk characterization are two key criteria in which MoE will rely in designating existing chemicals subject to registration, which it will do by ministerial decree.[20] The MoE has expressed its intent to select approximately 2,000 existing chemicals believed to pose hazards for registration under K-REACH. At the time of this writing, 800 substances have been submitted to the competent authority for consideration for inclusion under the first registration deadline. Of these, it is expected that 500 substances will be selected.

Various data regarding the information of the manufacturer or importer, chemical substance identification, use, classification and labeling, physical and chemical properties, inherent hazard, and guidance on safe use plus the results of tests conducted by approved testing laboratories, must be submitted in the registration dossier for new chemicals and existing chemicals at one ton or greater per year.[21] Those registrations for substances at 100 or more tons per year must be accompanied by a risk assessment (hazard and exposure assessments with safety confirmation), exposure details, and estimated volumes. South Korean testing laboratories approved for these purposes will be designated by ministerial decree. Foreign testing laboratories may be used if testing is completed to comply with the Organization for Economic Cooperation and Development's Good Laboratory Practice (GLP). Whether or not data generated in foreign laboratories approved by other government authorities will be acceptable under the regulation remains unclear. Low volume new chemicals (less than one ton per year until the end of 2020) will benefit from a simplified data submission. Such submissions include the following components: information of the manufacturer or importer, chemical substance identification, use, classification and labeling, and physical and chemical properties. After 2020, the requirements will be strengthened to 0.1 ton per year beginning in 2021 with an increase in the data requirements.

Registration of chemical substances between ten and 100 tons per year will have different submission dates, as shown in table 1.

---

20. *Id.* art. 9.
21. *Id.* art. 14.

TABLE 1. TONNAGE SUBMISSION DATES

TONNAGE	SUBMISSION DATE OF RISK ASSESSMENT
≥100 tons per year	January 1, 2015
≥70 to 100 tons per year	January 1, 2017
≥50 to 70 tons per year	January 1, 2018
≥20 to 50 tons per year	January 1, 2019
≥10 to 20 tons per year	January 1, 2020[22]

Substances exempt from registration are similar to those exempt under other provisions of K-REACH: chemicals imported in machinery; chemicals imported in machinery devices for test applications; chemicals contained in solid form product that is not released during use; and certain substances that required a "Confirmation of Registration Exemption" from the MoE; chemicals manufactured at ten tons or less per year for the purpose of export only; manufactured/imported substances for the purpose of research and studies; and substances for tests and nonisolated intermediates.

Evaluation

The MoE will evaluate chemicals that are manufactured or imported in quantities greater than ten tons per year. The evaluations will be staggered with chemicals manufactured or imported in the greatest amounts evaluated first in 2015 and low threshold chemicals last, with assessments of the ten tons or more a year band beginning in 2020. Based on the results of the evaluation, the MoE will designate chemicals as toxic.[23] Substances are likely to be designated as hazardous, authorized, and restricted/banned. Designation criteria are further clarified below. Further details are expected to be provided in forthcoming ministerial decrees.

Hazardous

The MoE will announce the results of its examination/assessment for those substances that are designated as hazardous. The announcement will include the name of the substance and toxicity. The designation

---

22 *Id.* at addendum, art. 1.
23. *Id.* arts. 18–20.

criteria are based on the results of the hazard examination, which determines whether the substance is toxic to human health and/or the environment. Substances that fall within this category and thus may be subject to Authorization and Restricted and Banned substances will be managed in accordance with the CCA.

Authorization

The MoE will designate chemicals that require authorization prior to manufacture, use, or import in South Korea based on the evaluation of chemicals described above.[24] Chemicals eligible for authorization are those that exhibit one of the following characteristics: (1) cause cancer, mutations, impairment of reproductive ability, disruption in endocrine system, or raise concerns about such effects; (2) accumulate in humans, animals, or plants in high concentrations and remain in the environment for a long period; and (3) may cause significant risks equal to or more than the chemicals included in (1) and (2) above.[25]

As of this writing, it remains unclear what standards the MoE will use to grant authorization. Under EU REACH, authorization is provided if the benefits of continued use of the substance outweigh its potential risks or if the potential risks are considered to be "adequately controlled." The MoE has stated that it will announce the name of the substance, the use details, and grace period, commonly referred to as the "sunset date" under EU REACH.

Restricted and Banned Chemicals

The MoE is also expected to publish a list of restricted or prohibited substances.[26] Chemicals on this list will be restricted from being manufactured, imported, sold, stored, or transported for certain purposes, or prohibited from being manufactured, imported, sold, stored, or transported for all purposes. The MoE will use certain criteria to make its findings. These include that the chemical: poses risks based on a "hazard examination and risk assessment"; is recognized as having risks by an international organization; is prohibited or restricted from

---

24. *Id.* art. 25.
25. *Id.*
26. *Id.* art. 27.

manufacturing or import by international treaty or other international convention; or a substitute has been identified, the chemical use is no longer risky due to the development of a new technology, or new scientific evidence has demonstrated the chemical is no longer a risk.[27]

Based on these criteria, it would appear the K-REACH restriction list may be somewhat broader than the EU REACH restriction list and broader than the substances of very high concern (SVHC) list maintained under REACH Annex XIV. SVHCs must have at least one of the following characteristics: carcinogenic, mutagenic, or toxic for reproduction (CMR) category 1 and 2; persistent, bioaccumulative, or toxic (PBT); very persistent and very bioaccumulative (vPvB); or an "equivalent level of concern."

Only Representative

Foreign manufacturers exporting chemical substances or products into South Korea may appoint an Only Representative.[28] Qualification for Only Representative status is to be decided by the MoE. An Only Representative's obligations include annual reporting of use and tonnage of chemical substances; registration; notification of products; and other activities as set out in the presidential decree. The Only Representative must report the appointment or release of the appointment by foreign manufacturers to the MoE.[29]

Supply Chain Communications

Any person who transfers a substance covered under article 10 (registration requirements) must prepare and provide the registration number, chemical name, hazard information, and safe use information to the downstream user. Most of this information is covered in a chemical's Safety Data Sheet. The downstream user then must provide levels of use, exposure information, sales volumes, and safe use information back to the manufacturer or importer. This step is intended to ensure a bilateral transfer of information that provides both parties with sufficient information to manage a chemical safely.[30] It is widely expected these supply chain communications will be significantly embellished through ministerial decrees.

---

27. Id. art. 26.
28. Id. art. 38.
29. Id.
30. Id. arts. 29–30.

Supply Chain Requirements Applicable to Product Manufacturers

Manufacturers and importers of articles also have notification requirements under K-REACH. The notification requirement applies when the content of a chemical in a product is greater than one ton per year or the concentration is greater than 0.1 percent.[31] If a product contains hazardous substances at ≥1 ton/yr, the manufacturer or importer must notify the MoE in advance. The substance name, content, hazard information, and use of the hazardous chemical in the substance must be provided. Exemptions to the notification (known as Exemption Confirmation of Notification) can be obtained for substances for which there is no exposure during normal use and for substances already registered under article 10. Articles with no releases are also exempt from this provision. Applications for Exemption Confirmations of Notification will be reviewed by the MoE, and applicants will be notified if they qualify for exemption.[32]

A product safety or risk assessment is required for any product deemed potentially hazardous and thus believed to pose a risk. The assessment must be performed by the MoE or a suitable trusted institution. Once completed, the results of the assessment will be published by the MoE and will include safety labeling and standards. These standards can include product sales bans and recall orders.[33] The MoE will also, under K-REACH, publish standards of product safety labeling for products believed to pose risks.[34]

These product and supply chain notification provisions extend well beyond what is required under the EU REACH program. Whether this is part of an Asian trend or merely the next logical progression in chemical product regulation is unclear. How the program unfolds also remains to be seen. MoE is granted significant authority under K-REACH to conduct risk assessments or require the product manufacturer, importer, or distributor to submit "necessary data" for a risk assessment.[35] Whether and under what circumstances MoE will deploy its new authority under K-REACH remains to be seen. Intuitively, products likely to receive the most scrutiny, and thus more likely to be subject to risk assessment and/or product recalls or bans,

---

31. *Id.* art. 32.
32. *Id.*
33. *Id.* art. 33.
34. *Id.* art. 34.
35. *Id.* art. 33.

are those marketed to consumers containing chemicals determined to be of concern.

Enforcement Provisions

The penal provisions in K-REACH are stiff. Administrative and criminal penalties apply and range from fines not exceeding 10 million won (approximately $10,000 USD) to prison labor of up to seven years.[36]

Trends

As with many countries, South Korea has stepped up considerably its chemicals management program to ensure that substances being placed on its market are safe. The burden of safe use is squarely on chemical manufacturers and importers using the EU's REACH as a model. As K-REACH is new, business interests will need to monitor its development carefully and be aware of how the program is implemented, interpreted, and enforced. As the supply chain communications and product regulation provisions are novel and have no precedent in EU REACH, these provisions in particular will be essential to monitor. Whether these supply chain and product provisions will become the template for other Asian countries and regions elsewhere in the world or unique to South Korea remains to be seen.

At the time of this writing, MoE and KCMA are continuing to ensure that industry support is forthcoming to achieve the target date of January 1, 2015, for implementation. MoE is supporting a Help Desk that provides a consultation system for domestic and foreign industries for implementation of K-REACH.

A pilot project is under way to identify potential operational concerns when conducting a joint registration. The pilot project is expected to be completed in May 2015. Seven substances have been identified for inclusion in the pilot project. The pilot project is a joint project between industry and the MoE. The information technology system designed to support K-REACH is not yet available and is expected to be implemented in phases. Industry continues to express

---

36. *Id.* arts. 49–54.

its concerns that the infrastructure may not sufficiently be implemented to support 2015 reporting requirements. There is no indication that the January 1, 2015, implementation date will be extended.

Resources

Unofficial English translation of K-REACH is available at http://www.lawbc.com/uploads/docs/English_translation_-_Korea_REACH__Act_on_Registration_and_Evaluation_e.pdf.

Complete text of the K-REACH in Korean is available at http://likms.assembly.go.kr/bill/jsp/BillDetail.jsp?bill_id=PRC_C1P3R0E4E3O0M1T7S0Q6W5E4O5U8E4.

8

Concluding Remarks

Lynn L. Bergeson

The preceding chapters illustrate the renaissance underway in the global management of chemicals. Some say the EU REACH program jump-started the movement. Others suggest the maturity of environmental management, improvements in technological and chemical detection, and growing awareness of chemical impacts on human health and the environment have coalesced and are the reason for the sharply increased amount of activity in the global management of chemicals. Whatever the reason, these developments are having and will continue to have a profound impact on the global business of chemicals. Legal practitioners, product stewards, scientists, regulatory affairs managers, and others in the business of chemicals will need to familiarize themselves with these developments, and stay on top of them to remain compliant and competitive.

As of this writing, there is no sign of diminished activity. As K-REACH in Korea ramps up, REACH in Europe enters a new phase of chemical authorization, and the United States continues to plod along with TSCA reform, governments around the world are setting their sights on a new era of chemical regulation. As noted, no one resource can possibly answer all questions all the time. We hope this Handbook has made rationalizing this "space" easier and equipped readers with a sufficient foundation to be able to formulate the right questions, and direct you to resources to answer them.

Glossary

AA: Alternative analysis; an evaluation and comparison of a priority product and one or more alternatives.

Active implantable medical device: Any active implantable medical device within the meaning of point (c) of Article 1(2) of Council Directive 90/385/EEC of June 20, 1990, on the approximation of the laws of the Member States relating to active implantable medical devices.

Active ingredient (AI): An ingredient in a pesticide product that prevents, destroys, repels, or mitigates any pest in a pesticide that is not a plant regulator, defoliant, desiccant, or nitrogen stabilizer. For plant regulators, the AI accelerates or retards the rate of growth or maturation, or alters the behavior, of plants and their products through physiological action. For defoliants, the AI causes leaves or foliage to fall from plants. For desiccants, the AI artificially accelerates the dying of plant tissue. For nitrogen stabilizers, the AI prevents or hinders the process of nitrification, denitrification, ammonia volatilization, or urease production through action affecting soil bacteria.

Acute toxicity: Propensity of a substance to have adverse effects from a single exposure or from multiple exposures in a short space of time.

Agrochemicals: A chemical, such as a fertilizer, hormone, fungicide, insecticide, or soil treatment, that improves the production of crops.

Allegation: Claim by any person that any chemical substance or mixture manufactured, imported, processed, distributed, or released into the environment has caused a "significant adverse reaction" on human health or the environment.[1]

Ames: Method that uses bacteria to test whether a given chemical can cause cancer.

Antimicrobial pesticide: A pesticide that is intended to "disinfect, sanitize, reduce, or mitigate growth or development of microbiological organisms" or "protect inanimate objects, industrial processes or

---

1. 40 C.F.R. § 717.3(a).

systems, surfaces, water, or other chemical substances from contamination, fouling, or deterioration caused by bacteria, viruses, fungi, protozoa, algae, or slime."[2]

Article: A manufactured item (1) that is formed to a specific shape or design during manufacture, (2) that has end-use function(s) dependent in whole or in part upon its shape or design during end use, and (3) that has either no change of chemical composition during its end use or only those changes of composition which have no commercial purpose separate from that of the article, except that fluids and particles are not considered articles regardless of shape or design.[3]

Article 26 inquiry: Pursuant to Article 26, every potential registrant of a non-phase-in substance or of a phase-in substance that was not preregistered must inquire of ECHA whether a registration has been submitted for the same substance and must provide information about substance identity and the anticipated data requirements. This has come to be known as an "Article 26 inquiry."

Assay: An analysis or examination.

Assemble: Fit, join, put, or otherwise bring together components to create, repair, refurbish, maintain, or make nonmaterial alterations to a consumer product.

Authoritative body: An agency or formally organized program or group that uses one of the methods set forth in the Proposition 65 regulations for the identification of chemicals, and that the Carcinogen Identification Committee has identified as having expertise in the identification of chemicals as causing cancer or the DARTIC has identified as having expertise in the identification of chemicals as causing reproductive toxicity.

Authorization dossier: Dossier submitted to apply for an authorization under REACH pursuant to Article 62 that includes:

1. Specific uses covered by the application;
2. A detailed CSR;
3. An analysis of alternatives (AoA) for the substances' and technologies' alternatives, description of their availability,

---

2. FIFRA § 2(mm), 7 U.S.C. § 136(mm).
3. *See* 40 C.F.R. § 720.3(c).

risk, technical and economic feasibility, and a summary of the current and planned R&D;

4. A substitution plan (SP): A planning for each use of the suggested actions, including an agenda of the substitution and/or R&D; and
5. A socioeconomic analysis (SEA): A comparison between the alternative's implementation costs and the benefits of the SVHC's elimination in terms of environment and human health.

Authorized representative: A natural or legal person established within the EU who has received a written mandate from a manufacturer to act on his behalf in relation to specified tasks.

BFI (Bona Fide Intent): Notice to EPA under section 5 to request written determination of the TSCA Inventory status for a chemical substance.

Bioaccumulative: Propensity of a substance to cause accumulation of substances, such as pesticides, or other organic chemicals in an organism.

Biochemical pesticide: A pesticide that is produced by a plant or animal and that operates through a nontoxic mechanism (e.g., insect pheromones, hormones, natural plant and insect regulators, and enzymes).

Biocidal Products Directive (BPD): EC Directive 98/8/EC was adopted on February 16, 1998, with a broad scope to identify and assess all active substances in biocidal products on the EU market at that time. These active substances were subsequently categorized into 23 different product-types, including disinfectants, nonagricultural pesticides, and antifouling coatings. The BPD, its inventory, and the substance-specific information generated for this regulation form the basis for regulation under the BPR.

Biocidal Products Regulation (BPR): EC Regulation 528/2012 revises the BPR and places specific requirements on all producers of active substances and biocidal products. The BPR simplifies authorization of biocidal products, introducing EU-wide authorizations and standardizing the process of mutual recognition. A key aspect of the BPR is the provision that all legal entities making active substances or biocidal products available on the EU market must, starting September 1, 2013, possess an approval for each active substance within that biocidal product as well as an authorization for the biocidal product itself. This provision levels the playing field for placing biocidal products on the market and is expected to promote the development of new, less hazardous active substances.

Biodegradation data: Data pertaining to capability of being decomposed by biological agents, especially bacteria.

Bulk form: To import in bulk form means to import a chemical substance (other than as part of a mixture or article) in any quantity, in any container used for purposes of transportation or containment, if the chemical substance has an end-use or commercial purpose separate from the container.[4]

CAS Registry Number: Chemical Abstracts Service (CAS) Registry Number, often referred to as CAS RNs, are universally used numbers to provide a unique identification for chemical substances. A division of the American Chemical Society, CAS is a company that works with the U.S. EPA to develop appropriate identifiers for chemical substances based on their unique molecular structure.

CBI: Confidential business information.

CC: Candidate chemical.

CDR (Chemical Data Reporting): Reporting obligation under section 8 to provide information on production volumes and process and use information on chemicals over the reporting threshold.[5]

Certified applicator: An individual who is certified to use, or supervise the use of, RUPs.

Certified limits: The upper and lower concentrations at which a pesticide's ingredients must be present as specified in a Confidential Statement of Formula.

Chemical Registration Center of the Ministry of Environmental Protection of China (CRC): The former CRC was responsible for the documentation and registration of all chemicals produced and developed in China. The center is now part of the Solid Waste and Chemical Management Center (SCC).

Chemical Safety Assessment (CSA): A CSA, documented in a CSR, is required for substances registered at ten tons per annum and for persistent, bioaccumulative, and toxic (PBT), very persistent and

---

4. Id. § 712.3(c).
5. See id. pt. 711.

very bioaccumulative (vPvB), and carcinogenic, mutagenic, or toxic to reproduction (CMR) substances registered at one ton per annum. The CSA is a risk assessment, the first step in developing which is identifying the substance's hazard profile by evaluating the available data against standardized hazard categorization criteria. Following the general provisions of REACH Annex I, the CSA includes a human health hazard assessment; a physicochemical hazard assessment; an environmental health assessment; and a PBT and a vPvB assessment. Where, based on these various assessments, the registrant concludes that the substance is either "dangerous" (in accordance with a referenced directive) or is either a PBT or a vPvB, the CSA must include both an exposure assessment and a risk characterization.

Chemical Safety Report (CSR): Report documenting CSA, conducted in accordance with paragraphs 2 to 7 and with Annex I for either a substance on its own, in a mixture, or in an article or group of substances.

Chemical substance: Any organic or inorganic substance of a particular molecular identity, including (1) any combination of such substances occurring in whole or in part as a result of a chemical reaction or occurring in nature, and (2) any element or uncombined radical.[6]

Chemicals: Elements, compounds, and substances obtained by causing artificial reactions therewith and those obtained by extracting or refining substances existing in nature.

China Entry-Exit Inspection and Quarantine (CIQ): Under Decree No. 591, CIQ, which is part of the general Administration of Quality Supervision, Inspection, and Quarantine (AQSIQ), enforces the import/export requirements.

Chromosomal test: Testing on an organized structure of deoxyribonucleic acid (DNA), protein, and ribonucleic acid found in cells.

CIC: Carcinogen Identification Committee.

Cite-all method: One of two ways in which to support a pesticide registration application, in which an applicant cites, and offers to pay for, all relevant data previously submitted to EPA that support its registration application.

---

6. TSCA § 3(2)(A), 15 U.S.C. § 2602(2)(A). *See also* 40 C.F.R. § 720.3(e).

Citizen's petition: Petition issued under section 21 to initiate a proceeding for the issuance, amendment, or repeal of a rule under TSCA section 4, 6, or 8, or an order issued under TSCA section 5(e) or 6(b)(2).[7]

COC: Chemicals of concern.

Company number: A number assigned to a particular entity that is part of the pesticide registration number assigned to each pesticide registered by that entity, and that also is part of the individual establishment number assigned to each pesticide-producing establishment owned by that entity.

Competent Authority (CA): Definition 1. Each EU Member State must appoint a governmental department to conduct and enforce the provisions of the BPR. This CA is responsible for evaluating applications for approval and authorization and for assessing the potential risks posed by the active substance in the use proposed. The CA is also responsible for enforcement of decisions made under the BPR (such as exclusion and substitution) and for enforcing that active substances are not used in uses that have not been approved. A key responsibility of each CA is to coordinate with other Member State CAs and the EC to review continuously the potential risks posed by active substances and make recommendations, either for their increased regulation and restriction, or alternatively, for less regulation and restriction in the case of less hazardous substances.

Definition 2. The authority or authorities or bodies established by the Member States to carry out the obligations arising from REACH.

Confidential Statement of Formula (CSF): A statement of the ingredients in a pesticide formulation, its concentration, its purpose, any associated impurities, and, for some ingredients, the certified limits using EPA Form 8570-4.

"Conformité Européenne" marking (CE marking): Marking by which the manufacturer indicates that the product is in conformity with the applicable requirements set out in EU harmonization legislation providing for its affixing.

Conservation of Clean Air and Water in Europe (CONCAWE): An association of oil companies established to focus on environmental, health, and safety issues relevant to the oil industry.

Council: Administrative Council on Toxics Use Reduction.

---

7. TSCA § 21(a), 15 U.S.C. § 2620(a).

Country of import: The "country where the goods are to be consumed, further processed, or manufactured, as known to the shipper at the time of exportation. If the exporter does not know the country of ultimate destination, the shipment is credited to the last country to which the exporter knows that the merchandise will be shipped."[8]

Dangerous substance: Liquid, gas, or solid that poses a risk to health or safety.

DARTIC: Developmental and Reproductive Toxicant Identification Committee.

Data call-in (DCI): A notice issued by EPA to registrants of pesticides with a particular active ingredient that additional data are needed to maintain their pesticide registration.

Data submitters list: "The current Agency list, entitled 'Pesticide Data Submitters by Chemical,' of persons who have submitted data to the Agency" and who wish to be notified and offered compensation for the use of those data.[9]

Decree: An authoritative order having the force of law.

Device: A category not subject to FIFRA section 3 registrations, including any instrument or contrivance, other than a firearm, intended for trapping, destroying, repelling, or mitigating any pest or other form of plant or animal life, except humans and microorganisms on or in living humans or other living animals (e.g., using physical or mechanical means to act against a pest without the incorporation of a substance or mixture of substances to act against the pest devices).

Discharge or release into water or onto or into land: Includes a discharge or release to air that is directly and immediately deposited into water or onto land.

Disruption in endocrine system: Effect of chemicals that at certain doses can interfere with the endocrine (or hormone) system in mammals.

Distribute or sell: A key factor in determining the applicability of many FIFRA requirements, meaning "to distribute, sell, offer for sale, hold for distribution, hold for sale, hold for shipment, ship, deliver for shipment, release for shipment, or receive and (having so received) deliver or offer to deliver."

---

8. GUIDE FOR CHEMICAL IMPORTERS/EXPORTERS, at 13.
9. 40 C.F.R. § 152.83.

Distributor: Definition 1. Any natural or legal person established within the European Community, including a retailer, who only stores and places on the market a substance, on its own or in a mixture, for third parties.

Definition 2. Any natural or legal person in the supply chain, other than the manufacturer or the importer, who makes an EEE available on the market. This definition does not prevent a distributor from being, at the same time, a producer.

Distributor product: A registered pesticide distributed and sold under the name and address of a person other than, or in addition to, the registrant.

Downstream user: A person (in the case of a corporation, it shall be limited to a corporation founded in Korea) who uses chemical substances or mixtures in the course of business activities, provided, however, that a person who manufactures, imports, or sells a chemical substance or mixture to consumers is excluded.

Downstream user: Any natural or legal person established within the European Community, other than the manufacturer or the importer, who uses a substance, either on its own or in a mixture, in the course of his industrial or professional activities. A distributor or a consumer is not a downstream user. A reimporter exempted pursuant to Article 2(7)(c) shall be regarded as a downstream user.

DTSC: Department of Toxic Substances Control.

EINECS: European Inventory of Existing Commercial Chemical Substances.

Electrical and electronic equipment (EEE): Equipment that is dependent on electric currents or electromagnetic fields in order to work properly and equipment for the generation, transfer, and measurement of such currents and fields and designed for use with a voltage rating not exceeding 1,000 volts for alternating current and 1,500 volts for direct current.

End-use product (EP): A pesticide product intended for use either as it is distributed or sold or after the user combines it with other substances, and not intended to be used to manufacture or formulate other pesticide products.

Enforceable consent agreement (ECA): Formal agreement among EPA, industry, and other interested parties on what testing must be conducted.

Establishment: Any facility where a pesticidal product, active ingredient (AI), or device is "produced." Establishments must be registered with EPA through the submission of an Application for Registration of Pesticide-Producing and Device-Producing Establishments (EPA Form 3540-8).

European Chemicals Agency (ECHA): The registration and authorization process is overseen by ECHA, which was established pursuant to REACH Article 75 and is composed of regulatory specialists and scientific experts from throughout the EU.

European Commission (EC): The EC was established in 1951 as a nine-member parliamentary body, reflecting the nine members of the EU at that time. It is now a 27-member (reflecting the current 27 EU Member States) executive body responsible for proposing legislation, implementing decisions, adopting legislation, and upholding EU treaties.

In its role of proposing and adopting legislation, the EC is the legal base of the BPR and is responsible for establishing the legal obligations for all Member States and legal entities within the EU.

European Economic Area: A free-trade area created in 1994 by an agreement between the European Free Trade Association, excluding Switzerland, and the European Union (EU).

European Inventory of Existing Commercial Chemical Substances (EINECS): EINECS lists all substances considered to be "existing," defined as those substances that were deemed to be on the EU market between January 1, 1971, and September 18, 1981.

European Union (EU): The economic and political union of the 27 European Member States.

Eurozone: The geographical area containing the countries that have adopted the euro as their single currency.

Existing chemicals subject to registration: Substances published by the MoE among existing chemical substances as necessary to register for hazard examination pursuant to Article 18 or risk assessment pursuant to Article 24 through the deliberation of the chemical substance evaluation committee, pursuant to Article 7.

Existing chemical substance: A chemical substance that falls under one of the following:

1. Chemical substances published by the MoE through consultation with the minister of employment and labor as chemical substances distributed for commercial uses in the country prior to February 2, 1991; or
2. Chemical substances published by the MoE as the substances clear the new chemical notification procedure under the previous Toxic Chemicals Control Act after February 2, 1991.

Existing chemicals: Definition 1. Chemical substances that are listed on the TSCA Inventory.

Definition 2. Chemicals listed on the Korea Existing Chemicals Inventory (KECI). These chemicals include those chemical substances that were commercially used in Korea prior to February 2, 1991, and were notified to MoE, or subsequently notified to MoE and listed on the KECI.

Existing stocks: Inventory of a pesticide currently in the United States that has been previously or is currently registered and that is packaged, labeled, and released for shipment prior to the date of a cancellation or suspension order or a registration amendment, including label changes.

Existing substances: Those substances listed on the IECSC published by the CRC.

Experimental use permit (EUP): A permit, granted after submission of an EUP application (EPA Form 8570-17), allowing a pesticide to be used to accumulate information needed to register the pesticide or a new use of the pesticide.

Export-only chemicals: Chemical substances manufactured or imported solely for export, provided the chemical substance is labeled to indicate that it is intended for export, and the chemical substance is distributed domestically only to persons who intend to export the chemical substance directly or process it solely for export.[10]

Exporter: The person who, as the principal party in interest in the export transaction, has the power and responsibility for determining and controlling the sending of the chemical substance or mixture to a destination out of the Customs territory of the United States.[11]

Expose: To cause to ingest, inhale, contact via body surfaces, or otherwise come into contact with a chemical listed on Proposition 65.

---

10. Id. § 720.3(s).
11. Id. § 707.63(b).

Exposure scenario: The set of conditions, including operational conditions and risk management measures, that describes how the substance is manufactured or used during its life cycle and how the manufacturer or importer controls, or recommends downstream users to control, exposures of humans and the environment. An exposure scenario may cover one specific process or use or several processes or uses as appropriate.

Food additives: Substances added to food to preserve flavor or enhance its taste and appearance.

Formulator's exemption: An exemption from the requirement to submit or cite data pertaining to a registered pesticide, or to offer to pay data compensation for those data, when a producer purchases the registered pesticide from another producer to formulate it into a pesticide for which the producer is seeking registration.

Full notification: This applies to all chemicals except those new chemical substances subject to reduced notification and polymer notification. The following data are required:
1. Application Form for the Hazard Review of Manufacture/Import of Chemical Substance.
2. Test reports for acute toxicity, genetic toxicity, degradation, and ecotoxicity.
3. The document regarding the main release channel to the environment and the estimated release amount.

Full study report: A complete and comprehensive description of the activity performed to generate the information. This covers the complete scientific paper as published in the literature describing the study performed or the full report prepared by the test house describing the study performed.

Globally Harmonized System of Classification and Labelling of Chemicals (GHS): Internationally agreed-upon system created by the United Nations in 1992 designed to generate consistent criteria for classification and labeling on a global level. The GHS was formed to establish a uniform global system of classification and labeling of chemicals to assist in trade. The standard reference document is known as the "Purple Book," and it has undergone several revisions. Countries may have adopted various versions of the Purple Book or may have adopted domestic modifications. It is important to reference each country's specific adoption of GHS.

Good Laboratory Practice (GLP): A quality system of management controls for research laboratories, and organizations to ensure the uniformity, consistency, reliability, reproducibility, quality, and integrity of chemicals. GLP helps to ensure that the data submitted are an accurate reflection of the results obtained during a study, thus making them reliable and reproducible when performing risk and safety assessments.

Harmonized standard: Standard adopted by one of the European standardization bodies listed in Annex I to Directive 98/34/EC of the European Parliament and of the Council of June 22, 1998, laying down a procedure for the provision of information in the field of technical standards and regulations and of rules on Information Society services on the basis of a request made by the Commission in accordance with Article 6 of Directive 98/34/EC.

Hazard: The intrinsic properties of chemicals that adversely affect human health or the environment, such as toxicity of chemicals.

Hazardous chemicals: Per Article 3 of Decree No. 591, hazardous chemicals are highly toxic substances and other chemicals that are toxic, corrosive, explosive, flammable, or combustion-supporting and can do harm to people, facilities, or the environment.

Hazardous substance: Toxic chemicals, authorization substances, restricted substances, prohibited substances, and other substances of hazards or risks or concerned with such hazards or risks.

Health and safety study: Any study of any effect of a chemical substance or mixture on health or the environment or on both, including underlying data and epidemiological studies, studies of occupational exposure to a chemical substance or mixture, toxicological, clinical, and ecological or other studies of a chemical substance or mixture, and any test performed under TSCA.[12]

High Production Volume (HPV) program: The U.S. HPV program defines HPV chemicals as organic nonpolymer chemicals supplied at or greater than 450 metric tons (1 million lbs.) in the United States.

Homogenous material: One material of uniform composition throughout or a material, consisting of a combination of materials, that cannot be disjointed or separated into different materials by mechanical actions such as unscrewing, cutting, crushing, grinding, and abrasive processes.

---

12. TSCA § 8(d), 15 U.S.C. § 2607(d); 40 C.F.R. § 716.3.

Identified use: A use of a substance on its own or in a mixture, or a use of a mixture, that is intended by an actor in the supply chain, including his own use, or that is made known to him in writing by an immediate downstream user.

Imminently hazardous chemical substance or mixture: Chemical substance or mixture that presents an imminent and unreasonable risk of serious or widespread injury to health or the environment and is likely to result in such injury before a final rule under section 6 can protect against the risk.[13]

Import: The physical introduction into the customs territory of the European Community.

Import certificate: Certification statement regarding a chemical substance's compliance with TSCA.

Import certificate, blanket certification: A single import certificate that is used to certify TSCA compliance for multiple shipments of the same chemical substances, including those contained in mixtures, or articles over a one-year period on a calendar year basis.[14]

Import certificate, negative certification: Certification statement that chemical substance(s) in imported shipments are exempt from TSCA, unless the chemical substance, mixture, or article is exempt from TSCA import certification requirements.

Import certificate, positive certification: Certification statement that chemical substance(s) in imported shipment comply with TSCA.

Import tolerance: A tolerance for a pesticide that is not registered in the United States but that may be imported into the United States as a residue on raw agricultural commodities or processed food.

Importer: Definition 1. A person, an enterprise, or an entity that brings in goods from a foreign country.

Definition 2. Under U.S. law, the U.S. Customs regulations define "importer" as "the person primarily liable for the payment of any duties on the merchandise, or an authorized agent acting on his behalf." TSCA regulations further provide that the importer may be:

1. The consignee, or
2. The importer of record, or

---

13. TSCA § 7(f), 15 U.S.C. § 2606(f).
14. GUIDE FOR CHEMICAL IMPORTERS/EXPORTERS, at 20–21.

3. The actual owner of the merchandise, if an actual owner's declaration and superseding bond has been filed in accordance with 19 C.F.R. § 141.20 of this chapter, or
4. The transferee of the merchandise, if the right to withdraw merchandise in a bonded warehouse has been transferred in accordance with subpart C of part 144 of 19 C.F.R. § 144.[15]

In vitro: In glass; made to occur outside the living organism in an artificial environment, such as a culture medium.

In vitro diagnostic medical device: In vitro diagnostic medical device within the meaning of point (b) of Article 1(2) of Directive 98/79/EC.

Industrial monitoring and control instruments: Monitoring and control instruments designed for exclusively industrial or professional use.

Inert ingredient: An ingredient that is intentionally added to a pesticide product, but that is not pesticidally active, such as a solvent, detergent, emulsifier, filler, carrier, or perfume.

Intermediate: A substance that is manufactured for and consumed in or used for chemical processing in order to be transformed into another substance.

International Uniform Chemical Information Database (IUCLID): Database designed to assist jurisdictions with the collection and sharing of relevant hazard data in chemical control and similar programs.

International Union of Pure and Applied Chemistry (IUPAC): IUPAC is commonly referred to when creating standardized chemical names for chemical regulation.

Inventory of Existing Chemical Substances in China (IECSC): A list of chemical substances considered to be existing in China as compiled by the MEP-SCC. The list is divided into two sections: the public section, which is free, and the confidential section, in which a query must be submitted to the MEP-SCC accompanied by a fee of approximately $100 USD.

Korea Existing Chemicals Inventory (KECI): Inventory containing those chemicals considered existing, as distinguished from those substances considered new.

---

15. 40 C.F.R. § 712.3(d).

Label and labeling: A label is the written, printed, or graphic matter on, or attached to, a pesticide or device or any of its containers or wrappers. The term "labeling" is broader, encompassing "label," but also including all other written, printed, or graphic matter accompanying the pesticide or device at any time or to which reference is made on the label or in literature accompanying the pesticide or device.

Large-scale fixed installation: Large-scale combination of several types of apparatus and, where applicable, other devices, which are assembled and installed by professionals, intended to be used permanently in a predefined and dedicated location, and deinstalled by professionals.

Large-scale stationary industrial tools: Large-scale assembly of machines, equipment, and/or components, functioning together for a specific application, permanently installed and deinstalled by professionals at a given place, and used and maintained by professionals in an industrial manufacturing facility or research and development facility.

Lead registrant: One registrant or importer acting with the agreement of other assenting registrants who submits joint registration dossier with all information, prior to submission of member dossiers.

Letter of access: A document granting the right to refer to the Lead Registrant dossier for the sole purpose of making a REACH registration for a specific substance, and to participate in the Joint Submission in ECHA REACH-IT.

LQTU (large quantity toxics user): Any toxics user who manufactures, processes, or otherwise uses any toxic or hazardous substance in an amount the same as or greater than the applicable threshold amount in a calendar year at a facility.

Manufacture: Definition 1. To produce or manufacture in the United States or import into the Customs territory of the United States.[16] This broad definition includes importers of chemical substances.[17]

Definition 2. To produce, prepare, import, or compound a toxic or hazardous substance; also means to produce a toxic or hazardous substance coincidentally during the manufacture, processing, use, or disposal of another substance or mixture of substances.

---

16. TSCA § 3(7), 15 U.S.C. § 2602(7); see also, e.g., 40 C.F.R. §§ 710.3(d), 720.3(q).
17. TSCA § 3(7), 15 U.S.C. § 2602(7) ("'manufacture' means to import into the customs territory of the United States (as defined in general note 2 of the Harmonized Tariff Schedule of the United States), produce, or manufacture").

Manufacturer: Definition 1. Any natural or legal person established within the European Community who manufactures a substance within the Community.

Definition 2. Any natural or legal person who manufactures an EEE or who has an EEE designed or manufactured and markets it under his name or trademark.

Manufacturing: Production or extraction of substances in the natural state.

Manufacturing use product (MUP): Any pesticide product that is not an end-use product, some of which consist only of a technical grade of active ingredient (TGAI); others contain the TGAI and inert ingredients, such as stabilizers or solvents.

Massachusetts DEP: Massachusetts Department of Environmental Protection.

Material Data Safety Sheet (MSDS): An MSDS provides workers and emergency personnel with procedures for handling or working with hazardous substances. The protocol includes information such as physical data, toxicity, health effects, first aid, reactivity, storage, disposal, protective equipment, instructions for safe use, and spill-handling procedures.

Me-too registration: A registration of a pesticide that is identical or substantially similar in composition and labeling to a currently registered pesticide, or that differs in composition and labeling from it only in ways that do not significantly increase the risk of unreasonable adverse effects on the environment, thus allowing the me-too registrant to rely in large part on data submitted for the previously registered product to support the me-too registration.

Member State: State that is a party to treaties of the EU and thereby subject to the privileges and obligations of EU membership.

Ministry of Environment (MoE): The South Korea branch of government charged with environmental protection. The tasks of MoE include enactment and amendment of environmental laws and regulations; introduction of environmental institutions; building up a framework structure for environmental administration; drafting and implementation of mid- to long-term comprehensive measures for environmental conservation; setting up standards for regulation; providing administrative and financial support for environmental management to local governments; inter-Korean environmental cooperation; and environmental cooperation with other countries.

Ministry of Environmental Protection (MEP): The Chinese ministry charged with developing and organizing the implementation of national policies and plans for environmental protection and preservation.

Mixture: Any combination of two or more chemical substances if the combination does not occur in nature and is not, in whole or in part, the result of a chemical reaction.[18]

Monomer: A substance that is capable of forming covalent bonds with a sequence of additional like or unlike molecules under the conditions of the relevant polymer-forming reaction used for the particular process.

Mutagenic: An agent, such as a chemical, ultraviolet light, or a radioactive element, that can induce or increase the frequency of mutation in an organism.

Mutation: The act or process of being altered or changed.

National Institute of Environmental Research (NIER): Korean agency charged with environmental research, education, international cooperation, and setting criteria for various pollutants.

National Measurement Office (NMO): An executive agency of the Department of Business, Innovation, and Skills seeking to provide a measurement infrastructure that supports innovation, facilitates fair competition, protects consumers, health, and the environment. NMO is the U.K. enforcement agency for RoHS 2.

National Registration Center of Chemicals (NRCC): Managed under SAWS, the NRCC is a comprehensive technical support agency that deals with dangerous chemicals in China. The center performs a wide range of tasks, such as hazardous chemical registration, emergency rescue in the case of chemical accidents, draft and revision of regulations and standards on safety management of chemicals, and subject research on chemical safety management, as well as relevant assessment, technical development, training, and consulting services.

Naturally occurring substance: Any chemical substance that is naturally occurring and that is (1) unprocessed; (2) processed only by manual, mechanical, or gravitational means; by dissolution in water; by flotation; or by heating solely to remove water; or (3) extracted from air by any means.[19]

---

18. TSCA § 3(8), 15 U.S.C. § 2602(8). *See also* 40 C.F.R. § 720.3(u).
19. *See* 40 C.F.R. § 710.4(b).

New chemical substances: Those substances not listed on the IECSC that must be notified (registered) prior to manufacture in or import into China if they are within the scope of Order No. 7.

New chemicals: Definition 1. Chemical substances that are not listed on the TSCA Inventory.

Definition 2. Chemicals excluding those listed in each of the following items:

1. Chemicals distributed for commercial purposes domestically before February 2, 1991, and announced by the MoE through consultation with the minister of labor on December 23, 1996; and
2. Chemicals that have undergone the examination of toxicity under the former provisions or the provisions of this Act after February 2, 1991, and were announced by the MoE.

Nitrogen stabilizer: Under FIFRA section 2(hh), pesticide "intended for preventing or hindering the process of nitrification, denitrification, ammonia volatilization, or urease production through action upon soil bacteria."

Non-phase-in substances: Substances that do not fulfill any of the criteria for phase-in substances. Normally, non-phase-in substances have not been manufactured, placed on the market, or used in the EU before June 1, 2008, unless they were notified under Directive 67/548/EEC (NONS). Potential manufacturers and importers of non-phase-in substances have to submit an inquiry to ECHA and subsequently register the substance in accordance with REACH before they can manufacture or import the substance.

Nonisolated intermediates: Any intermediate that is not intentionally removed from the equipment in which it is manufactured, including the reaction vessel and ancillary equipment to the reaction vessel, but not including tanks or other vessels in which the chemical substance is stored.[20]

Nonroad mobile machinery made available exclusively for professional use: Machinery, with an on-board power source, the operation of which requires either mobility or continuous or semicontinuous movement between a succession of fixed working locations while working, and is made available exclusively for professional use.

---

20. *Id.* § 710.3(d).

Notice of Arrival: A form (EPA Form 3540-1) listing the pesticides' brand names and active ingredients and the names and addresses of the importer and customs broker that must be submitted to EPA by an importer of pesticides and approved by EPA before the pesticide may enter the United States.

Notice of commencement (NOC): Notice required under section 5 of TSCA within 30 calendar days of the date a new chemical substance is first manufactured or imported for nonexempt commercial purposes. The chemical substance is considered to be on the TSCA Inventory and an existing chemical as soon as a complete NOC is received by EPA.

OEHHA: Office of Environmental Health Hazard Assessment.

Offer to pay: A requirement that an applicant relying upon data already submitted to EPA provide to such data owners an offer to pay compensation for such reliance.

Only Representative (OR): Under Article 8, "A natural or legal person established outside the Community who manufactures a substance on its own, in preparations or in articles, formulates a preparation or produces an article that is imported into the Community may by mutual agreement appoint a natural or legal person established in the Community to fulfil, as his only representative, the obligations on importers under this Title." Appointment of an OR also enables the non-EU entity to maintain the confidentiality of product compositions, volume, uses, and downstream user details (client list) from an EU importer, who in many instances may be a competitor.

OTA: Office of Technical Assistance and Technology.

Otherwise use or other use: Any use of a toxic substance that is not covered by the terms "manufacture" or "process"; relabeling or redistributing a container of a toxic substance where no repackaging of the toxic substance occurs does not constitute use or processing of the toxic substance.

Per year: Per calendar year, unless stated otherwise. For phase-in substances that have been imported or manufactured for at least three consecutive years, quantities per year shall be calculated on the basis of the average production or import volumes for the three preceding calendar years.

Persistent, bioaccumulative, and toxic (PBT): Chemicals that are toxic, persist in the environment, and bioaccumulate in food chains and thus pose risks to human health and ecosystems. The biggest concerns about PBTs are that they transfer rather easily among air, water, and land, and span boundaries of programs, geography, and generations.

Person in the course of doing business: Excludes any person employing fewer than ten employees in his or her business; local, state, and federal departments and agencies; and any entity in its operation of a public water system.

Person: Any natural or juridical person including any individual, corporation, partnership, or association, any state or political subdivision thereof, or any municipality, any interstate body and any department, agency, or instrumentality of the federal government.[21]

Pest: Under FIFRA section 2(t), "(1) any insect, rodent, nematode, fungus, weed, or (2) any other form of terrestrial or aquatic plant or animal life or virus, bacteria, or other micro-organism (except viruses, bacteria, or other micro-organisms on or in living man or other living animals) which [EPA] declares to be a pest. . . ."

Pesticide: Under FIFRA section 2(u), "[a]ny substance or mixture of substances intended for preventing, destroying, repelling, or mitigating any pest," with certain specified exceptions. Pesticides must be registered under FIFRA.

Pesticide chemical residue: A residue on or in a raw agricultural commodity or processed food of a pesticide chemical or any other added substance that is present on or in the commodity or food primarily as a result of the metabolism or other degradation of a pesticide chemical.

Pesticide product: For registration purposes, a pesticide in the particular form, including the composition, packaging, and labeling, in which the pesticide is, or is intended to be, distributed or sold. A pesticide product includes any physical apparatus used to deliver or apply the pesticide if it is distributed or sold with the pesticide.

Phase-in substance: A substance that meets at least one of the following criteria:

1. It is listed in the European Inventory of Existing Commercial Chemical Substances (EINECS).
2. It was manufactured in the European Community, or in the countries acceding to the European Union on January 1,

---

21. 710.3(d).

1995, May 1, 2004, or January 1, 2007, but not placed on the market by the manufacturer or importer, at least once in the 15 years before the entry into force of this regulation, provided the manufacturer or importer has documentary evidence of this.
3. It was placed on the market in the European Community, or in the countries acceding to the European Union on January 1, 1995, May 1, 2004, or January 1, 2007, by the manufacturer or importer before the entry into force of this regulation and it was considered as having been notified in accordance with the first indent of Article 8(1) of Directive 67/548/EEC in the version of Article 8(1) resulting from the amendment effected by Directive 79/831/EEC, but it does not meet the definition of a polymer as set out in this regulation, provided the manufacturer or importer has documentary evidence of this, including proof that the substance was placed on the market by any manufacturer or importer between September 18, 1981, and October 31, 1993, inclusive.

Photovoltaic panels: Panels capable of producing a voltage when exposed to radiant energy, especially light.

Physical-chemical: Relating to both physical and chemical properties.

Place into the stream of commerce in California: Sell, offer for sale, distribute, supply, or manufacture in or for use in California as a finished product or as a component in an assembled product.

Placing on the market: Supplying or making available, whether in return for payment or free of charge, to a third party. Import shall be deemed to be placing on the market.

Poisonous substances: Chemicals with toxicity, which are designated and announced by the MoE in accordance with the standards prescribed by presidential decree.

Polymer: A chemical substance consisting of molecules characterized by the sequence of one or more types of monomer units and comprising a simple weight majority of molecules containing at least three monomer units that are covalently bound to at least one other monomer unit or other reactant and that consists of less than a simple weight majority of molecules of the same molecular weight. Such molecules must be distributed over a range of molecular weights wherein differences in the molecular weight are primarily attributable to differences in the number of monomer units. In the context of this definition, "sequence" means that the monomer units under consideration are covalently bound to one another and form a continuous string within the molecule, uninterrupted by units other than monomer units.[22]

PP: Priority products.

Preliminary Assessment Information Rule (PAIR): Rule issued by EPA under section 8 to require manufacturers and importers of certain chemical substances listed on the TSCA Inventory to submit a one-time report on (1) the quantities of chemical substances manufactured, imported, used as a reactant, used in industry and consumer products, or lost to the environment; and (2) worker exposure.[23]

Premanufacture notice (PMN): Notice required under section 5 of TSCA for anyone who plans to manufacture or import a new chemical substance for a nonexempt commercial purpose.

Process: The preparation of a chemical substance or mixture, after its manufacture, for distribution in commerce—(1) in the same form or physical state as, or in a different form or physical state from, that in which it was received by the person so preparing such substance or mixture, or (2) as part of an article containing the chemical substance or mixture.[24]

Produce: Under FIFRA section 2(w), "to manufacture, prepare, compound, propagate, or process any pesticide or device or active ingredient used in producing a pesticide."

Producer: Any natural or legal person who, irrespective of the selling technique used, including distance communication within the meaning of Directive 97/7/EC of the European Parliament and of the Council of May 20, 1997, on the protection of consumers in respect of distance contracts:

1. Is established in a Member State and manufactures EEE under his own name or trademark, or has EEE designed or manufactured and markets it under his name or trademark within the territory of that Member State;
2. Is established in a Member State and resells within the territory of that Member State, under his own name or

---

22. Id. § 723.250(b).
23. See id. pt. 712, subpt. B.
24. TSCA § 3(10), 15 U.S.C. § 2602(10); 40 C.F.R. § 720.3(aa). See also EPA, Question and Answer Summary: EPA Seminar on Industry Obligations under TSCA (June 10, 1986), at 1–2.

trademark, equipment produced by other suppliers, a reseller not being regarded as the "producer" if the brand of the producer appears on the equipment, as provided for in point (1);
3. Is established in a Member State and places on the market of that Member State, on a professional basis, EEE from a third country or from another Member State; or
4. Sells EEE by means of distance communication directly to private households or to users other than private households in a Member State, and is established in another Member State or in a third country.

Whoever exclusively provides financing under or pursuant to any finance agreement shall not be deemed to be a "producer" unless he also acts as a producer within the meaning of points (1) to (4).

Product: Each of the following that is likely to result in the exposure of chemicals to consumers as final consumer goods or parts thereof and accessories:
1. Product consisting of a mixture.
2. Product of which the chemicals are not released during the course of uses and perform a certain function in a particular solid form.

Product and process-orientated research and development (PPORD): Any scientific development related to product development or the further development of a substance, on its own, in mixtures, or in articles in the course of which pilot plant or production trials are used to develop the production process and/or to test the fields of application of the substance.

Product-type: Under the BPR, 22 product-types are used to categorize the use of the active substances in biocidal products.

Prohibited substance: Substance published by the MoE as being recognized for its high risks through consultation with the head of the relevant central administrative agency, pursuant to Article 27, and deliberation of the chemical substance evaluation committee, pursuant to Article 7, to prohibit from manufacturing, import, sale, warehouse storage, transportation, or use for all applications.

Proposition 65: Safe Drinking Water and Toxic Enforcement Act of 1986.

Raw agricultural commodity (RAC): Under FFDCA section 201, section 321(r), "any food in its raw or natural state, including all fruits that are washed, colored, or otherwise treated in their unpeeled natural form prior to marketing."

Recall: Any measure aimed at achieving the return of a product that has already been made available to the end user.

Recovery: Any operation the principal result of which is waste serving a useful purpose by replacing other materials that would otherwise have been used to fulfill a particular function, or waste being prepared to fulfill that function, in the plant or in the wider economy.

Recycling: Any recovery operation by which waste materials are reprocessed into products, materials, or substances whether for the original or other purposes. It includes the reprocessing of organic material but does not include energy recovery and the reprocessing into materials that are to be used as fuels or for backfilling operations.

Registrant: The manufacturer or the importer of a substance or the producer or importer of an article submitting a registration for a substance.

Registration certificates: An official document stating that a person or company has provided all the necessary information for an official or government record.

Registration dossier: The registrant of a substance needs to compile all the required information in a registration dossier, which consists of two main components:

1. A technical dossier, always required for all substances subject to the registration obligations; and
2. A CSR, required if the registrant manufactures or imports a substance in quantities of ten metric tons or more per year.

The registration dossier has to be prepared using the IUCLID 5 software application. IUCLID 5 implements the Harmonized Templates developed by the OECD and it is compatible with other chemical legislations around the world. Once the dossier has been created with IUCLID 5, it has to be submitted to ECHA through REACH-IT.

Registration, Evaluation, Authorization, and Restriction of Chemicals (REACH): EU Regulation 1907/2006, which regulates industrial chemicals in the EU.

Regulation for Biocidal Products (R4BP): This is an Internet-based system for submission of biocide applications and communication from the CA, EC, or ECHA.

Reregistration eligibility decision (RED): A document summarizing the findings of EPA's reregistration review for individual chemical cases—either individual substances or related groups of substances—which reflects EPA's risk assessment and risk management decisions for the pesticides at issue.

Research and development (R&D) substances: Substances manufactured for noncommercial R&D work, or manufactured in small quantities solely for commercial R&D.

Responsible entity: The manufacturer, importer, assembler, or retailer of a consumer product.

Restricted substance: A substance published by the MoE as being recognized for its high risks in using a specific application and published by the MoE through consultation with the head of the relevant central administrative agency, pursuant to Article 27, and in deliberation with the chemical substance evaluation committee, pursuant to Article 7, to prohibit from manufacturing, import, sale, warehouse storage, transportation, or use for such application.

Restricted use pesticide (RUP): A pesticide that EPA finds may generally cause, without additional regulatory restrictions, unreasonable adverse effects on the environment, including injury to the applicator of the pesticide, when applied in accordance with its directions for use, warnings, and cautions, and for one of the uses for which it is registered, or in accordance with a widespread and commonly recognized practice.

Restriction: Any condition for or prohibition of the manufacture, use, or placing on the market.

Retailer: A person to whom a consumer product is delivered or sold for purposes of sale or distribution by the person to a consumer.

Reuse: Any operation by which products or components that are not waste are used again for the same purpose for which they were conceived.

RMB: Renminbi, the official currency of the People's Republic of China.

Robust study summary: A detailed summary of the objectives, methods, results, and conclusions of a full study report providing sufficient information to make an independent assessment of the study minimizing the need to consult the full study report.

RoHS: Restriction of the use of certain hazardous substances in electrical and electronic equipment.

RoHS 2: Directive 2011/65/EU of the European Parliament and of the Council of June 8, 2011, on the restriction of the use of certain hazardous substances in electrical and electronic equipment.

SAB: Science Advisory Board.

Safe harbor level: Exposures and discharges that fall below the no significant risk levels for chemicals listed as causing cancer and maximum allowable dose levels for chemicals listed as causing reproductive toxicity; exposures and discharges below the safe harbor levels are exempt from the requirements of Proposition 65.

Safe use determination: A written statement, issued by OEHHA, that interprets and applies Proposition 65 requirements to a specific set of facts.

Safety Data Sheet: A document requiring certain information about a chemical substance's potential hazards to provide workers and emergency personnel with procedures for safely managing the substance. The protocol includes information such as physical data, toxicity, health effects, first aid, reactivity, storage, disposal, protective equipment, instructions for safe use, and spill-handling procedures.

Sale: An act to put chemical substances, mixtures, or products on the market.

Scientific research and development: Any scientific experimentation, analysis, or chemical research carried out under controlled conditions in a volume less than one metric ton per year.

Scientific research record (SRR) notification: Intended to accommodate the use of new chemicals for research and development purposes only at a volume of <100 kg/year.

SCPR: Safer Consumer Products Regulations.

"Self-executing" exemption: An exemption that does not require EPA approval. Instead, once a manufacturer or importer determines that one of the self-executing exemptions applies, the new chemical substance may be manufactured or imported without first submitting a PMN, so long as the company complies with any recordkeeping or other requirements for the particular exemption.

Signal word: A word indicating a pesticide's toxicity, based on the toxicity category into which EPA has placed the pesticide, that must appear on the front of a pesticide label (e.g., pesticides in Toxicity Category I must bear the signal word "Danger," and if the pesticide product is in

Toxicity Category I because of oral, inhalation, or dermal toxicology, it must bear the word "Poison" in red letters on a distinctly contrasting background and the skull and crossbones symbol. Pesticides in Toxicity Category II must bear the signal word "Warning," and those in Toxicity Categories III and IV must bear the signal word "Caution.")

Significant adverse reaction: "[R]eactions that may indicate a substantial impairment of normal activities, or long-lasting or irreversible damage to health or the environment."[25]

Site: For importers, a "site" for reporting purposes is the operating unit within the company that is directly responsible for importing the chemical substance and controlling the transaction, in some cases this may refer to a company's headquarters in the United States.[26]

Skin sensitization data: Data pertaining to an allergic reaction to a particular irritant that results in the development of skin inflammation and itchiness.

SNUN (significant new use notice): Notice required by manufacturers of chemical substances subject to SNURs, which allows EPA the opportunity to review and if necessary prevent or limit potentially adverse exposure to, or effects from, the new use of the substance.

SNUR (Significant New Use Rule): Rule issued to limit use or production of a chemical substance, when EPA determines that a chemical substance use under certain conditions may pose an unreasonable risk or the chemical substance may be produced in substantial quantities and will either enter the environment in substantial quantities or may result in significant or substantial human exposure.

Solid Waste and Chemical Management Center (SCC): In 2013, the Competent Authority for new chemical notification activities became the SCC, which also addresses solid waste, chemical, and pollution management.

SQTU: Small quantity toxics user; any toxics user who is not a large quantity toxics user.

State Administration of Work Safety (SAWS): An agency directly under the State Council created to supervise and regulate the risks to occupational safety and health. Additionally, SAWS is the working

---

25. 40 C.F.R. § 717.3(i).
26. *Id.* § 711.3.

body of the office of the State Council Work Safety Commission and manages the State Administration of Coal Mine Safety.

State's qualified experts: The Carcinogen Identification Committee and Developmental and Reproductive Toxicant Identification Committee.

Stop sale, use, or removal order (SSURO): An order issued by EPA that prohibits the sale, use, or removal of a pesticide or device.

Study summary: A summary of the objectives, methods, results, and conclusions of a full study report providing sufficient information to make an assessment of the relevance of the study.

Substance: A chemical element and its compounds in the natural state or obtained by any manufacturing process, including any additive necessary to preserve its stability and any impurity deriving from the process used, but excluding any solvent that may be separated without affecting the stability of the substance or changing its composition.

Substance Information Exchange Forum (SIEF): A SIEF is composed of all preregistrants and late preregistrants of a substance that have an interest in the substance. Active members will typically join together to ensure that successful registration of a substance, including technical dossier preparation activities, is completed. Where participants in a SIEF or other arrangement cannot reach agreement on cost sharing, REACH provides an avenue for ECHA to become involved in the process, although this occurs infrequently.

Substances of very high concern (SVHC): Substances that may have serious and often irreversible effects on human health and the environment. If a substance is identified as an SVHC, it will be added to the Candidate List for eventual inclusion in the Authorization List.

Substances under observation: Chemicals feared to have toxicity, and designated and announced by the MoE in accordance with the standards prescribed by presidential decree.

Substantial production: Threshold for EPA finding in order to issue testing requirement under section 4 of TSCA, which, under EPA policy, is production of quantities of 1 million pounds or more annually.[27]

Substantial quantities that will enter the environment: Threshold for EPA finding in order to issue testing under section 4. EPA has stated that it may make the requisite finding whenever a chemical has been

---

27. 58 Fed. Reg. 28,736 (May 14, 1993).

released to the environment in quantities equal to at least 10 percent of its total production, or 1 million pounds per year, whichever is lower.[28]

Supplemental distribution: The distribution or sale of a registered product under the name and address of a person other than (or in addition to) the registrant.

Technical dossier: Registration documentation including information on:

1. The identity of the substance;
2. Manufacture and use of the substance;
3. The classification and labeling of the substance;
4. Guidance on its safe use;
5. (Robust) study summaries of the information on intrinsic properties;
6. Proposals for further testing, if relevant; and
7. For substances registered in quantities between one and ten metric tons, the technical dossier also contains exposure information for the substance (main use categories, types of uses, significant routes of exposure).

Technical grade of active ingredient (TGAI): Under EPA regulations, "a material containing an active ingredient: (1) [w]hich contains no inert ingredient, other than one used for purification of the active ingredient; and (2) [w]hich is produced on a commercial or pilot-plant production scale (whether or not it is ever held for sale)."[29]

Technical specification: Document that prescribes technical requirements to be fulfilled by a product, process, or service.

Third-party representative (TPR): The TPR acts upon the instruction of the appointee company and represents the company's interest in discussions with other supply chain actors. The TPR does not adopt the obligations of the registrant as the OR does but provides a level of confidentiality in the registrant's interactions with other legal entities.

Tolerance: The amount of a pesticide residue that legally may be present on a raw agricultural commodity, established pursuant to FFDCA section 408, or that legally may be present in or on a processed food under the terms of a food additive regulation established pursuant to FFDCA section 409.

---

28. Id. at 28,746.
29. 40 C.F.R. § 158.300.

Toxicity: The unique nature of a chemical that adversely affects human health or the environment, such as the toxicity of chemicals.

Toxicity category: A classification given to a pesticide based on the highest hazard shown by the pesticide based on the pesticide's oral and dermal lethal dose, the lethal inhalation concentration, eye effects, and skin effects, which then determines the signal word and precautionary statement that must appear on the pesticide's label, as well as the personal protective equipment required for the pesticide under the Worker Protection Standards. There are four toxicity categories, with Toxicity I being the most toxic and Toxicity IV being the least.

Toxics user: A person who owns or operates a facility that manufactures, processes, or otherwise uses any toxic or hazardous substance that is classified in SIC Codes 10 to 14, inclusive, 20 to 40, inclusive, 44 to 51, inclusive, 72, 73, 75, or 76 or the corresponding NAICS codes.

Trade secret: Any formula, plan, pattern, process, production data, device, information, or compilation of information used in a toxics user's business that provides the toxics user an opportunity to obtain an advantage over competitors who do not know or use it.

Treated article: An article that is exempt from FIFRA section 3 registration requirements because it is an article or substance treated with, or containing, a pesticide to protect the integrity of the article or substance itself.

TSCA Confidential Inventory: A list of all existing chemical substances whose identity has been claimed as confidential business information (CBI).[30]

TSCA Public Inventory: A list of all existing chemical substances whose identity has not been claimed as confidential business information (CBI).[31]

TUR: Toxics use reduction; in-plant changes in production processes or raw materials that reduce, avoid, or eliminate the use of toxic or hazardous substances or generation of hazardous by-products per unit of product.

TURA: Toxics Use Reduction Act.

TURI: Toxics Use Reduction Institute.

---

30. *Id.* § 720.25(b)(1).
31. *Id.*

Unknown, of Variable Composition, or of Biological Origin (UVCB): A chemical substance that cannot be identified by spectral or analytical means, has a variable composition (such as a reaction mass), or is of biological origin, any of which may affect the profile of the substance.

U.S. Environmental Protection Agency (U.S. EPA): The agency of the U.S. federal government responsible for protecting human health and the environment from industrial activities.

Use: Any processing, formulation, consumption, storage, keeping, treatment, filling into containers, transfer from one container to another, mixing, production of an article, or any other utilization.

User segment: A set of no fewer than five toxics users who use a similar production unit.

Warning: May be provided by general means such as consumer product labels, notices in mailings to water customers, posted notices, or notices in public news media.

Waste: Any substance or object that the holder discards or intends or is required to discard.

Waste from electrical and electronic equipment (WEEE): Electrical or electronic equipment that is waste within the meaning of Article 3(1) of Directive 2008/98/EC, including all components, subassemblies, and consumables that are part of the product at the time of discarding.

WEEE 2: Directive 2012/19/EU of the European Parliament and of the Council of July 4, 2012, on waste electrical and electronic equipment.

Index

AA. *See* Alternatives analyses (AA)
AAL. *See* Allowable absolute limit (AAL)
ABNT. *See* Brazilian Association of Technical Standards (Associação Brasileira de Normas Técnicas)
Active ingredients, 85, 88
Active substances, 311, 317, 320–321, 324–328
Act on Registration and Evaluation, Authorization, and Restriction of Chemical Substances (K-REACH)
 authorization, 371
 enforcement and penalties, 374
 evaluation, 370
 hazardous substances, 370–371
 notification, registration, approval requirements, and compliance, 367–368
 notification requirements, 14
 only representative, 372
 overview, 13–14, 366–374
 REACH and, 291–292, 362, 367, 371, 372, 373, 374
 registration of chemical substances, 368–370
 resources, 375
 restricted and banned chemicals, 371–372
 supply chain requirements applicable to manufacturers, 373–374
 trends, 374–375
AD. *See* Antimicrobials Division (AD)

Administración Nacional de Medicamentos, Alimentos y Technologia Médica (ANMAT). *See* National Administration for Drugs, Food, and Medical Technology (ANMAT)
Administration of Quality Supervision, Inspection, and Quarantine (AQSIQ), 356
Administrative Council on Toxics Use Reduction (Council), 168
Administrative monetary penalty (AMP), 214
Adverse health/environmental reactions, allegations of, 47–49, 54
Agência Nacional de Transportes Aquaviários (ANTAQ). *See* Brazilian Agency of Waterway Transportation (Agência Nacional de Transportes Aquaviários)
Agência Nacional de Transportes Terrestres (ANTT). *See* Brazilian Agency of Terrestrial Transportation (Agência Nacional de Transportes Terrestres)
Agência Nacional de Vigilância Sanitária (ANVISA). *See* Sanitary Surveillance Agency (Agência Nacional de Vigilância Sanitária)
AHERA. *See* Asbestos Hazard Emergency Response Act (AHERA)

Allowable absolute limit (AAL), 237
Alternatives analyses (AA), 157–159, 160, 162
American Arbitration Association, 116
AMP. *See* Administrative monetary penalty (AMP)
Andean Community of Nations (Comunidad Andina), 8, 9, 239, 242
ANMAT. *See* National Administration for Drugs, Food, and Medical Technology (ANMAT)
ANSI Z400.1, 240
ANTAQ. *See* Brazilian Agency of Waterway Transportation (Agência Nacional de Transportes Aquaviários)
Anteproyecto de Reglamento de Almacenamiento de Sustancias Químicas Peligrosas. *See* Draft Regulation for the Storage of Hazardous Chemicals (Anteproyecto de Reglamento de Almacenamiento de Sustancias Químicas Peligrosas)
Antimicrobials Division (AD), 98–99
ANTT. *See* Brazilian Agency of Terrestrial Transportation (Agência Nacional de Transportes Terrestres)
ANVISA. *See* Sanitary Surveillance Agency (Agência Nacional de Vigilância Sanitária)
Approval requirements, 25–37
AQSIQ. *See* Administration of Quality Supervision, Inspection, and Quarantine (AQSIQ)
Arbitration, 116
Argentina, 8, 242
  notification, registration, approval requirements, and compliance, 234–235
  regulation of substances/products, 235

Argentine Ministry of the Economy, 235
Article 47, Uruguayan Constitution, 246
Articles, 22
Asbestos Hazard Emergency Response Act (AHERA), 19
Asia, 11–12, 339–375
Associação Brasileira de Normas Técnicas (ABNT). *See* Brazilian Association of Technical Standards (Associação Brasileira de Normas Técnicas)
Atomic Energy Act, 20

Basel Convention, 242
Biochemicals, 192
Biocidal products, 312, 318–319, 322–323
Biocidal Products Directive (BPD), 10, 309, 310, 323, 334
Biocidal Products Regulation (BPR)
  basic provisions, 310
  confidential and trade secret information, 332–334
  enforcement and penalties, 334–335
  executive summary, 309–310
  imports and exports, 332
  inventories or other listing procedures, 320
  issues, 335–338
  notification, registration, approval requirements, and compliance, 317–323
  overview, 10–11, 310
  regulation of substances/products, 323–327
  reporting and recordkeeping requirements, 331
  resources, 338–339
  scope of, 310–317
  standard of review/burdens of proof, 319

substances/products subject to
and/or exempt from, 310–317
test data development, 327–331
Biocides, 309–339
Biopesticides and Pollution Prevention
Division (BPPD), 98–99
Blanket import certificates, 69–70
Bolivia, 8, 239
BPD. *See* Biocidal Products
Directive (BPD)
BPPD. *See* Biopesticides and Pollution
Prevention Division (BPPD)
BPR. *See* Biocidal Products
Regulation (BPR)
Brazil, 8, 242
enforcement and penalties, 255
overview, 251
regulation of substances/
products, 251–254
trends, 255
Brazilian Agency of Terrestrial
Transportation (Agência
Nacional de Transportes
Terrestres), 253
Brazilian Agency of Waterway
Transportation (Agência
Nacional de Transportes
Aquaviários), 253
Brazilian Association of Technical
Standards (Associação
Brasileira de Normas
Técnicas), 254
Brazilian Institute of the Environment
and Renewable Natural
Resources (Instituto Brasileiro do
Meio Ambiente e dos Recursos
Naturais Renováveis), 253
Brazilian Standard NBR
14725:2009, 254
Business issues
CEPA 1999, 216–218
FIFRA, 135–136
Proposition 65, 151–152
SCPR, 166

TSCA, 78–79
TURA, 175
By-products, 21, 27

California
Green Chemistry Initiative, 5
Safe Drinking Water and Toxic
Enforcement Act of 1986,
5, 144–153
Safer Consumer Products
Regulations, 5, 153–167
California Administrative Procedure
Act, 148
California Department of Toxic
Substances Control (DTSC),
153–154, 156–164, 166
California Office of Environmental
Health Hazard Assessment
(OEHHA), 143, 149, 151, 152
CAN. *See* Andean Community of
Nations (Comunidad Andina)
Canada
CEPA 1999, 5–6, 179–219
Pollutant Release Inventory
program, 347
Canadian Environmental
Protection Act, 1999
(CEPA 1999)
activities subject to, 186–187
basic provisions, 180–181
business issues, 216–218
confidential and trade secret
information, 210–212
enforcement and penalties, 212–216
Environment Canada/Health
Canada review and
response, 193–196
executive summary, 179–181
fundamental aspects of law, 7
guiding principles, 189–191
imports and exports, 206–210
inherently toxic, 7
inventories or other listing
procedures, 180, 191–193

414 Index

Canadian Environmental
 Protection Act (*continued*)
 notification, registration,
  approval requirements, and
  compliance, 188–196
 overview, 179
 persons subject to, 187–188
 regulation of substances/
  products, 196–202
 resources, 219
 scope of, 181–188
 standard of review/burdens of
  proof, 188–191
 test data development, 203–204
 toxic, 7
 TSCA and, 6–7
Cancellations, pesticide products,
 5, 107–111
Candidate chemicals (CC), 154,
 156–162, 164
Carcinogenic, mutagenic, or toxic to
 reproduction substances (CMR
 substances), 269, 271, 292, 372
Carcinogen Identification Committee
 (CIC), 147–148
CAS. *See* Chemical Abstracts
 Service (CAS)
CASRN. *See* Chemical
 Abstracts Service Registry
 Number (CASRN)
CBI. *See* Confidential business
 information (CBI)
CC. *See* Candidate chemicals (CC)
CDR. *See* Chemical Data
 Reporting (CDR)
CE markings. *See* "Conformité
 Européenne" (CE markings)
Center for Evaluation of Risks
 to Human Reproduction
 (CERHR), 148, 153
Central America, 7–9
CERCLA. *See* Comprehensive
 Environmental Response,
 Compensation, and Liability
 Act (CERCLA)

CERHR. *See* Center for Evaluation
 of Risks to Human
 Reproduction (CERHR)
Chemical Abstracts Service (CAS),
 26, 285, 366
Chemical Abstracts Service Registry
 Number (CASRN), 26
Chemical Assessment and
 Management Program, 38
Chemical bans
 BPR provisions, 325
 CEPA 1999 provisions, 201–202
 FIFRA provisions, 107–111
 Proposition 65 provisions, 160–162
 TSCA provisions, 40–41
Chemical Data Reporting (CDR), 21,
 45–47, 53
Chemical Information Exchange
 Network (CIEN), 222
Chemical Management Plan (CMP),
 181, 218
Chemical nomenclature, 80
Chemical Registration Center
 (CRC), 12, 348, 351, 358–359
Chemical Safety Assessment
 (CSA), 268–269
Chemical Safety Improvement
 Act, 3
Chemical Safety Report (CSR),
 268–269, 272, 281
Chemicals of concern
 (COC), 154, 156–162,
 163, 164, 165
Chemical substance
 notifications, 349–352
Chemical substances, 2–3, 368–370
Chemical Substances Instruction
 (Instrucción de Substancias
 Químicas), 229
Children's products, 177
Chile, 8
 imports and exports, 239
 notification, registration,
  approval requirements, and
  compliance, 236–237

regulation of substances/
products, 237–238
reporting and recordkeeping
requirements, 238–239
China
chemical regulation, 13
enforcement and penalties, 356–358
imports and exports, 356
issues, 358–359
management of chemicals in, 13
MEP Order No. 7, 12, 291, 342,
348–352, 353, 354
Order No. 7, 291
overview, 341–343
resources, 359
risk assessment, 354–355
SAWS Order No. 53, 341, 347
scope of laws, 344–353
State Council Decree No. 591, 13,
341, 344–347, 356
test data development, 355
China Entry-Exit Inspection and
Quarantine (CIQ), 356
China REACH, 348
CIC. *See* Carcinogen Identification
Committee (CIC)
CICOPLAFEST. *See* Comisión
Intersecretarial para el Control
del Proceso y Uso de Plaguicidas,
Fertilizantes y Sustancias
Tóxicas (CICOPLAFEST)
CIEN. *See* Chemical Information
Exchange Network (CIEN)
CIQ. *See* China Entry-Exit Inspection
and Quarantine (CIQ)
Circular No. 2/C 152 (1982), 239
Citizens' petitions, 77–78
Civil penalties
Brazil, 254–255
CEPA 1999 provisions, 214
FIFRA provisions, 132–133
Proposition 65 provisions, 150–151
REACH provisions, 284
TSCA provisions, 74
TURA provisions, 174–175

Classification, Labeling, and
Packaging (CLP), 289
CLP. *See* Classification, Labeling, and
Packaging (CLP)
CMP. *See* Chemical Management
Plan (CMP)
CMR substances. *See* Carcinogenic,
mutagenic, or toxic to
reproduction substances
(CMR substances)
COC. *See* Chemicals of
concern (COC)
Colombia, 8
imports and exports, 241
overview, 239–240
regulation of substances/
products, 240–241
Comisión Intersecretarial para el
Control del Proceso y Uso
de Plaguicidas, Fertilizantes
y Sustancias Tóxicas
(CICOPLAFEST), 227,
228, 229
Comisión Nacional de Seguridad
Química. *See* National
Chemical Safety Commission
(Comisión Nacional de
Seguridad Química)
Committee on Technical Barriers to
Trade, 250
Community Decision 602, 241
Compliance
TSCA import requirements, 66–67
TSCA section 5, 37
TSCA section 8(d), 50
Comprehensive Environmental
Response, Compensation,
and Liability Act
(CERCLA), 167–168
Comunidad Andina. *See* Andean
Community of Nations
(Comunidad Andina)
CONCAWE. *See* Conservation
of Clean Air and Water in
Europe (CONCAWE)

Confidential business information (CBI)
  BPR provisions, 332–334
  CEPA 1999 provisions, 181, 210–212
  FIFRA provisions, 128–130
  Proposition 65 and, 149
  REACH provisions, 262, 281–283
  SCPR provisions, 165
  TSCA provisions, 25, 64
  TURA provisions, 173–174
Confidential inventory list, 25–27
Confidential Statement of Formula (CSF), 89
"Conformité Européenne" (CE markings), 305, 306–307
Consent orders, 36
Conservation of Clean Air and Water in Europe (CONCAWE), 336
Costa Rica, 8
Cost sharing, 288–289
Council. *See* Administrative Council on Toxics Use Reduction (Council)
COVENIN 2226, 244
COVENIN 2670, 244
COVENIN 3059, 243, 244
COVENIN 3060, 244
COVENIN 3061, 243
COVENIN norms, 245
CRC. *See* Chemical Registration Center (CRC)
Criminal penalties
  Brazil, 255
  CEPA 1999 provisions, 215–216
  FIFRA provisions, 133–134
  REACH provisions, 284
  TSCA provisions, 75
CSA. *See* Chemical Safety Assessment (CSA)
CSF. *See* Confidential Statement of Formula (CSF)
CSR. *See* Chemical Safety Report (CSR)
Customs Shipments Procedure, 70

DARTIC. *See* Developmental and Reproductive Toxicant Identification Committee (DARTIC)
Data and cost sharing, 288–289, 337
Data protection, 337
Data sharing and compensation
  BPR provisions, 328–331
  CEPA 1999 provisions, 204
  FIFRA provisions, 4–5, 114–118
  Proposition 65 provisions, 163
  REACH provisions, 261–262, 276–279
  SCPR provisions, 163
  TSCA provisions, 43–44
Decree Law No. 2, 225
Decree No. 7.505/2011, 250
Decree No. 14.390/92, 250
Decree No. 78, 236
Decree No. 144 of July 26, 1985, 238
Decree No. 307/009, 247
Decree No. 351/79 (Feb. 5, 1979), 235
Decree No. 406/988, 246
Decree No. 594/2003, 237, 238
Decree No. 725, 237
Decree No. 2635, 244
Decree No. 21919, 250
Derived No Effect Levels (DNEL), 354
Developmental and Reproductive Toxicant Identification Committee (DARTIC), 148
Devices, 87
Digital records, 138–139
Directive 67/548/EEC, 235
Distributor obligations, 305–306
DNEL. *See* Derived No Effect Levels (DNEL)
Domestic Substance List (DSL), 6, 179, 180, 191–192, 194–195, 216, 217
Draft Regulation for the Storage of Hazardous Chemicals (Anteproyecto de Reglamento de Almacenamiento de Sustancias Químicas Peligrosas), 238

DSL. *See* Domestic Substance
    List (DSL)
DTSC. *See* California Department
    of Toxic Substances
    Control (DTSC)
Dukakis, Michael, 167
DuPont. *See* E. I. du Pont de Nemours
    and Company

EAB. *See* Environmental Appeals
    Board (EAB)
ECHA. *See* European Chemicals
    Agency (ECHA)
ECLN Regulations. *See* Export Control
    List Notification Regulations
    (ECLN Regulations)
Ecolabel program, 338
Ecosystem approach, 190
EC Regulation 1049/2001, 332
EC Regulation 1896/2000, 338
Ecuador, 8, 239
EEA. *See* Environmental Enforcement
    Act (EEA)
E. I. du Pont de Nemours and
    Company, 57
EINECS. *See* European Inventory of
    Existing Commercial Chemical
    Substances (EINECS)
Electronic labels, 138–139
Electronic reporting requirement, 82
Elementis Chromium, Inc., 57
Emergency Planning and
    Community Right to Know
    Act (EPCRA), 167–168,
    171, 175, 347
Emergency suspension, pesticide
    products, 112
Endocrine effect testing, 140–141
End-use product (EUP), 97
Enforcement and penalties
    BPR provisions, 334–335
    Brazil, 254–255
    FIFRA provisions, 130–135
    Proposition 65 provisions, 149–151

REACH provisions, 283–284
RoHS provisions, 306
TSCA provisions, 74–77
TURA provisions, 174–175
Uruguay, 247–248
WEEE provisions, 306
Environmental Appeals Board
    (EAB), 57
Environmental contamination,
    reporting emergency incidents
    of, 57–59
Environmental Enforcement Act
    (EEA), 214, 215
Environmental Fate and Effects
    Division, 99
Environmental protection
    alternative measures
    (EPAMs), 215–216
Environmental protection compliance
    orders (EPCO), 213
Environmental Registry, 217
Environmental Violations
    Administrative Monetary
    Penalties Act (EVAMPA), 214
Environment Canada, 6, 180, 181,
    184, 188–196, 204–205
EPA. *See* U.S. Environmental
    Protection Agency (EPA),
    FIFRA provisions; U.S.
    Environmental Protection
    Agency (EPA), Proposition
    65 provisions; U.S.
    Environmental Protection
    Agency (EPA), TSCA
    provisions; U.S. Environmental
    Protection Agency (EPA),
    TURA provisions
EPAMs. *See* Environmental protection
    alternative measures (EPAMs)
EPCO. *See* Environmental protection
    compliance orders (EPCO)
EPCRA. *See* Emergency Planning
    and Community Right to
    Know Act (EPCRA)

418 Index

ESECL Regulations. *See* Export of Substances on the Export Control List Regulations (ESECL Regulations)
ESURC Regulations. *See* Export of Substances under the Rotterdam Convention Regulations (ESURC Regulations)
EU. *See* European Union (EU)
EUP. *See* End-use product (EUP)
European Chemicals Agency (ECHA)
    BPR provisions, 11, 316, 317, 320–323, 336
    REACH provisions, 9, 260, 264, 267–271, 274–282, 292
European Economic Area, 9, 259
European Inventory of Existing Commercial Chemical Substances (EINECS), 267, 320
European Petroleum Refiners Association, 336
European Union (EU)
    BPD, 10, 309, 310, 323, 334
    BPR, 10–11, 309–339
    REACH, 9–10, 259–294
    RoHS, 10, 295–308
    WEEE, 10, 295–308
EVAMPA. *See* Environmental Violations Administrative Monetary Penalties Act (EVAMPA)
E-waste exports, 242
Exclusion, 325–326
Executive Decree 24867, 223
Executive Decree 26805-S, 223
Executive Decree No. 305, 224
Executive Decree No. 640, 224
Executive Decree No. 28113-S, 223
Existing substances, 25, 203
Experimental use permit, 86, 87–88, 90
Export Control List, 207–209
Export Control List Notification Regulations (ECLN Regulations), 207

Export of Substances on the Export Control List Regulations (ESECL Regulations), 207, 209
Export of Substances under the Rotterdam Convention Regulations (ESURC Regulations), 207
Export-only substances, 186
Exports
    BPR provisions, 332
    CEPA 199 provisions, 206–210
    Chile, 239
    Colombia, 241
    FIFRA provisions, 125–127
    Peru, 249
    REACH provisions, 281
    TSCA provisions, 21–22, 62–65

Failure to Comply List, 165–166
FDA. *See* U.S. Food and Drug Administration (FDA)
Federal Decree No. 2,657/1998, 254
Federal Decree No. 3,029/1999, 253
Federal Decree No. 4,074/2002, 253
Federal Decree No. 4,262/2002, 252
Federal Food, Drug, and Cosmetic Act (FFDCA), 4, 20, 83, 92, 95, 101
Federal Insecticide, Fungicide, and Rodenticide Act (FIFRA)
    active ingredients, 85, 88
    activities subject to, 89–90
    authority to regulate inventory/listed substances or products, 101
    basic provisions, 83–84
    business issues, 135–136
    cancellations, 108–111
    confidential and trade secret information, 128–130
    data compensation, 4–5
    digital records/electronic labels, 138–139
    emergency suspension, 112
    enforcement and penalties, 130–135
    EPA review and response, 98–101

executive summary, 83–84
imports and exports, 125–128
inert ingredients, 88
inventories/other listing
    procedures, 96–98
notification/registration/approval
    requirements, 92–93
overview, 4–5, 83
persons subject to, 90–92
pesticide registration, 85
practical tips, 136–138
prioritization and risk
    assessments, 102–104
provision exemptions, 95
REACH provisions, 260
reduced risk prioritization, 103–104
registration review cycle, 138
registrations, 93–95
regulation of substances/
    products, 101–112
reporting and recordkeeping
    requirements, 118–125
resources, 141
risk assessment, 4
scope of, 84–92
standard of review/burdens of
    proof, 93–96
substances exempt from TSCA, 20
substances/products exempt
    from, 85–88
substances/products subject to
    and/or/exempt from, 84–85
tolerances, 95–96
trends, 138–141
21st-Century Toxicology, 140–141
use restrictions, 104–107
Federal Labor Law and Work Safety
    Regulation (Ley Federal del
    Trabajo y Seguridad en el
    Trabajo), 229
Federal Law No. 7,802/1989, 252, 253
Federal Law No. 9,782/1999, 252–253
Federal Law No. 10,357/2001, 252
FFDCA. *See* Federal Food, Drug, and
    Cosmetic Act (FFDCA)

FIFRA. *See* Federal Insecticide,
    Fungicide, and Rodenticide
    Act (FIFRA)
File searches, 50–51
FONDONORMA. *See* Fondo para la
    Normalización y Certificación de
    la Calidad (FONDONORMA)
Fondo para la Normalización y
    Certificación de la Calidad
    (FONDONORMA), 245
Food Quality Protection Act (FQPA),
    4, 83, 92, 95, 138
Formally required mechanism, 149
FQPA. *See* Food Quality Protection
    Act (FQPA)

General Environmental Law
    No. 41, 224
General Health Law, Law 5395, 222
General Health Law, Law 26842
    (Ley General de Salud No.
    26842), 248
General Law of the Environment, 246
Generic chemical names, 212
Globally Harmonized System of
    Classification and Labelling of
    Chemicals (GHS), 8, 9, 140, 223,
    238–239, 241, 248, 254, 289, 344
GLP. *See* Good Laboratory
    Practice (GLP)
Good Laboratory Practice (GLP),
    113, 366, 369
Green Chemistry Initiative, 5, 153
Guidance documents, 357–358

Hazardous chemicals, 250
Hazardous substances, 239, 307, 370–371
Hazardous waste exports, 242
Health and safety studies reporting,
    49–51, 54–55
Health Canada, 6, 180, 181, 184,
    188–196, 204–205
Health Effects Division, 99
High Production Volume Chemicals
    Program, 38, 181, 337

Higiene y Seguridad en el Trabajo, 235
Honduras, 226–227

IARC. *See* International Agency for Research on Cancer (IARC)
IBAMA. *See* Brazilian Institute of the Environment and Renewable Natural Resources (Instituto Brasileiro do Meio Ambiente e dos Recursos Naturais Renováveis)
IECSC. *See* Inventory of Existing Chemical Substances Produced or Imported in China (IECSC)
Illegal drugs, 252
IMO. *See* International Maritime Organization (IMO)
Imports
   BPR provisions, 332
   CEPA 199 provisions, 206–207, 210
   Chile, 239
   Colombia, 241
   FIFRA provisions, 89–91, 125–128
   Peru, 249
   REACH provisions, 281
   TSCA provisions, 66–71
   WEEE provisions, 305
Impurities, 27, 183
Impurity, 21
Incidental reaction products, 183
Inert ingredients, 88, 98
Inherently toxic, 7
Inspections, 75–76
Instituto Brasileiro do Meio Ambiente e dos Recursos Naturais Renováveis (IBAMA). *See* Brazilian Institute of the Environment and Renewable Natural Resources (Instituto Brasileiro do Meio Ambiente e dos Recursos Naturais Renováveis)
Instituto Nacional de Tecnología, Normalización y Metrologia (INTN), 249
Instituto Nacional de Vigilancia de Medicamentos y Alimentos (INVIMA). *See* National Institute for the Surveillance of Food and Medicines (Instituto Nacional de Vigilancia de Medicamentos y Alimentos)
Instrucción de Substancias Químicas. *See* Chemical Substances Instruction (Instrucción de Substancias Químicas)
Interagency Testing Committee (ITC), 43
Intergovernmental cooperation, 190
International Agency for Research on Cancer (IARC), 148, 235
International Convention for the Control and Management of Ships' Ballast Water and Sediments, 314
International Maritime Organization (IMO), 254
International Organization for Standardization (ISO), 320
International Uniform Chemical Information Database (IUCLID), 268, 279, 287
International Union of Pure and Applied Chemistry (IUPAC), 282, 283, 320
INTN. *See* Instituto Nacional de Tecnología, Normalización y Metrologia (INTN)
Inventory of Existing Chemical Substances Produced or Imported in China (IECSC), 12, 348–349, 351, 353
INVIMA. *See* National Institute for the Surveillance of Food and Medicines (Instituto Nacional de Vigilancia de Medicamentos y Alimentos)
ISO. *See* International Organization for Standardization (ISO)
ISO 11014-1, 223

Index 421

ITC. *See* Interagency Testing Committee (ITC)
IUCLID. *See* International Uniform Chemical Information Database (IUCLID)
IUPAC. *See* International Union of Pure and Applied Chemistry (IUPAC)

KCMA. *See* Korea Chemicals Management Association (KCMA)
KECI. *See* Korea Existing Chemicals Inventory (KECI)
Korea Chemicals Management Association (KCMA), 14, 364
Korea Existing Chemicals Inventory (KECI), 14, 364, 365
Korea REACH. *See* Act on Registration and Evaluation, Authorization, and Restriction of Chemical Substances (K-REACH)
K-REACH. *See* Act on Registration and Evaluation, Authorization, and Restriction of Chemical Substances (K-REACH)
Kyoto Protocol, 226, 246

Labels and labeling, 91–92
Labor Code mechanism, 147, 149
La Ley Orgánica contra el Tráfico Illicito y el Consumo de Sustancias Estupefacientes y Psicotrópicas. *See* Organic Law against Illicit Trafficking and Consumption of Narcotic and Psychotropic Substances (La Ley Orgánica contra el Tráfico Illicito y el Consumo de Sustancias Estupefacientes y Psicotrópicas)
Land transportation of hazardous cargoes, 253
Large quantity toxics users (LQTUs), 170, 171–173
Law 9, 240, 241
Law 18.164 (1982), 239

Law No. 55/1993, 240, 241
Law on Dangerous Substances, Materials, and Wastes (Ley sobre Sustancias, Materiales y Desechos Peligrosos), 243
Lead-Based Paint Exposure Reduction Act, 19
Letter of access (LOA), 270, 277, 288, 311, 316, 317, 329, 330
Ley Federal del Trabajo y Seguridad en el Trabajo. *See* Federal Labor Law and Work Safety Regulation (Ley Federal del Trabajo y Seguridad en el Trabajo)
Ley General de Salud No. 26842. *See* General Health Law, Law 26842 (Ley General de Salud No. 26842)
Ley sobre Sustancias, Materiales y Desechos Peligrosos. *See* Law on Dangerous Substances, Materials, and Wastes (Ley sobre Sustancias, Materiales y Desechos Peligrosos)
Listed substances/products, 324
LOA. *See* Letter of access (LOA)
LoREX. *See* Low release and low exposure exemptions (LoREX)
Los Solvents Orgánicos Nocivos para la Salud. *See* Organic Solvents Harmful to Health (Los Solvents Orgánicos Nocivos para la Salud)
Low release and low exposure exemptions (LoREX), 31
Low volume exemptions (LVE), 31
Low volume substances, 183
LQTUs. *See* Large quantity toxics users (LQTUs)
LVE. *See* Low volume exemptions (LVE)

Maine, 5, 177
Maine Center for Disease Control and Prevention, 177

Maine DEP. *See* Maine Department of Environmental Protection (Maine DEP)
Maine Department of Environmental Protection (Maine DEP), 177
Maine Department of Health and Human Services, 177
Manufacture, 24
Manufactured items, 182
Manufacturers
   BPR provisions, 316
   FIFRA provisions, 90–91
   K-REACH provisions, 368–369, 373–374
   REACH provisions, 265
   RoHS provisions, 301–303
   TSCA provisions, 24
   WEEE provisions, 302–303
Manufacturing use product (MUP), 97
MAPA. *See* Ministry of Agriculture, Livestock, and Food Supply (Ministério da Agricultura, Pecuária e Abastecimento)
Masked Name Regulations, 211
Masked names, 181
Massachusetts, 5, 144, 167–176
Massachusetts DEP. *See* Massachusetts Department of Environmental Protection (Massachusetts DEP)
Massachusetts Department of Environmental Protection (Massachusetts DEP), 170–176
Material Safety Data Sheet (MSDS), 346, 352
MEA. *See* Multilateral environmental agreements (MEA)
MEP. *See* Ministry of Environmental Protection (MEP)
MEP – Chemical Registration Center (MEP-CRC), 343
MEP-CRC. *See* MEP – Chemical Registration Center (MEP-CRC)
MEP-SCC. *See* MEP – Solid Waste and Chemical Management Center (MEP-SCC)

MEP – Solid Waste and Chemical Management Center (MEP-SCC), 343, 348–349, 351, 352, 358–359
Mercado Común del Sur. *See* Mercosur (Mercado Común del Sur)
Mercosur (Mercado Común del Sur), 8, 9, 242
Mexico, 7–9, 227–231
Microbial products of biotechnology, 35
MINAET. *See* Ministério de Salud, the Ministério de Ambiente, Energía y Telecomunicaciones (MINAET)
Ministério da Agricultura, Pecuária e Abastecimento (MAPA). *See* Ministry of Agriculture, Livestock, and Food Supply (Ministério da Agricultura, Pecuária e Abastecimento)
Ministerio de Salud. *See* Ministry of Health (Ministerio de Salud)
Ministerio de Salud Pública y Bienestar Social. *See* Ministry of Public Health and Social Welfare (Ministerio de Salud Pública y Bienestar Social); National Food and Nutrition Institute (Ministerio de Salud Pública y Bienestar Social)
Ministério de Salud, the Ministério de Ambiente, Energía y Telecomunicaciones (MINAET), 223
Ministerio de Salud y Servicios Sanitarios. *See* Ministry of Health and Health Services (Ministerio de Salud y Servicios Sanitarios)
Ministry of Agriculture, Livestock, and Food Supply (Ministério da Agricultura, Pecuária e Abastecimento), 253
Ministry of Environmental Protection (MEP)
   Chemical Registration Center, 12

Index 423

Order No. 7, 12, 291, 342, 348–352, 353, 354
Order No. 57, 346
Ministry of Environment (MoE), 363, 365, 367–374
Ministry of Health and Health Services (Ministerio de Salud y Servicios Sanitarios), 237, 239
Ministry of Health (Ministerio de Salud), 241
Ministry of Public Health and Social Welfare (Ministerio de Salud Pública y Bienestar Social), 250
Ministry of Transportation (MOT), 342, 343
Minnesota, 5, 177–178
Minnesota Department of Health, 177
Minnesota Pollution Control Agency, 177
Mixtures, 182
MoE. *See* Ministry of Environment (MoE)
Montreal Protocol, 226, 246
MOT. *See* Ministry of Transportation (MOT)
MSDS. *See* Material Safety Data Sheet (MSDS)
Multilateral environmental agreements (MEA), 242
MUP. *See* Manufacturing use product (MUP)
Mutual recognition, 319, 323

Nanomaterials, 184–185, 313
Nanotechnology, 184
National Administration for Drugs, Food, and Medical Technology (ANMAT), 234–235
National Chemical Safety Commission (Comisión Nacional de Seguridad Química), 250
National Customs Service (SNA), 236
National Food and Nutrition Institute (Ministerio de Salud Pública y Bienestar Social), 250

National Institute for Occupational Safety and Health (NIOSH), 148
National Institute for the Surveillance of Food and Medicines (Instituto Nacional de Vigilancia de Medicamentos y Alimentos), 240
National People's Congress (NPC), 342, 343
National Registration Center of Chemicals (NRCC), 345–346
National Registry of Chemical Precursors (Registro Nacional de Precursores Químicos), 235
National Security Policy for Chemicals, 238
National standards, 190
National Toxicology Program (NTP), 148, 153
Naturally occurring substances, 23
NDSL. *See* Nondomestic Substance List (NDSL)
Negative import certificates, 68–69
New chemical notifications
    FIFRA provisions, 20
    MEP-SCC review of, 352
    TCCA provisions, 364
    TSCA provisions, 27–30
New substance notification (NSN), 6, 180, 203
New Substance Notification Regulations (NSN Regulations), 179, 186, 191–192, 193
New substances, 25
NGOs. *See* Nongovernmental organizations (NGOs)
NIOSH. *See* National Institute for Occupational Safety and Health (NIOSH)
NOAEL. *See* No observed adverse effect level (NOAEL)
NOC. *See* Notice of commencement (NOC)
NOEQ. *See* Notice of excess quantity (NOEQ)

NOM. *See* Official Mexican Standards (Normas Oficial Mexicana)
NOMI. *See* Notice of manufacture or import (NOMI)
Non-Customs Courier Shipments Procedure, 70
Nondomestic Substance List (NDSL), 6, 180, 191–192
Nongovernmental organizations (NGOs), 270
Nonisolated intermediates, 21, 27
No observed adverse effect level (NOAEL), 95
Norma Chilena No. 2245 (2003), 238
Normas Oficial Mexicana. *See* Official Mexican Standards (Normas Oficial Mexicana)
Normas sobre el Control de Recuperation de Materiales Peligrosos y Manejo de Residuos Peligrosos. *See* Norms on the Control of Recovery of Hazardous Materials and Handling of Hazardous Waste (Normas sobre el Control de Recuperation de Materiales Peligrosos y Manejo de Residuos Peligrosos)
Norma Técnica Colombiana 4335 (NTC 4335), 240
Norms on the Control of Recovery of Hazardous Materials and Handling of Hazardous Waste (Normas sobre el Control de Recuperation de Materiales Peligrosos y Manejo de Residuos Peligrosos), 244
Notice of arrival, 90, 91
Notice of commencement (NOC), 3, 25, 34, 194–195
Notice of excess quantity (NOEQ), 194, 195, 216
Notice of manufacture or import (NOMI), 194, 195, 216
Notification requirements, 14, 25

NPC. *See* National People's Congress (NPC)
NRCC. *See* National Registration Center of Chemicals (NRCC)
NSN. *See* New substance notification (NSN)
NSN Regulations. *See* New Substance Notification Regulations (NSN Regulations)
NTC 4335. *See* Norma Técnica Colombiana 4335 (NTC 4335)
NTP. *See* National Toxicology Program (NTP)

Occupational exposure limits (OEL), 237
Occupational Safety and Health Administration (OSHA), 49, 140, 238
OECD. *See* Organization for Economic Co-operation and Development (OECD)
OEHHA. *See* California Office of Environmental Health Hazard Assessment (OEHHA)
OEL. *See* Occupational exposure limits (OEL)
Office of Health Assessment and Translation (OHAT), 148–149, 153
Office of Pesticide Programs (OPP), 98, 136, 139
Office of Technical Assistance and Technology (OTA), 170
Office of Technical Coordination for the Management of Chemical Substances (OTCMCS), 223
Official Mexican Standards (Normas Oficial Mexicana), 229, 230
OHAT. *See* Office of Health Assessment and Translation (OHAT)
Only Representative (OR), 265, 283, 290, 372
OPP. *See* Office of Pesticide Programs (OPP)

OR. *See* Only Representative (OR)
Organic Law against Illicit Trafficking and Consumption of Narcotic and Psychotropic Substances (La Ley Orgánica contra el Tráfico Illicito y el Consumo de Sustancias Estupefacientes y Psicotrópicas), 244
Organic Solvents Harmful to Health (Los Solvents Orgánicos Nocivos para la Salud), 238
Organization for Economic Co-operation and Development (OECD), 286
OSHA. *See* Occupational Safety and Health Administration (OSHA)
OTA. *See* Office of Technical Assistance and Technology (OTA)
OTCMCS. *See* Office of Technical Coordination for the Management of Chemical Substances (OTCMCS)

PAIR. *See* Preliminary Assessment Information Rule (PAIR)
Panama, 8, 224–226
Paraguay, 8, 242, 249–251
PBT. *See* Persistent, bioaccumulative, and toxic (PBT)
PCBs. *See* Polychlorinated biphenyl (PCBs)
Perfluorooctanoic acid (PFOA), 57
Persistent, bioaccumulative, and toxic (PBT), 168, 269, 271, 292, 325, 327, 372
Persistent organic pollutants (POPs), 242
Peru, 8, 239, 248–249
Pesticide Registration Improvement Act of 2003 (PRIA), 99, 139
Pesticide Registration Improvement Extension Act of 2012 (PRIA 3), 99, 100, 139
Pesticide residues, 90, 92, 98

Pesticides
    Brazilian regulations, 253
    Chilean regulations, 236
    Chinese regulations, 348, 353, 356
    Costa Rican regulations, 223
    FIFRA provisions, 83–141
    Mexican regulations, 227, 229
    Uruguayan regulations, 246
    Venezuelan regulations, 242
Pests, 84, 86
PFOA. *See* Perfluorooctanoic acid (PFOA)
Phase-in substances, 267
PIC. *See* Prior Informed Consent (PIC)
PMN. *See* Premanufacture notice (PMN)
Pollutant Emissions Register program, 347
Pollutant release and transfer registry (PRTR), 347
Pollutant Release Inventory program, 347
"Polluter pays" principle, 190
Pollution prevention, 190
Polychlorinated biphenyl (PCBs), 41
Polymers
    CEPA 1999 provisions, 183, 192
    EPA exemptions, 28–29
    EPA regulations, 22
    TCCA provisions, 365
POPs. *See* Persistent organic pollutants (POPs)
Positive import certificates, 67–68
Postchemical notification obligations, 352
PPORD. *See* Product and process-orientated research and development (PPORD)
PPs. *See* Priority Products (PPs)
Precautionary principle, 190
Predicted No Effect Concentrations (PNEC), 354
Preliminary Assessment Information Rule (PAIR), 44–45, 52–53

426 *Index*

Premanufacture notice (PMN), 3, 23, 25, 27, 28, 33–34, 35, 180
Preregistrations, 286–287
PRIA. *See* Pesticide Registration Improvement Act of 2003 (PRIA)
PRIA 3. *See* Pesticide Registration Improvement Extension Act of 2012 (PRIA 3)
Prior Informed Consent (PIC), 242
Prioritization
 BPR provisions, 324
 CEPA 1999 provisions, 196–199
 Colombian regulations, 240
 FIFRA provisions, 102–104
 Proposition 65 provisions, 157–159
 TSCA provisions, 38
Priority Products (PPs), 157–159
Priority Substances List (PSL), 198–199
Processing aids, 314–315
Processors of chemical substances, 24
Product and process-orientated research and development (PPORD), 264, 283
Product bans
 BPR provisions, 325
 CEPA 1999 provisions, 201–202
 FIFRA provisions, 5, 107–111
 Proposition 65 provisions, 160–162
 TSCA provisions, 40–41
Prohibition, 202
Proposed Regulatory Framework for Nanomaterials under the Canadian Environmental Protection Act, 1999, 184
Proposition 65. *See* Safe Drinking Water and Toxic Enforcement Act of 1986 (Proposition 65)
Proteins, 183
PRTR. *See* Pollutant release and transfer registry (PRTR)
PSL. *See* Priority Substances List (PSL)
Public inventory list, 25–27

Qualified experts mechanism, 149
Quantitative structural-activity relationships (QSAR), 275

R&D. *See* Research and development (R&D)
R&D exemption. *See* Research and development exemption (R&D exemption)
R&D substances. *See* Research and development substances (R&D substances)
RAC. *See* Raw agricultural commodities (RAC)
Raw agricultural commodities (RAC), 92–93
RCR. *See* Risk characterization ratios (RCR)
RCRA. *See* Resource Conservation and Recovery Act (RCRA)
RCRA Toxicity Characteristic Limits, 58
REACH. *See* Registration, Evaluation, Authorization, and Restriction of Chemicals (REACH)
REACH-IT, 279, 287
Recordkeeping requirements
 BPR provisions, 331
 CEPA 1999 provisions, 205
 FIFRA provisions, 87, 90, 120–123
 Proposition 65 provisions, 164
 TSCA provisions, 52–55, 61–62, 65, 71
 TURA provisions, 173
Reduced risk prioritization, 103–104
Register of Importers and Exporters of Ozone-Depleting Substances, 236
Registration Division (RD), 98–99
Registration, Evaluation, Authorization, and Restriction of Chemicals (REACH), 374
 activities subject to, 264
 authorization, 262
 basic provisions, 260–262
 BPR provisions, 335–336, 337

burden of proof, 261
business issues, 285–293
CEPA 1999 and, 6
confidential and trade secret
  information, 281–283
ECHA fees and review, 270–271
enforcement and penalties, 283–285
executive summary, 259–262
guidance and interpretation
  shortfalls, 287–288
imports and exports, 281
K-REACH and, 361–362, 367,
  372, 373, 374
persons/entities subject to, 265–266
registration dossier and chemical
  safety report, 268–269
registration requirements,
  261, 266–268
registration software, 279–280
regulation of substances/
  products, 271–274
resources, 293–294
restrictions, 262, 273–274
scope of, 262–266
test data development, 274–281
Registration review, 4
Registro Nacional de Precursores
  Químicos. *See* National
  Registry of Chemical Precursors
  (Registro Nacional de
  Precursores Químicos)
Reglamento de la Ley General de
  Salud en Materia de Control
  Sanitario de Actividades,
  Establecimientos, Productos
  y Servicios. *See* Regulation
  under the General Health
  Law Regarding the Sanitary
  Control of Activities, Facilities,
  Products, and Services
  (Reglamento de la Ley General
  de Salud en Materia de Control
  Sanitario de Actividades,
  Establecimientos, Productos
  y Servicios)

Reglamento General Técnico de
  Seguridad, Higiene y Medicina
  en el Trabajo, 250
Regular notification, 351
Regulation No. 229, 254
Regulation of Health Conditions and
  Basic Environmental Workplaces
  (Regulación de las Condiciones
  de Salud y Lugares de Trabajo
  Básico del Medio Ambiente), 238
Regulation on Basic Sanitary and
  Environmental Conditions
  in Workplaces (Reglamento
  sobre Condiciones Sanitarias
  y Ambientales Básicas en los
  Lugares de Trabajo), 237
Regulation under the General Health
  Law Regarding the Sanitary
  Control of Activities, Facilities,
  Products, and Services
  (Reglamento de la Ley General
  de Salud en Materia de Control
  Sanitario de Actividades,
  Establecimientos, Productos y
  Servicios), 229
Reporting and recordkeeping
  requirements
  BPR provisions, 331
  CEPA 1999 provisions, 204–206
  Chilean SDSs, 238–239
  FIFRA provisions, 120–125
  Proposition 65 provisions, 163–164
  SCPR provisions, 163–164
  TSCA provisions,
    44–62, 82
  TURA provisions, 171–173
Uruguay, 247
Requirements for Registration and
  Authorization of Operators
  of Dangerous Substances,
  Materials and Waste
  (Requisitos para el Registro y
  Autorización de Operadores
  de Sustancias Peligrosas,
  Materiales y Residuos), 244

Research and development exemption (R&D exemption), 29–30
Research and development (R&D), 185, 192–193, 216, 312–313
Research and development substances (R&D substances), 22
Resolution No. 2,239/2011, 253–254
Resolution No. 40, 244
Resolution No. 420/2004, 253
Resolution No. 709/98, 234–235
Resource Conservation and Recovery Act (RCRA), 58
Restricted use pesticides (RUPs), 92
Restriction of Hazardous Substances (RoHS)
  basic provisions, 296–297
  executive summary, 295–298
  exemptions, 297–298
  manufacturer obligations, 301–303
  notification, registration, approval requirements, and compliance, 301–303
  overview, 10, 296
  regulation of substances/products, 306–307
  resources, 308
  scope and implementation deadlines of, 298–300
Risk, 7
Risk assessment
  BPR provisions, 324
  CEPA 1999 provisions, 196–199
  Colombian regulations, 240
  FIFRA provisions, 102–104
  Proposition 65 provisions, 157–159
  TSCA provisions, 38
Risk characterization ratios (RCR), 354
Risk management, 80–81
Risk notification
  CEPA 1999 provisions, 205–206
  FIFRA provisions, 123–125
  Proposition 65 provisions, 164
  TSCA provisions, 55–62
RoHS. *See* Restriction of Hazardous Substances (RoHS)
Romney, Mitt, 167

Rotterdam Convention, 242, 348, 356
RUPs. *See* Restricted use pesticides (RUPs)

SAB. *See* Science Advisory Board (SAB)
Safe Drinking Water and Toxic Enforcement Act of 1986 (Proposition 65), 5, 143
  activities subject to, 145–146
  business issues, 151–152
  confidential and trade secret information, 149
  enforcement and penalties, 149–151
  listing procedures, 147–148
  notification, registration, approval requirements, and compliance, 147–149
  OEHHA review and response, 148–149
  overview, 144
  persons subject to, 146
  practical tips, 152–153
  resources, 153
  scope of, 145–146
  standard of review/burdens of proof, 147
  substances subject to, 145
Safer Consumer Products Regulations (SCPR), 5, 143
  business issues, 166
  confidential and trade secret information, 165
  enforcement and penalties, 165–166
  Failure to Comply List, 165–166
  notification, registration, approval requirements, and compliance, 156–157
  overview, 153–154
  regulation of substances/products, 157–162
  reporting and recordkeeping requirements, 163–164
  resources, 166
  scope of, 154–156
  test data development, 162–163

Safety Data Sheets (SDS)
   BPR provisions, 331
   Central and South American countries, 8
   Chilean regulations, 236, 238
   Chinese regulations, 346, 352
   Colombian regulations, 240
   Costa Rica regulations, 223–224
   Panamanian regulations, 225
   Paraguayan regulations, 250
   Peruvian regulations, 249
   REACH provisions, 269, 283, 290
   Venezuelan regulations, 242, 244
Sanitary Surveillance Agency (Agência Nacional de Vigilância Sanitária), 252–253
SAWS Order No. 53. *See* State Administration of Work Safety Order No. 53 (SAWS Order No. 53)
Schedule 1, 200–202
Schwarzenegger, Arnold, 154
Science Advisory Board (SAB), 168, 171
Science-based decision making, 191
Scientific research record (SSR), 349–350
SCPR. *See* Safer Consumer Products Regulations (SCPR)
SDS. *See* Safety Data Sheets (SDS)
Seizure, 76–77
Servicio Nacional de Aduanas (SNA). *See* National Customs Service (SNA)
Settlements, 74–75
SFF. *See* SIEF Formation Facilitators (SFF)
SIEF. *See* Substance Information Exchange Forum (SIEF)
SIEF Formation Facilitators (SFF), 287
Significant New Activity (SNAc), 6, 181, 195, 199–200, 217
Significant New Use Rule (SNUR), 26, 33, 38–39, 80–81
Simplified notification (general), 350

Simplified notification (special), 350
SIP. *See* Substance Identification Profile (SIP)
Site-limited intermediate substance, 185–186
Small- and medium-sized enterprises (SMEs), 270, 272
Small quantity toxics users (SQTUs), 170
SNA. *See* National Customs Service (SNA)
SNAc. *See* Significant New Activity (SNAc)
SNUR. *See* Significant New Use Rule (SNUR)
South America, 7–9
   Argentina, 8, 234–235
   Brazil, 8, 251–255
   Chile, 8, 236–239
   Colombia, 8, 239–241
   executive summary, 233–234
   Paraguay, 8, 249–251
   Peru, 8, 248–249
   Uruguay, 8, 246–248
   Venezuela, 9, 242–245
South Korea
   K-REACH, 13–14, 362, 366–375
   overview, 361–362
   TCCA, 13–14, 362–366
SQTUs. *See* Small quantity toxics users (SQTUs)
SSURO. *See* Stop sale, use, or removal orders (SSURO)
State Administration of Work Safety Order No. 53 (SAWS Order No. 53), 341, 347, 352
State Council Decree No. 344, 344
State Council Decree No. 591, 13, 341, 344–347, 356
State laws
   California, 5, 143, 144–166
   Maine, 177
   Massachusetts, 5–6, 167–176
   Minnesota, 177–178
   Washington, 176–177
Stockholm Convention, 226, 242, 246

Stop sale, use, or removal orders (SSURO), 134–135
Subpoena power, 76
Substance Identification Profile (SIP), 271
Substance identity, 285–286
Substance Information Exchange Forum (SIEF), 262, 276–279, 280, 286–288, 290
Substances, 181–184
Substance sameness, 335–336
Substances occurring in nature, 183
Substances of very high concern (SVHC), 271–273, 289, 292, 372
Substances subject to other acts, 183
Substantial risk information, 56–57
Substitution, 326
Supply chain communication, 289–291, 372, 374
Supreme Decree No. 1-84-ITI/IND, 249
Suspension, pesticide products, 111–112
Sustainable development, 189–190
SVHC. *See* Substances of very high concern (SVHC)

TCCA. *See* Toxic Chemicals Control Act (TCCA)
Technical equivalence, 335–336
Technically qualified individual (TQI), 29, 30
Technical Standard No. G 050, 249
Temporary permissible limit (TPL), 237
Test data development
    BPR provisions, 327–331
    CEPA 1999 provisions, 203–204
    China, 355
    FIFRA provisions, 112–114
    Proposition 65 provisions, 162–163
    REACH provisions, 274–281
    TCCA provisions, 366
    TSCA provisions, 41–44
Test marketing exemptions (TME), 31–32
Third-party representative (TPR), 265, 283, 290

3M Company, 57, 132
Time-weighted average (TWA), 237
TME. *See* Test marketing exemptions (TME)
Tolerances, for pesticides, 92, 95–96
Toxic, 7, 180–181, 194, 217
Toxic Chemicals Control Act (TCCA)
    K-REACH and, 367
    new chemical notification requirements exemptions, 365
    new chemicals notification, 365
    notification, registration, approval requirements, and compliance, 364
    overview, 13, 361, 362–363
    resources, 366
    scope of, 363–366
    test data development, 366
Toxic Free Kids Act, 177
Toxicity
    CEPA 1999 provisions, 6, 180
    TSCA provisions, 6, 180
Toxics Information Clearinghouse, 153
Toxic Substances Control Act (TSCA)
    activities subject to, 23
    allegations of significant adverse health or environmental reactions, 47–49
    authority to require testing/testing triggers, 41–42
    awareness, 79
    basic provisions, 2, 18
    business issues, 78–79
    CEPA 1999 and, 6–7
    Chemical Data Reporting, 45–47
    chemical or product bans, 40–41
    citizens' petitions, 77–78
    confidential and trade secret information, 72–73
    confidential business information, 64
    definition of chemical substances, 2–3
    enforcement and penalties, 74–77, 132

EPA approval requirement
  exemptions, 30–32
executive summary, 17–18
file searches, 50–51
generic chemical names, 212
identifying chemical substances
  and mixtures subject for
  export, 62–63
imports and exports, 62–71
incidents of environmental
  contamination, 57–59
inspections, 75–76
inventories or other listing
  procedures, 33–35
MEP Order No. 7, 354
new chemical notification
  requirements exemptions, 27–30
new technologies, 79–80
notice of commencement, 195
notification, approval
  requirements, and
  compliance, 25–37
persons subject to, 23–24
Preliminary Assessment
  Information Rule, 44–45
prioritization and risk
  assessment, 38
processors of chemical
  substances, 24
products that must be tested/
  implementing test
  requirements, 42–43
provisions exemptions, 20–23
REACH provisions, 260
recordkeeping, 32
reform, 79
regulation of substances, 37–41
reporting and recordkeeping
  requirements, 44–62
reporting exemptions, 60–61
resources, 82
risk, 7
risk notification, 55–62
scope of, 19–24
Section 5(e) consent orders, 36
Section 5(f) orders, 36–37
Section 8(d) report preparation/
  submission to EPA, 51–52
Section 8(e) report preparation/
  submission to EPA, 59–60
Section 12(b) export notice
  preparation/submission to
  EPA, 64
Section 16 Settlements, 74–75
SNAcs and Significant New Use
  Rule under, 181
standard of review/burdens of
  proof, 32–33
statutory overview, 17
submission of copies of health and
  safety studies, 51
submission of lists of health and
  safety studies, 51
substances exempt from, 19–23
substances subject to, 19
test data development, 41–44
trends, 79–82
use restrictions, 38–40
Toxics Use Reduction Act (TURA),
  5, 144
  activities subject to, 169
  business issues, 175
  confidential and trade secret
    information, 173–174
  enforcement and penalties, 174–175
  overview, 167
  persons subject to, 169–171
  practical tips, 176
  regulation of substances/
    products, 171
  reporting and recordkeeping
    requirements, 171–173
  resources, 176
  scope of, 167–171
  substances/products subject to
    and/or exempt from, 167–168
Toxics Use Reduction Institute
  (TURI), 168, 171, 176
Toxics use reduction plans (TUR
  plans), 167, 172–173, 175

432 *Index*

Toys, 177
TPL. *See* Temporary permissible limit (TPL)
TPR. *See* Third-party representative (TPR)
TQI. *See* Technically qualified individual (TQI)
Trade secret information
   BPR provisions, 332–334
   CEPA 1999, 210–212
   FIFRA provisions, 128–130
   Proposition 65 provisions, 149
   REACH provisions, 281–283
   SCPR provisions, 165
   TSCA provisions, 72–73
   TURA provisions, 173–174
Transient reaction intermediates, 183
Transit of hazardous products in port facilities, 253
Transit of substances into Canada, 183
Transparency, 81–82
Treated articles, 312
Trends
   BPR, 337–338
   Brazil, 255
   CEPA 1999, 218
   FIFRA, 138–141
   Mexico, 230–231
   Paraguay, 251
   REACH, 291–293
   state laws, 176–178
   TSCA, 79–82
   Venezuelan, 245
TSCA. *See* Toxic Substances Control Act (TSCA)
TSCA Chemical Substance Inventory (TSCA Inventory)
   CEPA 1999 and, 216
   determining inventory status, 26–27
   exemptions from certain TSCA provisions, 20
   new and existing substances, 25
   Nondomestic Substance List and, 191
   public and confidential lists, 25–26
   role of, 3
TSCA Work Plan Chemicals, 38, 81
TURA. *See* Toxics Use Reduction Act (TURA)
TURI. *See* Toxics Use Reduction Institute (TURI)
TUR plans. *See* Toxics use reduction plans (TUR plans)
TWA. *See* Time-weighted average (TWA)
21st-Century Toxicology, 140–141

U.N. Committee of Experts on the GHS, 241
UNEP. *See* United Nations Environment Programme (UNEP)
United Nations Environment Programme (UNEP), 222
United Nations Framework Convention on Climate Change, 246
United Nations Institute for Training and Research-International Labour Organization (UNITAR-ILO), 245
United States
   FIFRA, 4–5, 83–141
   state laws, 5–6, 143–178
   TSCA, 2–3, 17–82
Unknown, of Variable Composition, or of Biological Origin (UVCB), 285, 336
Unlawful activities
   CEPA 1999 provisions, 212–213
   FIFRA provisions, 130
   MEP Order No. 7, 357
   Proposition 65 provisions, 149
   REACH provisions, 284
   TSCA provisions, 74
Uruguay, 8, 242
   enforcement and penalties, 247
   overview, 246
   reporting and recordkeeping requirements, 247
U.S. Customs Service, 69, 70, 127

U.S. Environmental Protection
  Agency (EPA), FIFRA
  provisions
 active ingredients in pesticides, 85
 authority to regulate inventory/listed
   substances or products, 101
 budget/resources, 139
 cancellations, 108–110
 definition of pesticide, 84
 device registration, 87
 device subject to, 87
 digital records/electronic
   labels, 138–139
 emergency suspension, 112
 FIFRA exemptions, 95
 imports and exports, 125–128
 inert ingredients in pesticides,
   88, 98
 international activities, 139–140
 inventories/other listing
   procedures for pesticides, 96–98
 labels and labeling, 91
 Office of Pesticide Programs,
   98–99, 136, 139
 pesticide registration, 4, 94–95, 136
 pesticide regulation, 83
 pesticide residue regulation,
   92, 98
 pesticide tolerance levels, 95–96
 prioritization and risk
   assessments, 102–104
 recordkeeping, 120–123
 reduced risk prioritization, 103–104
 registration review, 4
 registration review cycle, 138
 regulatory control options
   available to, 5
 report on amount of pesticide
   distributed, 89
 reports to be submitted, 118–120
 review and response, 98–101
 risk notification, 123–125
 stop sale orders, 134–135
 suspension, 111–112
 test data development, 112–118
 use restrictions, 5, 104–107

U.S. Environmental Protection
  Agency (EPA), Proposition 65
  provisions, 148
U.S. Environmental Protection Agency
  (EPA), TSCA provisions
 actions, 79
 allegations of significant adverse
   health or environmental
   reactions, 54
 authority to regulate inventory/listed
   substances, 37–38
 authority to require testing/testing
   triggers, 41–42
 business issues, 78–79
 chemical or product bans, 40–41
 citizens' petitions, 77–78
 confidential and trade secret
   information, 72–73
 data sharing and compensation, 43
 enforcement and penalties, 74–77
 exemptions from certain TSCA
   provisions, 20–23
 exemptions requiring
   approval, 30–32
 focus on chemical
   nomenclature, 80
 incidents of environmental
   contamination, 57–59
 inspections, 75–76
 microbial products of
   biotechnology, 35
 naturally occurring substances
   regulations, 23
 new technologies, 79–80
 NOC preparation and submission
   to, 34
 polymer exemption, 28
 polymer regulations, 22
 premanufacture notice, 34
 recordkeeping, 32
 reporting exemptions, 60–61
 reports to be submitted, 44–50
 resources, 82
 review/response for chemicals
   added to inventory, 3, 35–37
 risk assessment, 38

U.S. Environmental Protection Agency
    (EPA) *(continued)*
  risk management, 80–81
  risk notification, 55–62
  Section 8(d) report
    preparation/submission, 51–52
  Section 8(e) report
    preparation/submission, 59–60
  Section 12(b) export notice
    preparation/submission, 64
  shipments without import
    certificates, 70–71
  standard of review/burdens of
    proof, 33
  substances subject to, 19
  substantial risk information, 56–57
  transparency actions, 81–82
  trends, 79–82
  TSCA awareness, 79
  TSCA Inventory status, 26–27
  use restrictions, 38–40
U.S. Environmental Protection
  Agency (EPA), TURA
  provisions, 171, 175
Use restrictions
  BPR provisions, 324–325
  CEPA 1999 provisions, 199–201
  FIFRA provisions, 5, 92, 104–107
  Honduran regulations, 227
  Proposition 65 provisions, 160–162
  TSCA provisions, 38–40
U.S. Food and Drug Administration
  (FDA), 69, 148
UVCB. *See* Unknown, of Variable
  Composition, or of Biological
  Origin (UVCB)

Venezuela, 9
  business issues, 244–245
  overview, 242
  regulation of substances/
    products, 243–244
Very persistent and very
  bioaccumulative (vPvB), 269,
  271, 325, 372

Virtual elimination, 190, 201–202
Voluntary cancellation, pesticide
  products, 5, 110–111
vPvB. *See* Very persistent and very
  bioaccumulative (vPvB)

Washington, 5, 176–177
Washington Department of
  Ecology, 176–177
Waste Electrical and Electronic
  Equipment (WEEE)
  basic provisions, 296–297
  containing hazardous
    substances, 307
  executive summary, 295–298
  exemptions, 297–298
  manufacturer obligations, 301–303
  notification, registration,
    approval requirements, and
    compliance, 301–303
  overview, 10, 296
  regulation of substances/
    products, 303–307
  resources, 308
  scope and implementation
    deadlines of, 298–300
  Venezuelan regulations, 242
  waste compliance, 303
Wastes, 182
WEEE. *See* Waste Electrical and
  Electronic Equipment (WEEE)
WHO. *See* World Health
  Organization (WHO)
Worker Protection Standard
  (WPS), 91
World Health Organization (WHO),
  225, 235
World Trade Organization (WTO),
  250, 293
WPS. *See* Worker Protection
  Standard (WPS)
WTO. *See* World Trade
  Organization (WTO)